Music Data
Mining

Chapman & Hall/CRC
Data Mining and Knowledge Discovery Series

SERIES EDITOR
Vipin Kumar
University of Minnesota
Department of Computer Science and Engineering
Minneapolis, Minnesota, U.S.A

AIMS AND SCOPE

This series aims to capture new developments and applications in data mining and knowledge discovery, while summarizing the computational tools and techniques useful in data analysis. This series encourages the integration of mathematical, statistical, and computational methods and techniques through the publication of a broad range of textbooks, reference works, and handbooks. The inclusion of concrete examples and applications is highly encouraged. The scope of the series includes, but is not limited to, titles in the areas of data mining and knowledge discovery methods and applications, modeling, algorithms, theory and foundations, data and knowledge visualization, data mining systems and tools, and privacy and security issues.

PUBLISHED TITLES

UNDERSTANDING COMPLEX DATASETS:
DATA MINING WITH MATRIX DECOMPOSITIONS
David Skillicorn

COMPUTATIONAL METHODS OF FEATURE SELECTION
Huan Liu and Hiroshi Motoda

CONSTRAINED CLUSTERING: ADVANCES IN
ALGORITHMS, THEORY, AND APPLICATIONS
Sugato Basu, Ian Davidson, and Kiri L. Wagstaff

KNOWLEDGE DISCOVERY FOR COUNTERTERRORISM
AND LAW ENFORCEMENT
David Skillicorn

MULTIMEDIA DATA MINING: A SYSTEMATIC
INTRODUCTION TO CONCEPTS AND THEORY
Zhongfei Zhang and Ruofei Zhang

NEXT GENERATION OF DATA MINING
Hillol Kargupta, Jiawei Han, Philip S. Yu,
Rajeev Motwani, and Vipin Kumar

DATA MINING FOR DESIGN AND MARKETING
Yukio Ohsawa and Katsutoshi Yada

THE TOP TEN ALGORITHMS IN DATA MINING
Xindong Wu and Vipin Kumar

GEOGRAPHIC DATA MINING AND
KNOWLEDGE DISCOVERY, SECOND EDITION
Harvey J. Miller and Jiawei Han

TEXT MINING: CLASSIFICATION, CLUSTERING, AND
APPLICATIONS
Ashok N. Srivastava and Mehran Sahami

BIOLOGICAL DATA MINING
Jake Y. Chen and Stefano Lonardi

INFORMATION DISCOVERY ON ELECTRONIC HEALTH
RECORDS
Vagelis Hristidis

TEMPORAL DATA MINING
Theophano Mitsa

RELATIONAL DATA CLUSTERING: MODELS,
ALGORITHMS, AND APPLICATIONS
Bo Long, Zhongfei Zhang, and Philip S. Yu

KNOWLEDGE DISCOVERY FROM DATA STREAMS
João Gama

STATISTICAL DATA MINING USING SAS APPLICATIONS,
SECOND EDITION
George Fernandez

INTRODUCTION TO PRIVACY-PRESERVING DATA
PUBLISHING: CONCEPTS AND TECHNIQUES
Benjamin C. M. Fung, Ke Wang, Ada Wai-Chee Fu, and
Philip S. Yu

HANDBOOK OF EDUCATIONAL DATA MINING
Cristóbal Romero, Sebastian Ventura,
Mykola Pechenizkiy, and Ryan S.J.d. Baker

DATA MINING WITH R: LEARNING WITH
CASE STUDIES
Luís Torgo

MINING SOFTWARE SPECIFICATIONS: METHODOLOGIES
AND APPLICATIONS
David Lo, Siau-Cheng Khoo, Jiawei Han, and Chao Liu

DATA CLUSTERING IN C++: AN OBJECT-ORIENTED
APPROACH
Guojun Gan

MUSIC DATA MINING
Tao Li, Mitsunori Ogihara, and George Tzanetakis

Music Data Mining

EDITED BY

Tao Li
Mitsunori Ogihara
George Tzanetakis

CRC Press
Taylor & Francis Group
Boca Raton   London   New York

CRC Press is an imprint of the
Taylor & Francis Group an **informa** business

A CHAPMAN & HALL BOOK

CRC Press
Taylor & Francis Group
6000 Broken Sound Parkway NW, Suite 300
Boca Raton, FL 33487-2742

© 2012 by Taylor & Francis Group, LLC
CRC Press is an imprint of Taylor & Francis Group, an Informa business

No claim to original U.S. Government works

Printed in the United States of America on acid-free paper
Version Date: 20110603

International Standard Book Number: 978-1-4398-3552-4 (Hardback)

Visit the Taylor & Francis Web site at
http://www.taylorandfrancis.com

and the CRC Press Web site at
http://www.crcpress.com

Contents

5 Mood and Emotional Classification   135

Mitsunori Ogihara and Youngmoo Kim

6   Zipf's Law, Power Laws, and Music Aesthetics   **169**

*Bill Manaris, Patrick Roos, Dwight Krehbiel, Thomas Zalonis, and
J.R. Armstrong*

11 Symbolic Data Mining in Musicology   327

Ian Knopke and Frauke Jürgensen

Index   347

List of Figures

List of Tables

Preface

During the last 10 years there has been a dramatic shift in how music is produced, distributed, and consumed. A combination of advances in digital storage, audio compression, as well as significant increases in network bandwidth have made digital music distribution a reality. Portable music players, computers, and smart phones frequently contain personal collections of thousands of music tracks. Digital stores in which users can purchase music contain millions of tracks that can be easily downloaded.

The research area of music information retrieval gradually evolved during this time period in order to address the challenge of effectively accessing and interacting with these increasing large collections of music and associated data such as styles, artists, lyrics, and music reviews. The algorithms and systems developed, frequently employ sophisticated signal processing and machine-learning techniques in their attempt to better capture the frequently elusive relevant music information.

The purpose of this book is to present a variety of approaches to utilizing data mining techniques in the context of music processing. Data mining is the process of extracting useful information from large amounts of data. The multifaceted nature of music information provides a wealth of opportunities for mining useful information and utilizing it to create novel ways of interaction with large music collections.

This book is mainly intended for researchers and graduate students in acoustics, computer science, electrical engineering, and music who are interested in learning about the state of the art in music data mining. It can also serve as a textbook in advanced courses. Learning about music data mining is challenging as it is an interdisciplinary field that requires familiarity with several research areas and the relevant literature is scattered in a variety of publication venues. We hope that this book will make the field easier to approach by providing both a good starting point for readers not familiar with the topic as well as a comprehensive reference for those working in the field.

Although the chapters of the book are mostly self-contained and can be read in any order, they have been grouped and ordered in a way that can provide a structured introduction to the topic. The first part of the book deals with fundamental topics. Chapter 1 consists of a survey of music data mining and the different tasks and algorithms that have been proposed in the literature. It serves as a framework for understanding and placing the subsequent chapters in context. One of the fundamental sources of information that can be used for music data mining is the actual audio signal. Extracting relevant audio features requires sophisticated signal processing techniques. Chapter 2 introduces audio signal processing and how it can be used to derive audio features that characterize different facets of musical information such as timbre, rhythm, and pitch content.

The second part of the book deals with classification of the important tasks of music data mining. Chapter 3 describes how a computational approach inspired by human auditory perception can be used for classification and retrieval tasks. There is much literature in instrument recognition, which is explored in Chapter 4. Listening to music can have a profound effect on our mood and emotions. A number of systems for mood and emotion classification have been proposed and are reviewed in Chapter 5. Chapter 6 explores connections between power laws and music aesthetics in the context of *Armonique*, a music discovery engine based on power-law metrics. The engine is evaluated through psychological experiments with human listeners connecting recommendations to human emotional and psychological responses.

Social aspects play an important role in understanding music and are the topic of the third part of the book. There is a large amount of information about music that is available on the Web and peer-to-peer networks. Chapter 7 describes how this information can be extracted and used either directly for music mining tasks or as a way of evaluating music mining algorithms. Tags are words provided by users to categorize information. They are increasingly utilized as a way of indexing images, music, and videos. Chapter 8 provides a thorough overview of how tags can be used in music data mining. Many music data mining algorithms require large amounts of human labeling to train supervised machine-learning models. Human computation games are multiplayer online games that help collect volunteer data in a truthful manner. By designing a game that is entertaining, players are willing to spend a huge amount of time playing the game, contributing massive amounts of data. Chapter 9 shows how human computation games have been used in music classification. A key challenge in data mining is capturing the music information that is important to human listeners.

The last part of the book deals with two more specialized topics of music data mining. Predicting hit songs before they become hits is a music mining task that easily captures the popular imagination. Claims that it has been solved are made frequently in the press but they are very hard to verify. Chapter 10 is a thoughtful and balanced exploration of hit song science from a variety of perspectives. Most of the music mining systems described in this book have as their target the average music listener. Musicology is the scholarly study of music. Chapter 11 (the last chapter of the book) shows how music data mining can be used in the specialized context of symbolic musicology.

Editing a book takes a lot of effort. We would like to thank all the contributors for their chapters (their contacts are found in a few pages after this Preface) as well as their help in reviewing and proofreading. We would like to thank the following people at Chapman & Hall/Taylor & Francis for their help and encouragement: Randi Cohen, Shashi Kumar, Sunil Nair, Jessica

Vakili, and Samantha White. Finally, we would like to express our gratitude to our family, Jing, Emi, Ellen, Erica, Tiffany, Panos, and Nikos, for their kind support.

MATLAB is a registered trademark of The MathWorks, Inc. For product information, please contact:

The MathWorks, Inc.
3 Apple Hill Drive
Natick, MA 01760-2098 USA
Tel: 508 647 7000
Fax: 508-647-7001
E-mail: info@mathworks.com
Web: www.mathworks.com

Tao Li, Mitsu Ogihara, and George Tzanetakis

List of Contributors

J.R. Armstrong
Computer Science Department
College of Charleston
66 George Street
Charleston, SC 29424
E-mail: armstrong@cs.cofc.edu

Jayme Garcia Arnal Barbedo
Department of Communications
State University of Campinas
UNICAMP C.P. 6101
CEP: 13.083-852
Campinas, SP
Brazil
E-mail: jbarbedo@gmail.com

Frauke Jürgensen
Department of Music
University of Aberdeen
MacRobert Building
Aberdeen, AB24 5UA
Scotland, United Kingdom
E-mail: f.jurgensen@abdn.ac.uk

Youngmoo E. Kim
Department of Electrical and
    Computer Engineering
Drexel University
3141 Chestnut Street
Philadelphia, PA 19104
E-mail: ykim@drexel.edu

Ian Knopke
British Broadcasting Corporation
Future Media Technology
BC4 C3 Broadcast Centre BBC
201 Wood Lane, White City
London, W12 7TP
England
E-mail: ian.knopke@gmail.com

Dwight Krehbiel
Department of Psychology
Bethel College
300 East 27th Street
North Newton, KS 67117
E-mail: krehbiel@bethelks.edu

Edith Law
Machine Learning Department
Carnegie Mellon University
5000 Forbes Avenue
Pittsburgh, PA 15217
E-mail: edith@cmu.edu

Lei Li
School of Computer Science
Florida International University
11200 SW 8th Street
Miami, FL 33199
E-mail: lli003@cs.fiu.edu

Tao Li
School of Computer Science
Florida International University
11200 SW 8th Street
Miami, FL 33199
E-mail: taoli@cs.fiu.edu

Richard F. Lyon
Google, Inc.
1600 Amphitheatre Parkway
Mountain View, CA 94043
E-mail: dicklyon@acm.org

Bill Manaris
Computer Science Department
College of Charleston
66 George Street
Charleston, SC 29424
E-mail: manaris@cs.cofc.edu

Steven R. Ness
Department of Computer Science
University of Victoria
ECS, Room 504
P.O. Box 3055, STN CSC
Victoria, BC V8W 3P6
Canada
E-mail: sness@sness.net

Mitsunori Ogihara
Department of Computer Science
University of Miami
Coral Gables, FL 33146
E-mail: ogihara@cs.miami.edu

François Pachet
Sony CSL-Paris
6, Rue Amyot
75005 Paris
France
E-mail: pachet@csl.sony.fr

Patrick Roos
Department of Computer Science
University of Maryland
A.V. Williams Building
College Park, MD 20742
E-mail: roos@cs.umd.edu

Markus Schedl
Department of Computational
    Perception
Johannes Kepler University
Altenberger Str. 69
A-4040 Linz
Austria
E-mail: markus.schedl@jku.at

Douglass Turnbull
Department of Computer Science
Ithaca College
953 Dansby Road
Ithaca, NY 14850
E-mail: dturnbull@ithaca.edu

George Tzanetakis
Department of Computer Science
University of Victoria
P.O. Box 3055, STN CSC
Victoria, BC V8W 3P6
Canada
E-mail: gtzan@cs.uvic.ca

Thomas C. Walters
Google, Inc.
1600 Amphitheatre Parkway
Mountain View, CA 94043
E-mail: thomaswalters@google.com

Thomas Zalonis
Computer Science Department
College of Charleston
66 George Street
Charleston, SC 29424
E-mail: zalonis@cs.cofc.edu

Part I

Fundamental Topics

1

Music Data Mining: An Introduction

Tao Li

Florida International University

Lei Li

Florida International University

CONTENTS

In the Internet age, a gigantic amount of music-related information is easily accessible. For example, music and artist information can be obtained from the artist and record company Web sites, song lyrics can be downloaded from the lyrics databases, and music reviews are available from various discussion forums, blogs, and online magazines.

As the amount of available music-related information increases, the challenges of organizing and analyzing such information become paramount. Recently, many data mining techniques have been used to perform various tasks (e.g., genre classification, emotion and mood detection, playlist generation, and music information retrieval) on music-related data sources. Data mining is a process of automatic extraction of novel, useful, and understandable patterns from a large collection of data. With the large amount of available data from various sources, music has been a natural application area for data mining techniques. In this chapter, we attempt to provide a review of music data mining by surveying various data mining techniques used in music analysis. The chapter also serves as a framework for understanding and placing the rest of the book chapters in context. The reader should be cautioned that music data mining is such a large research area that truly comprehensive surveys are almost impossible, and thus, our overview may be a little eclectic. An interested reader is encouraged to consult with other articles for further reading, in particular, Jensen [50, 90]. In addition, one can visit the Web page: http://users.cis.fiu.edu/~lli003/Music/music.html, where a comprehensive survey on music data mining is provided and is updated constantly.

1.1 Music Data Sources

Table 1.1 briefly summarizes various music-related data sources, describing different aspects of music. We also list some popular Web sites, from which music data sources can be obtained. These data sources provide abundant information related to music from different perspectives. To better understand the data characteristics for music data mining, we give a brief introduction to various music data sources below.

Data Sources	Examples (Web Sites)
Music Metadata	*All Music Guide, FreeDB, WikiMusicGuide*
Acoustic Features	*Ballroom*
Lyrics	*Lyrics, Smartlyrics, AZlyrics*
Music Reviews	*Metacritic, Guypetersreviews, Rollingstone*
Social Tags	*Last.fm*
User Profiles and Playlists	*Musicmobs, Art of the Mix, Mixlister*
MIDI Files	*MIDIDB, IFNIMIDI*
Music Scores	*Music-scores*

Table 1.1
Various Music Data Sources

Music Metadata: Music metadata contains various information describing specific music recordings. Generally speaking, many music file formats support a structure known as *ID3*, which is designed for storing actual audio data music metadata, such as artist name, track title, music description, and album title. Thus, metadata can be extracted with little effort from the ID3 data format. Also, music metadata can be obtained from an online music metadatabase through application programming interfaces (APIs) running on them. These databases and their APIs are used by the majority of music listening software for the purpose of providing information about the tracks to the user. Some well-known music metadatabase applications, for example, *All Music Guide* and *FreeDB*, provide flexible platforms for music enthusiasts to search, upload, and manage music metadata.

Acoustic Features: Music acoustic features include any acoustic properties of an audio sound that may be recorded and analyzed. For example, when a symphonic orchestra is playing Beethoven's *9th Symphony*, each musical instrument, with the exception of some percussions, produces different periodic vibrations. In other words, the sounds produced by musical instruments are the result of the combination of different frequencies. Some basic acoustic features [90] are listed in Table 1.2.

Lyrics: Lyrics are a set of words that make up a song in a textual format. In general, the meaning of the content underlying the lyrics might be explicit or implicit. Most lyrics have specific meanings, describing the artist's emotion, religious belief, or representing themes of times, beautiful natural scenery, and so on. Some lyrics might contain a set of words, from which we cannot easily deduce any specific meanings. The analysis of the correlation between lyrics and other music information may help us understand the intuition of the artists. On the Internet, there are a couple of Web sites offering music lyrics searching services, for example, *SmartLyrics* and *AZLyrics*.

Music Reviews: Music reviews represent a rich resource for examining the ways that music fans describe their music preferences and possible impact of those preferences. With the popularity of the Internet, an ever-increasing number of music fans join the music society and describe their attitudes toward music pieces. Online reviews can be surprisingly detailed, covering not only the reviewers' personal opinions but also important background and contextual information about the music and musicians under discussion [47].

Music Social Tags: Music social tags are a collection of textual information that annotate different music items, such as albums, songs, artists, and so on. Social tags are created by public tagging of music fans. Captured in these tags is a great deal of information including music genre, emotion, instrumentation, and quality, or a simple description for the purpose of retrieval. Music social tags are typically used to facilitate searching for songs,

Acoustic Features	Description
Pitch	Related to the perception of the fundamental frequency of a sound; range from low or deep to high or acute sounds.
Intensity	Related to the amplitude, of the vibration; textual labels for intensity range from soft to loud.
Timbre	Defined as the sound characteristics that allow listeners to perceive as different two sounds with same pitch and same intensity.
Tempo	The speed at which a musical work is played, or expected to be played, by performers. The tempo is usually measured in beats per minute.
Orchestration	Due to the composers and performers' choices in selecting which musical instruments are to be employed to play the different voices, chords, and percussive sounds of a musical work.
Acoustics	A specialization on some characteristics of timbre, including the contribution of room acoustics, background noise, audio postprocessing, filtering, and equalization.
Rhythm	Related to the periodic repetition, with possible small variants, of a temporal pattern of onsets alone. Different rhythms can be perceived at the same time in the case of polyrhythmic music.
Melody	A sequence of tones with a similar timbre that have a recognizable pitch within a small frequency range.
Harmony	The organization, along the time axis, of simultaneous sounds with a recognizable pitch.

Table 1.2
Different Acoustic Features

exploring for new songs, finding similar music recordings, and finding other listeners with similar interests [62]. An illustrative example of well-known online music social tagging systems is *Last.fm*, which provides plenty of music tags through public tagging activities.

User Profiles and Playlists: User profile represents the user's preference to music information, for example, what kind of songs one is interested in, which artist one likes. Playlist, or also called listening history, refers to the list of music pieces that one prefers or has listened to. Traditionally, user profiles and playlists are stored in music applications, which can only be accessed by a single user. With the popularity of cyberspace, more and more music listeners share their music preference online. Their user profiles and playlists are stored and managed in the online music databases, which are open to all the Internet users. Some popular online music applications, for example, *playlist.com*, provide services of creating user profiles and playlists, and sharing them on social networks.

MIDI Files: MIDI, an abbreviation for musical instrument digital interface, is a criterion adopted by the electronic music industry for controlling devices, such as synthesizers and sound cards, that emit music. At minimum, a MIDI representation of a sound includes values for the sound's pitch, length, and volume. It can also include additional characteristics, such as attack and delay times. The MIDI standard is supported by most synthesizers, so sounds created on one synthesizer can be played and manipulated on another synthesizer. Some free MIDI file databases provide online MIDI searching services, such as *MIDIDB* and *IFNIMIDI*.

Music Scores: Music score refers to a handwritten or printed form of musical notation, which uses a five-line staff to represent a piece of music work. The music scores are used in playing music pieces, for example, when a pianist plays a famous piano music. In the field of music data mining, some researchers focus on music score matching, score following and score alignment, to estimate the correspondence between audio data and symbolic score [25]. Some popular music score Web sites (e.g., *music-scores.com*), provide music score downloading services.

These different types of data sources represent various characteristics of music data. Music data mining aims to discover useful information and inherent features of these data sources by taking advantage of various data mining techniques. In the following, we first give a brief introduction to traditional data mining tasks, and subsequently present music-related data mining tasks.

1.2 An Introduction to Data Mining

Data mining is the nontrivial extraction of implicit, previously unknown, and potentially useful information from a large collection of data. The data mining process usually consists of an iterative sequence of the following steps: *data management, data preprocessing, mining,* and *postprocessing* [67]. The four-component framework provides us with a simple systematic language for understanding the data mining process.

Data management is closely related to the implementation of data mining systems. Although many research papers do not explicitly elaborate on data management, it should be noted that data management can be extremely important in practical implementations. Data preprocessing is an important step to ensure the data format and quality as well as to improve the efficiency and ease of the mining process. For music data mining, especially when dealing with acoustic signals, feature extraction where the numeric features are extracted from the signals plays a critical role in the mining process. In the mining step, various data mining algorithms are applied to perform the data mining tasks. There are many different data mining tasks such as data visualization, association mining, classification, clustering, and similarity search. Various algorithms have been proposed to carry out these tasks. Finally, the postprocessing step is needed to refine and evaluate the knowledge derived from the mining step. Since postprocessing mainly concerns the nontechnical work such as documentation and evaluation, we then focus our attention on the first three components and will briefly review data mining in these components.

1.2.1   Data Management

Data management concerns the specific mechanism and structures of how the data are accessed, stored, and managed. In music data mining, data management focuses on music data quality management, involving data cleansing, data integration, data indexing, and so forth.

Data Cleansing: Data cleansing refers to "cleaning" the data by filling in missing values, smoothing noisy data, identifying or removing outliers, and resolving inconsistencies [44]. For example, in music databases, the "artists" value might be missing; we might need to set a default value for the missing data for further analysis.

Data Integration: Data integration is the procedure of combining data obtained from different data sources and providing users with an integrated and unified view of such data [64]. This process plays a significant role in music data, for example, when performing genre classification using both acoustic features and lyrics data.

Data Indexing: Data indexing refers to the problem of storing and arranging a database of objects so that they can be efficiently searched for on the basis of their content. Particularly for music data, data indexing aims at facilitating efficient content music management [19]. Due to the very nature of music data, indexing solutions are needed to efficiently support similarity search, where the similarity of two objects is usually defined by some expert of the domain and can vary depending on the specific application. Peculiar features of music data indexing are the intrinsic high-dimensional nature of the data to be organized, and the complexity of similarity criteria that are used to compare objects.

1.2.2 Data Preprocessing

Data preprocessing describes any type of processing performed on raw data to prepare it for another processing procedure. Commonly used as a preliminary data mining practice, data preprocessing transforms the data into a format that will be more easily and effectively processed. Data preprocessing includes data sampling, dimensionality reduction, feature extraction, feature selection, discretization, transformation, and so forth.

Data Sampling: Data sampling can be regarded as a data reduction technique since it allows a large data set to be represented by a much smaller random sample (or subset) of the data [44]. An advantage of sampling for data reduction is that the cost of obtaining a sample is proportional to the size of the sample. Hence, sampling complexity is potentially sub-linear to the size of the data. For acoustic data, data sampling refers to measuring the audio signals at a finite set of discrete times, since a digital system such as a computer cannot directly represent a continuous audio signal.

Dimensionality Reduction: Dimensionality reduction is an important step in data mining since many types of data analysis become significantly harder as the dimensionality of the data increases, which is known as the *curse of dimensionality*. Dimensionality reduction can eliminate irrelevant features and reduce noise, which leads to a more understandable model involving fewer attributes. In addition, dimensionality reduction may allow the data to be more easily visualized. The reduction of dimensionality by selecting attributes that are a subset of the old is know as *feature selection*, which will be discussed below. Some of the most common approaches for dimensionality reduction, particularly for continuous data, use techniques from linear algebra to project the data from a high-dimensional space into a lower-dimensional space, for example, *Principal Component Analysis* (PCA) [113].

Feature Extraction: Feature extraction refers to simplifying the amount of resources required to describe a large set of data accurately. For music data, feature extraction involves low-level musical feature extraction (e.g., acoustic features) and high-level features of musical feature extraction (e.g., music keys). An overview of feature extraction problems and techniques is given in Chapter 2.

Feature Selection: The purpose of feature selection is to reduce the data set size by removing irrelevant or redundant attributes (or dimensions). It is in some sense a direct form of dimensionality reduction. The goal of feature selection is to find a minimum set of attributes such that the resulting probability distribution of the data classes is as close as possible to the original distribution obtained using all features [44]. Feature selection can significantly improve the comprehensibility of the resulting classifier models and often build a model that generalizes better to unseen points. Further, it is often the case that finding the correct subset of predictive features is an important issue in its own right. In music data mining, feature selection is integrated with feature extraction in terms of selecting the appropriate feature for further analysis.

Discretization: Discretization is used to reduce the number of values for a given continuous attribute by dividing the range of the attribute into intervals. As with feature selection, discretization is performed in a way that satisfies a criterion that is thought to have a relationship to good performance for the data mining task being considered. Typically, discretization is applied to attributes that are used in classification or association analysis [113]. In music data mining, discretization refers to breaking the music pieces down into relatively simpler and smaller parts, and the way these parts fit together and interact with each other is then examined.

Transformation: Variable transformation refers to a transformation that is applied to all the values of a variable. In other words, for each object, the transformation is applied to the value of the variable for that object. For example, if only the magnitude of a variable is important, then the values of the variable can be transformed by taking the absolute value [113]. For acoustic data, a transformation consists of any operations or processes that might be applied to a musical variable (usually a set or tone row in 12-tone music, or a melody or chord progression in tonal music) in composition, performance, or analysis. For example, we can utilize *fast Fourier transform* or *wavelet transform* to transform continuous acoustic data to discrete frequency representation.

1.2.3   Data Mining Tasks and Algorithms

The cycle of data and knowledge mining comprises various analysis steps, each step focusing on a different aspect or task. Traditionally, data mining tasks involve data visualization, association mining, sequence mining, classification, clustering, similarity search, and so forth. In the following, we will briefly describe these tasks along with the techniques used to tackle these tasks.

1.2.3.1   Data Visualization

Data visualization is a fundamental and effective approach for displaying information in a graphic, tabular, or other visual format [113]. The goal of visualization is to provide visual interpretations for the information being considered, and therefore, the analysts can easily capture the relationship between data or the tendency of the data evolution. Successful visualization requires that the data (information) be converted into a visual format so that the characteristics of the data and the relationships among data items or attributes can be analyzed or reported. For music data visual techniques, for example, graphs, tables, and wave patterns, are often the preferred format used to explain the music social networks, music metadata, and acoustic properties.

1.2.3.2 Association Mining

Association mining, the task of detecting correlations among different items in a data set, has received considerable attention in the last few decades, particularly since the publication of the AIS and *a priori* algorithms [2, 3]. Initially, researchers on association mining were largely motivated by the analysis of market basket data, the results of which allowed companies and merchants to more fully understand customer purchasing behavior and as a result, better rescale the market quotient. For instance, an insurance company, by finding a strong correlation between two policies, A and B, of the form $A \Rightarrow B$, indicating that customers that held policy A were also likely to hold policy B, could more efficiently target the marketing of policy B through marketing to those clients that held policy A but not B. In effect, the rule represents knowledge about purchasing behavior [17]. Another example is to find music song patterns. Many music fans have their own playlists, in which music songs they are interested in are organized by personalized patterns. Music recommendation can be achieved by mining association patterns based on song co-occurrence.

1.2.3.3 Sequence Mining

Sequence mining is the task to find patterns that are presented in a certain number of data instances. The instances consist of sequences of elements. The detected patterns are expressed in terms of subsequences of the data sequences and impose an order, that is, the order of the elements of the pattern should be respected in all instances where it appears. The pattern is considered to be frequent if it appears in a number of instances above a given threshold value, usually defined by the user [102].

These patterns may represent valuable information, for example, about the customers behavior when analyzing supermarket transactions, or how a Web site should be prepared when analyzing the Web site log files, or when analyzing genomic or proteomic data in order to find frequent patterns which can provide some biological insights [33]. For symbolic data, a typical example of sequence mining is to recognize a complex chord from MIDI guitar sequences [107].

1.2.3.4 Classification

Classification, which is the task of assigning objects to one of several predefined categories, is a pervasive problem that encompasses many diverse applications. Examples include detecting spam e-mail messages based upon the message header and content, classifying songs into different music genres based on acoustic features or some other music information, and categorizing galaxies based on their shapes [113]. For music data, typical classification tasks include music genre classification, artist/singer classification, mood detection, instrument recognition, and so forth.

A classification technique (also called a *classifier*) is a systematic approach to building classification models from an input data set. Common techniques include decision tree classifiers, rule-based classifiers, neural networks, support vector machines, and naïve Bayes classifiers [57]. Each of these techniques employs a specific learning algorithm to identify a classification model that best fits the relationship between the attribute set and class label of the input data. The model generated by a learning algorithm should both fit the input data well and correctly predict the class labels of records it has never seen before. Therefore, a key objective of the learning algorithm is to build models with good generalization capability, that is, models that accurately predict the class labels of previously unknown records [113].

1.2.3.5   Clustering

The problem of clustering data arises in many disciplines and has a wide range of applications. Intuitively, clustering is the problem of partitioning a finite set of points in a multidimensional space into classes (called clusters) so that (i) the points belonging to the same class are similar and (ii) the points belonging to different classes are dissimilar. The clustering problem has been studied extensively in machine learning, databases, and statistics from various perspectives and with various approaches and focuses [66]. In music data mining, clustering involves building clusters of music tracks in a collection of popular music, identifying groups of users with different music interests, constructing music tag hierarchy, and so forth.

1.2.3.6   Similarity Search

Similarity search is an important technique in a broad range of applications. To capture the similarity of complex domain-specific objects, the feature extraction is typically applied. The feature extraction aims at transforming characteristic object properties into feature values. Examples of such properties are the position and velocity of a spatial object, relationships between points on the face of a person such as the eyes, nose, mouth, and so forth. The extracted values of features can be interpreted as a vector in a multidimensional vector space. This vector space is usually denoted as feature space [97]. The most important characteristic of a feature space is that whenever two of the complex, application-specific objects are similar, the associated feature vectors have a small distance according to an appropriate distance function (e.g., the Euclidean distance). In other words, two similar, domain-specific objects should be transformed to two feature vectors that are close to each other with respect to the appropriate distance function. In contrast to similar objects, the feature vectors of dissimilar objects should be far away from each other.

Thus, the similarity search is naturally translated into a neighborhood query in the feature space [97].

Similarity search is a typical task in music information retrieval. Searching for a musical work given an approximate description of one or more of other music works is the prototype task for a music search system, and in fact it is simply addressed as a similarity search. Later, we will briefly introduce the task of similarity search in music information retrieval.

1.3 Music Data Mining

Music plays an important role in the everyday life for many people, and with digitalization, large music data collections are formed and tend to be accumulated further by music enthusiasts. This has led to music collections—not only on the shelf in form of audio or video records and CDs—but also on the hard drive and on the Internet, to grow beyond what previously was physically possible. It has become impossible for humans to keep track of music and the relations between different songs, and this fact naturally calls for data mining and machine-learning techniques to assist in the navigation within the music world [50]. Here, we review various music data mining tasks and approaches. A brief overview of these tasks and representative publications is described in Table 1.3.

1.3.1 Overview

Data mining strategies are often built on two major issues: what kind of data and what kind of tasks. The same applies to music data mining.

⋆ *What Kind of Data?*

A music data collection consists of various data types, as shown in Table 1.1. For example, it consists of music audio files, metadata such as title and artist, and sometimes even play statistics. Different analysis and experiments are conducted on such data representations based on various music data mining tasks.

⋆ *What Kind of Tasks?*

Music data mining involves methods for various tasks, for example, genre classification, artist/singer identification, mood/emotion detection, instrument recognition, music similarity search, music summarization and visualization, and so forth. Different music data mining tasks focus on different data sources, and try to explore different aspects of data sources. For example, music genre classification aims at automatically classifying music signals into a single unique class by taking advantage of computational analysis of music feature representations [70]; mood/emotion detection tries to identify the mood/emotion represented in a music piece by virtue of acoustic features or other aspects of music data (see Chapter 5).

1.3.2   Music Data Management

It is customary for music listeners to store part of, if not all of, their music in their own computers, partly because music stored in computers is quite often easier to access than music stored in "shoe boxes." Transferring music data from their originally recorded format to computer accessible formats, such as MP3, a process that involves gathering and storing metadata. Transfer software usually uses external databases to conduct this process. Unfortunately, the data obtained from such databases often contains errors and offers multiple entries for the same album. The idea of creating a unified digital multimedia database has been proposed [26]. A digital library supports effective interaction among knowledge producers, librarians, and information and knowledge seekers. The subsequent problem of a digital library is how to efficiently store and arrange music data records so that music fans can quickly find the music resources of their interest.

Music Indexing: A challenge in music data management is how to utilize data indexing techniques based on different aspects of the data itself. For music data, content and various acoustic features can be applied to facilitate efficient music management. For example, Shen et al. present a novel approach for generating small but comprehensive music descriptors to provide services of efficient content music data accessing and retrieval [110]. Unlike approaches that rely on low-level spectral features adapted from speech analysis technology, their approach integrates human music perception to enhance the accuracy of the retrieval and classification process via PCA and neural networks. There are other techniques focusing on indexing music data. For instance, Crampes et al. present an innovative integrated visual approach for indexing music and for automatically composing personalized playlists for radios or chain stores [24]. Specifically, they index music titles with artistic criteria based on visual cues, and propose an integrated visual dynamic environment to assist the user when indexing music titles and editing the resulting playlists. Rauber et al. have proposed a system that automatically organizes a music collection according to the perceived sound similarity resembling genres or styles of music [99]. In their approach, audio signals are processed according to psychoacoustic models to obtain a time-invariant representation of its characteristics. Subsequent clustering provides an intuitive interface where similar pieces of music are grouped together on a map display. In this book, Chapter 8 provides a thorough investigation on music indexing with tags.

Data Mining Tasks	Music Applications	References
Data Management	Music Database Indexing	[24] [99] [110]
Association Mining	Music Association Mining	[55] [61] [74] [125]
Sequence Mining	Music Sequence Mining	[8] [21] [41] [43] [93] [107]
Classification	Audio Classification	[35] [124] [132]
	Genre/Type Classification	[4] [27] [42] [63] [72] [65] [70] [82] [84] [85] [86] [89] [101] [122] [123]
	Artist Classification	[10] [54]
	Singer Identification	[39] [53] [75] [109] [118] [131]
	Mood Detection	[46] [60] [68] [76] [119] [129]
	Instrument Recognition	[7] [16] [31] [32] [45] [58] [105]
Clustering	Clustering	[14] [20] [49] [52] [73] [78] [95] [96] [117]
Similarity Search	Similarity Search	[5] [11] [22] [28] [38] [71] [80] [87] [91] [100] [106] [111]
Summarization	Music Summarization	[22] [23] [59] [79] [94] [108] [126] [127]
Data Visualization	Single-Music Visualization	[1] [18] [36] [37] [48] [104]
	Multimusic Visualization	[13] [81] [92] [116] [121]

Table 1.3
Music Data Mining Tasks

1.3.3   Music Visualization

Visualization of music can be divided into two categories: (i) the ones that focus on visualizing the metadata content or acoustic content of single music documents; and (ii) the ones that aim at representing complete music collections for showing the correlations among different music pieces, or grouping music pieces into different clusters based on their pair-wise similarities. The former type is motivated by the requirement of a casual user, when the user skims through a music CD recording before listening to it carefully in order to roughly capture the main idea or the music style of the music documents. The latter is based on an idea that a spatial organization of the music collection will help the users find particular songs that they are interested in, because they can remember the position in the visual representation and they can be aided by the presence of similar songs near the searched one.

Individual Music Visualization: There are various approaches to single music visualization, most of which take advantage of music acoustic features for representing music recording. For example, Adli, Imieliński, and Swami state that symbolic analysis of music in MIDI can provide more accurate information about the musical aspects like tonality, melody line, and rhythm with less computational requirements if compared to the analysis in audio files. Also, visualizations based on MIDI files can create visual patterns closely related to musical context as the musical information can be explicitly or implicitly obtained [1]. Chen et al. [18] propose an emotion-based music player which synchronizes visualization (photos) with music based on the emotions evoked by auditory stimulus of music and visual content of visualization. Another example of music visualization for single music records is the piano roll view [104], which proposes a new signal processing technique that provides a piano roll-like display of a given polyphonic music signal with a simple transform in spectral domain. There are some other approaches, such as the self-similarity [36], the plot of the waveform [37], and the spectrogram [48]. Any representation has positive aspects and drawbacks, depending on the dimensions carried by the music form it is related to, and on the ability to capture relevant features. Representations can be oriented toward a global representation or local characteristics.

Music Collection Visualization: Visualization of a collection of musical records is usually based on the concept of similarity. Actually, the problem of a graphical representation, normally based on bidimensional plots, is typical of many areas of data analysis. Techniques such as Multidimensional Scaling and Principal Component Analysis are well known for representing a complex and multidimensional set of data when a distance measure—such as the musical similarity—can be computed between the elements or when the elements are mapped to points in a high-dimensional space. The application of bidimensional visualization techniques to music collections has to be carried out considering that the visualization will be given to nonexpert users, rather than to data analysts, who need a simple and appealing representation of the data.

One example of system for graphical representation of audio collection is Marsyas3D [121], which includes a variety of alternative 2D and 3D representations of elements in the collection. In particular, Principal Component Analysis is used to reduce the parameter space that describes the timbre in order to obtain either a bidimensional or tridimensional representation. Another example is the Sonic Browser, which is an application for browsing audio collections [13] that provides the user with multiple views, including a bidimensional scatterplot of audio objects, where the coordinates of each point depend on attributes of the data set, and a graphical tree representation, where the tree is depicted with the root at the center and the leaves over a circle. Sonic Radar, presented by Lübbers [81], is based on the idea that only a few objects, called prototype songs, can be presented to the user. Each prototype song is obtained through clustering the collection with a k-means algorithm and extracting the song that is closer to the cluster center. Prototype songs are plotted on a circle around a standpoint. In addition, Torrens et al. propose different graphical visualization views and their associated features to allow users to better organize their personal music libraries and also to ease selection later [116]. Pampalk et al. [92] present a system with islands of music that facilitates exploration of music libraries without requiring manual genre classification. Given pieces of music in raw audio format, they estimate their perceived sound similarities based on psychoacoustic models. Subsequently, the pieces are organized on a two-dimensional map so that similar pieces are located close to each other.

1.3.4 Music Information Retrieval

Music Information Retrieval (MIR) is an emerging research area devoted to fulfill users' music information needs. As it will be seen, despite the emphasis on retrieval of its name, MIR encompasses a number of different approaches aiming at music management, easy access, and enjoyment. Most of the research work on MIR, of the proposed techniques, and of the developed systems are content based [90]. The main idea underlying content-based approaches is that a document can be described by a set of features that are directly computed from its content. In general, content-based access to multimedia data requires specific methodologies that have to be tailored to each particular medium. Yet, the core information retrieval (IR) techniques, which are based on statistics and probability theories, may be more generally employed outside the textual case, because the underlying models are likely to describe fundamental characteristics being shared by different media, languages, and application domains [51].

A great variety of different methods for content-based searching in music scores and audio data have been proposed and implemented in research prototypes and commercial systems. Besides the limited and well-defined task of identifying recordings, for which audio fingerprinting techniques work well, it is hard to tell which methods should be further pursued. This underlines the importance of a TREC-like series of comparisons for algorithms (such as EvalFest/MIREX at ISMIR) for searching audio recordings and symbolic music notation. Audio and symbolic methods are useful for different tasks. For example, identification of instances of recordings must be based on audio data, while works are best identified based on a symbolic representation. For determining the genre of a given piece of music, approaches based on audio look promising, but symbolic methods might work as well. The interested reader can get a brief overview of different content-based music information retrieval systems from the article by Typke, Wiering, and Veltkamp [120].

Music Similarity Search: In the field of music data mining, similarity search refers to searching for music sound files similar to a given music sound file. In principle, searching can be carried out on any dimension. For instance, the user could provide an example of the timbre—or of the sound—that the user is looking for, or describe the particular structure of a song, and then the music search system will search for similar music works based the information given by the user.

The similarity search processes can be divided into feature extraction and query processing [71]. For feature extraction, the detailed procedure or instruction is introduced in Chapter 2. After feature extraction, music pieces can be represented based on the extracted features. In the step of query processing, the main task is to employ a proper similarity measure to calculate the similarity between the given music work and the candidate music works. A variety of existing similarity measures and distance functions have previously been examined in this context, spanning from simple Euclidean and Mahalanobis distances in feature space to information theoretic measures like the Earth Mover's Distance and Kullback-Leibler [11]. Regardless of the final measure, a major trend in the music retrieval community has been to use a density model of the features (often timbral space defined by Mel-frequency cepstral coefficients [MFCC]) [98]. The main task of comparing two models has then been handled in different ways and is obviously an interesting and challenging task.

The objective of similarity search is to find music sound files similar to a music sound file given as input. Music classification based on genre and style is naturally the form of a hierarchy. Similarity can be used to group sounds together at any node in the hierarchies. The use of sound signals for similarity is justified by an observation that audio signals (digital or analog) of music belonging to the same genre share certain characteristics, because they are composed of similar types of instruments, having similar rhythmic patterns, and similar pitch distributions [30].

The problem with finding sound files similar to a given sound file has been extensively studied during the last decade. Logan and Salomon propose the use of MFCC to define similarity [80]. Nam and Berger propose the use of timbral features (spectral centroids, short-term energy function, and zero-crossing) for similarity testing [87]. Cooper and Foote study the use of self-similarity to summary music signals [22]. Subsequently, they use this summarization for retrieving music files [38]. Rauber, Pampalk, and Merkl study a hierarchical approach in retrieving similar music sounds [100]. Schnitzer et al. rescale the divergence and use a modified FastMap implementation to accelerate nearest-neighbor queries [106]. Slaney, Weinberger, and White learn embeddings so that the pair-wise Euclidean distance between two songs reflects semantic dissimilarity [111]. Deliège, Chua, and Pedersen perform feature extraction in a two-step process that allows distributed computations while respecting copyright laws [28]. Li and Ogihara investigate the use of acoustic-based features for music information retrieval [71]. For similarity search, the distance between two sound files is defined to be the Euclidean distance of their normalized representations. Pampalk, Flexer, and Widmer present an approach to improve audio-based music similarity and genre classification [91]. Berenzweig et al. examine both acoustic and subjective approaches for calculating similarity between artists, comparing their performance on a common database of 400 popular artists [11]. Aucouturier and Pachet introduce a timbral similarity measure for comparing music titles based on a Gaussian model of cepstral coefficients [5].

1.3.5 Association Mining

As discussed in Section 1.2.3.2, association mining refers to detecting correlations among different items in a data set. Specifically in music data mining, association mining can be divided into three different categories: (i) detecting associations among different acoustic features. For example, Xiao et al. use statistic models to investigate the association between timbre and tempo and then use timbral information to improve the performance of tempo estimation [125]; and (ii) detecting associations among music and other document formats. For instance, Knopke [55] measures the similarity between the public text visible on a Web page and the linked sound files, the name of which is normally unseen by the user. Liao, Wang, and Zhang use a dual-wing harmonium model to learn and represent the underlying association patterns between music and video clips in professional MTV [74]; (iii) detecting associations among music features and other music aspects, for example, emotions. An illustrative example of research related to this category is the work by Kuo et al. [61], which investigates the music feature extraction and modifies the affinity graph for association discovery between emotions and music features. Such an affinity graph can provide insight for music recommendations.

1.3.6 Sequence Mining

For music data, sequence mining mainly aims to detect patterns in sequences, such as chord sequences. There are relatively few publications related to music sequence mining tasks. The main contribution of sequence mining is in the area of music transcription. When transcribing audio pieces, different types of errors might be introduced in the transcription version, such as segmentation errors, substitution errors, time alignment errors, and so on. To better control the error rate of transcription, researchers try to explore the feasibility of applying sequence mining into transcribing procedures. For example, Gillet and Richard discuss two postprocessing methods for drum transcription systems, which aim to model typical properties of drum sequences [41]. Both methods operate on a symbolic representation of the sequence, which is obtained by quantizing the onsets of drum strokes on an optimal tatum grid, and by fusing the posterior probabilities produced by the drum transcription system. Clarisse et al. propose a new system for automatic transcription of vocal sequences into a sequence of pitch and duration pairs [21]. Guo and Siegelmann use an algorithm for Time-Warped Longest Common Subsequence (T-WLCS) to deal with singing errors involving rhythmic distortions [43]. Other applications of sequence mining for music data include chord sequence detection [8, 107] and exploring music structure [93].

1.3.7   Classification

Classification is an important issue within music data mining tasks. Various classification problems have emerged during the recent decades. Some researchers focus on classifying music from audio pieces, whereas others are engaged in categorizing music works into different groups. The most general classification focuses on music genre/style classification. In addition, there are some other classification tasks, such as artist/singer classification, mood/emotion classification, instrument classification, and so on. In the following, we will provide a brief overview on different classification tasks. Table 1.4 briefly summarizes different classification tasks.

Audio Classification: The term *audio classification* has been traditionally used to describe a particular task in the fields of speech and video processing, where the main goal is to identify and label the audio in three different classes: speech, music, and environmental sound. This coarse classification can be used for aiding video segmentation and deciding where to apply automatic speech recognition. The refinement of the classification with a second step, where music signals are labeled with a number of predefined classes, has been presented by Wang et al. [132]. This is one of the first articles that uses hidden Markov models as a tool for MIR. An early work on audio classification by Wold et al. [124] was aimed at retrieving simple music signals via a set of semantic labels, in particular, musical instruments. The approach is based on the combination of segmentation techniques with automatic separation of different sources and the parameter extraction. The classification based on the particular orchestration is still an open problem with complex polyphonic performances.

An important issue in audio classification is the amount of audio data required for achieving good classification rates [35]. This problem has many

Tasks	Techniques	References
Audio Classification	Tree-Based Quantization	[35]
	Covariance Matrix	[124]
	Hidden Markov Model	[132]
Genre Classification	Bayesian Model	[27]
	Decision Tree	[4]
	Hidden Markov Model	[101]
	Statistical Pattern Recognition	[122]
	Wavelet Transformation	[42] [72]
	SVM	[65] [86]
	Taxonomy	[70]
	Multilabeling Classification	[82] [123]
	Neural Networks	[63] [84] [85]
Artist Classification	Singer Voice	[10]
	Text Categorization	[54]
Singer Identification	Gaussian Mixture Model	[39] [53] [109] [118] [131]
	KNN	[75]
Mood Detection	SVM on Text Features	[46]
	Multilabel Classification	[68] [119]
	Fuzzy Classifier	[129]
	Gaussian Mixture Model	[76]
Instrument Recognition	Statistical Model	[31] [32]
	Neural Networks	[58]
	Prior Knowledge	[105]
	Taxonomy	[45]

Table 1.4
Music Classification

aspects. First, the amount of data needed is strictly related to the computational complexity of the algorithms, which usually are at least linear with the number of audio samples. Second, perceptual studies have shown that even untrained listeners are quite good at classifying audio data with very short excerpts (less than 1 sec). Finally, in a query-by-example paradigm, where the examples have to be digitally recorded by users, it is likely that users will not be able to record a large part of audio.

Genre Classification: A particular aspect of music record classification is genre classification. The problem is to correctly label an unknown recording of a song with a music genre. Labels can be hierarchically organized in the collection of genres and subgenres. Labeling can be used to enrich the musical document with high-level metadata or to organize a music collection. Genre classification is still biased by Western music, and thus genres are the ones typically found in Western music stores. Some attempts have been made to extend the approach to other cultures (for example, Norowi, Doraisamy, and Wirza [89]), genre classification has been carried for traditional Indian musical forms together with Western genres.

Tzanetakis and Cook were among the first to introduce the problem of music classification [122]. The data set used in this work covered just a few classes and had some bias toward classical music and jazz. The lack of agreement in genre selection is a typical problem of music classification, because the relevance of the different categories is extremely subjective, as well as the categories themselves. These problems are faced also by human classifiers that try to accomplish the same task, and in fact it has been reported that college students achieved no more than about 70% of classification accuracy when listening to three seconds of audio (listening to a longer excerpt did not improve the performances) [122]. The automatic classification is based on three different feature sets, related to rhythmic, pitch, and timbre features. As also highlighted in subsequent works, rhythm seems to play an important role for the classification.

The features used as content descriptors are normally the ones related to timbre. This choice depends on the fact that approaches try to classify short excerpts of an audio recording, where middle-term features such as melody and harmony are not captured. Common music processing approaches compute the Mel-frequency cepstral coefficients (MFCCs), while the use of wavelet transform is exploited by Grimaldi, Cunningham, and Kokaram [42] and Ogihara and Li [72]. Systems on genre classification are normally trained with a set of labeled audio excerpts, and classification is carried out using different techniques and models from the classification literature. Various feature extraction methods have been used [4, 63, 72, 84, 85, 101, 122]. For classification, the Gaussian Mixture Model (GMM) [86] has been classically used but Support Vector Machines (SVMs) [65, 123] and Bayesian methods [27] are becoming more popular. Some other aspects of music, such as taxonomies [70], can be applied to tackling the genre classification problem. Advanced data mining techniques, such as multilabel classification [82, 123] and ensemble methods, can also be used to provide effective approaches for this problem. The basic idea of ensemble methods for genre classification is as follows: individual classification is performed using smaller segments of the input music signals followed by aggregation of individual classification results using combination rules (e.g., majority voting).

Mood and Emotion Detection/Classification: Music mood/emotion describe the inherent emotional meaning of a music clip. Mood/emotion-based annotation is helpful in music understanding, music search, and some music-related applications. One common opinion objecting to mood/emotion detection is that the emotional meaning of music is subjective and it depends on many factors including culture. Music psychologists now agree that culture is of great importance in people's mood response to music, as well as other factors including education and previous experiences. Krumhansl [60] points out that musical sounds might inherently have emotional meaning. For example, some music patterns represent contentment or relaxation, while others make an individual feel anxious or frantic.

There is some recent research on music mood/emotion detection and classification. Liu, Hu, and Zhang present a hierarchical framework to automate the task of mood detection from acoustic music data, by following some music psychological theories in Western cultures [76]. Yang, Liu, and Chen consider a different approach to music emotion classification [129]. For each music segment, the approach determines how likely it is that the song segment belongs to an emotion class. Two fuzzy classifiers are adopted to provide the measurement of the emotion strength. Hu, Downie, and Ehmann investigate the usefulness of text features in music mood classification on 18 mood categories derived from user tags [46].

In addition, some advanced data mining techniques are applied to music mood/emotion detection and classification, for example, multilabel classification. Li and Ogihara cast the emotion detection problem as a multilabel classification problem, where the music sounds are classified into multiple classes simultaneously [68]. The automated detection of emotion in music has been modeled as a multilabel classification task, where a piece of music may belong to more than one class [119]. Chapter 5 presents a detailed study on various aspects of mood and emotion classification.

Instrument Recognition and Classification: The need for automatic classification of sounds arises in different contexts. For example, in biology it aims at identifying species' given animal calls, in medicine it aims at detecting abnormal conditions of vital organs, in mechanical engineering it aims at recognizing machine-failure conditions, in defense it aims at detecting sounds of engine approaching, and in multimedia it aims at classifying video scenes. In this section, we focus on describing sound effects in the case of music, which means description calls for deriving indexes in order to locate melodic patterns, harmonic or rhythmic structures, and so forth [45].

Music instrument recognition (see Chapter 4) and classification are very difficult tasks that are far from being solved. The practical utility for musical instrument classification is twofold:

1. To provide labels for monophonic recordings, for "sound samples" inside sample libraries, or for new patches to be created with a given synthesizer.

2. To provide indexes for locating the main instruments that are included in a musical mixture (for example, one might want to locate a saxophone "solo" in the middle of a song).

The first problem is easier to solve than the second one, and it seems clearly solvable given the current state of the art. The second is tougher, and it is not clear if research done on solving the first one may help. Common sense dictates that a reasonable approach to the second problem would be the initial separation of the sounds corresponding to the different sound sources, followed by the segmentation and classification on those separated tracks. Techniques for source separation cannot yet provide satisfactory solutions although some promising approaches have been developed [7, 16]. As a consequence, research on music instrument classification has concentrated on working with isolated sounds under the assumption that separation and segmentation have been previously performed. This implies the use of a sound sample collection (usually isolated notes) consisting of different instrument families and classes.

Most publications on music instrument recognition and classification focus on analyzing acoustic features of sounds in the music pieces. For example, Eronen and Klapuri present a system for musical instrument recognition that takes advantage of a wide set of features to model the temporal and spectral characteristics of sounds [31]. Instrument classification process has been shown as a three-layer process consisting of pitch extraction, parameterization, and pattern recognition [58]. Sandvold, Gouyon, and Herrera present a feature-based sound modeling approach that combines general, prior knowledge about the sound characteristics of percussion instrument families (general models) with on-the-fly acquired knowledge of recording-specific sounds (localized models) [105]. Essid, Richard, and David utilize statistical pattern recognition techniques to tackle the problem in the context of solo musical phrases [32].

Artist Classification: The term *artist classification* refers to classifying musicians as the predefined artist label given a music document. Traditionally, artist classification is performed based on acoustic features or singer voice. For instance, Berenzweig, Ellis, and Lawrence present that automatically-located singing segments form a more reliable basis for classification than using the entire track, suggesting that the singer's voice is more stable across different performances, compositions, and transformations due to audio engineering techniques rather than the instrumental background [10]. An alternative approach to artist classification is to utilize text categorization techniques to classify artists. Knees, Pampalk, and Widmer retrieve and analyze Web pages ranked by search engines to describe artists in terms of word occurrences on related pages [54].

Singer Identification: Automated singer identification is important in organizing, browsing, and retrieving data in large music collections due to numerous potential applications including music indexing and retrieval, copyright management, and music recommendation systems. The development of singer identification enables the effective management of music databases based on "singer similarity." With this technology, songs performed by a particular singer can be automatically clustered for easy management or exploration, as described by Shen et al. [109].

Several approaches have been proposed to take advantage of statistical models or machine-learning techniques for automatic singer classification/identification [53, 75, 118, 131]. In general, these methods consist of two main steps: singer characteristic modeling based on solo voice and class label identification. In the singer characteristic modeling step, acoustic signal information is extracted to represent the music. Then specific mechanisms (e.g., a statistical model and machine-learning algorithms are constructed to assign songs to one of the predefined singer categories based on their extracted acoustic features. In addition, Shen et al. use multiple low-level features extracted from both vocal and nonvocal music segments to enhance the identification process with a hybrid architecture and build profiles of individual singer characteristics based on statistical mixture models [109]. Fujihara et al. describe a method for automatic singer identification from polyphonic musical audio signals including sounds of various instruments [39].

1.3.8 Clustering

Clustering, as introduced in Section 1.2.3.5, is the task of separating a collection of data into different groups based on some criteria. Specifically in music data mining, clustering aims at dividing a collection of music data into groups of similar objects based on their pair-wise similarities without predefined class labels.

There are several notable publications on music data clustering. An interrecording distance metric that characterizes diversity of pitch distribution together with harmonic center of music pieces has been introduced to measure dissimilarities among musical features, based on chroma-based features extracted from acoustic signals [78]. Camacho tracks the pitch strength trace of the signal, determining clusters of pitch and unpitched sound based on the criterion of the local maximization of the distance between the centroids [14]. Tsai, Rodgers, and Wang examine the feasibility of unsupervised clustering of acoustic data of songs based on the singer's voice characteristics, which are extracted via vocal segment detection followed by solo vocal signal modeling [117]. Li, Ogihara, and Zhu propose a clustering algorithm that integrates features from both lyrics and acoustic data sources to perform bimodal learning [73]. In order to reduce the dimensionality of music features, Jehan proposes a perceptually grounded model for describing music as a sequence of labeled sound segments, for reducing data complexity as well as for compressing audio [49]. In addition, some simple data representation formats are introduced when performing clustering music collections. An illustrative example is introduced by Pienimäki and Lemström, who propose a novel automatic analysis method based on paradigmatic and surface level similarity of music represented in symbolic form [96]. Kameoka, Nishimoto, and Sagayama decompose the energy patterns diffused in time frequency space, that is, a time series of power spectrum, into distinct clusters such that each of them is originated from a single sound stream [52]. Peng et al. propose an approach based on the generalized constraint of a clustering algorithm by incorporating the constraints for grouping music by "similar" artists [95]. Cilibrasi, Vitányi, and Wolf apply compression-based method to the clustering of pieces of music [20].

1.3.9 Music Summarization

Creation of a concise and informative extraction that best summarizes an original digital content is another challenge in music data mining and is extremely important in large-scale information organization and understanding [108]. Recently, most of the music summarization for commercial use has been manually generated from the original music recordings. However, since a large volume of digital content has been made publicly available in various media, in particular, the Internet, efficient approaches to automatic music summarization are increasingly in demand.

Like text summarization, music summarization aims to determine the most general and salient themes of a given music piece that may be used as a representative of the music and readily recognized by a listener. Automatic music summarization can be applied to music indexing, content-based music retrieval, and Web-based music delivery [127]. Several research methods are proposed in automatic music summarization. A music summarization system [59] was developed on MIDI format, which utilized the repetition nature of MIDI compositions to automatically recognize the main melody theme segment of a given piece of music and generate a music summary. Unfortunately, MIDI is a symbolic score used by synthesizers and is totally different from sampled audio format such as Waveform audio file format (WAV), which is highly unstructured, and thus, MIDI summarization methods cannot be applied to real music summarization. A music acoustic data summarization system developed by Logan and Chu [79] uses MFCCs to parameterize each piece of music. Based on the MFCC features, either a cross-entropy measure or Hidden Markov Model (HMM) is used to discover the song structure. Then heuristics are applied to extract key phrases in terms of this structure. This summarization method is suitable for certain genres of music such as rock or folk music, but it is less applicable to classical music. MFCCs were also used as features in the work by Cooper and Foote [22, 23]. They use a two-dimensional (2-D) similarity matrix to represent music structure and generate a music summary. A limitation to this approach is that it does not always produce an intuitive summarization. Peeters, La Burthe, and Rodet [94] propose a multipass approach to generate music summaries. Xu, Maddage, and Shao [126] propose effective algorithms to automatically classify and summarize music content. Support vector machines are applied to classify music into pure music and vocal music by learning from training data. Both for pure vocal music, a number of features are extracted to characterize the music content, respectively. Based on calculated features, a clustering algorithm is applied to structure the music content. Finally, a music summary is created based on clustering and domain knowledge related to pure vocal music.

1.3.10 Advanced Music Data Mining Tasks

In the wake of the increasing popularity of music and the avalanche of various music applications and softwares, the research directions of music data mining tend to be diverse. Specifically, advanced data mining techniques based on different learning metrics have emerged in the music data mining community. A couple of learning tasks, involving multitask learning, multiinstance learning, multilabel classification and so on, are introduced in this section.

Multitask: Multitask learning (MTL) [15] has attracted significant interests in the data mining and machine-learning community during the last decade [6, 12, 115, 128]. Many of the research publications on multitask learning have explored ideas in Bayesian hierarchical modeling [40], and such approaches have been successfully applied to information retrieval [12] and computer vision [115]. For music data, multitask learning has comprehensive applications. For example, Ni et al. [88] employ a nonparametric Bayesian approach [114] for multitask learning in which the number of states is not fixed *a priori*; the model is termed an *infinite HMM* (iHMM). To learn multiple iHMMs simultaneously, one for each sequential data set, the base distributions of the iHMMs may be drawn from a nested Dirichlet Process (nDP) [103], thus allowing intertask clustering.

Multiple-Instance: Multiple-instance learning (MIL) [29] trains classifiers from lightly supervised data, for example, labeled collections of items, known as *bags*, rather than labeled items. Particularly, in music data mining, there are many high quality sources of metadata about musical information such as *Last.fm*, the *All Music Guide*, *Pandora.com*, and so forth. However, each source provides metadata only at certain granularities, that is, it describes the music only at certain scales. For music data, clip (part of tracks)-level classifiers can be used to refine descriptions from one granularity to finer granularities, for example using audio classifiers trained on descriptions of artists to infer descriptions of albums, tracks, or clips. This metadata refinement problem is an MIL problem. Some publications are related to this research area, for instance, Mandel and Ellis [83] formulated a number of

music information related multiple-instance learning tasks and evaluated the
mixed-integer SVM (mi-SVM) and Multiple-Instance Learning via Embedded
Instance Selection (MILES) algorithms on them.

Multilabel: In machine learning, multilabel classification is the special
case within classification of assigning one of several class labels to an input
object. Unlike the better understood problem of binary classification, which
requires discerning between the two given classes, the multilabel one is a more
complex and less researched problem. Recently, multilabel classification has
been increasingly applied to the music categorization problem. For example,
Wang et al. [123] propose a multilabel music style classification approach,
called *Hypergraph-integrated Support Vector Machine* (HiSVM), which can in-
tegrate both music content and music tags for automatic music style classification. Li and Ogihara [68] cast the emotion detection problem as a multilabel
classification problem, where the music sounds are classified into multiple
classes simultaneously.

Semisupervised Learning: Semisupervised learning is a type of
machine-learning technique that makes use of both labeled and unlabeled data
for training—typically a small amount of labeled data with a large amount of
unlabeled data. Semisupervised learning falls between unsupervised learning
(without any labeled training data) and supervised learning (with completely
labeled training data). Many machine-learning researchers have found that unlabeled data, when used in conjunction with a small amount of labeled data,
can produce considerable improvement in learning accuracy. The acquisition
of labeled data for a learning problem often requires a skilled human agent to
manually classify training examples. The cost associated with the labeling process thus may render a fully labeled training set infeasible, whereas acquisition
of unlabeled data is relatively inexpensive. In such situations, semisupervised
learning can be of great practical value. In music data mining, semisupervised
learning can be applied to the task of classifying music metadata. For instance, Li and Ogihara [69] study the problem of identifying "similar" artists
using both lyrics and acoustic data. The approach uses a small set of labeled
samples for the seed labeling to build classifiers that improve themselves using
unlabeled data. You and Dannenberg [130] explore the use of machine learning
to improve onset detection functions. To solve the problem of training data,
they use a semisupervised learning technique combined with score alignment.
The result of alignment, is an estimate of the onset time of every note in the
MIDI file, and these estimates are improved by iteratively applying our onset
detector and then retraining on the new data.

Tensor-Based Learning: A *tensor* is a multidimensional array. More
formally, an N-way or Nth-order tensor is an element of the tensor product
of N vector spaces, each of which has its own coordinate system. Decompositions of higher-order tensors (e.g., N-way arrays with $N \geq 3$) have applications
in psychometrics, chemometrics, signal processing, numerical linear algebra,
computer vision, numerical analysis, data mining, neuroscience, graph analysis, and elsewhere [56]. Particularly, in data mining, tensor decomposition and

factorization have comprehensive applications. For example, Liu et al. [77] propose a text representation model, the Tensor Space Model (TSM), which models the text by multilinear algebraic high-order tensor instead of the traditional vector. Sun, Tao, and Faloutsos [112] introduce the dynamic tensor analysis (DTA) method to summarize high-order or high-dimensional data and reveal the hidden correlations among data. In music data mining, tensor analysis has its own applications in different aspects. Benetos and Kotropoulos [9] propose an automatic music genre classification system using tensor representations, where each recording is represented by a feature matrix over time. An algorithm which performs shifted nonnegative tensor factorization is presented to separate harmonic instruments from multichannel recordings, extending shifted nonnegative matrix factorization to the multichannel case (see FitzGerald, Cranitch, and Coyle [34]).

1.4 Conclusion

The term *music data mining* encompasses a number of different research and development activities that have the common denominator of being related to music data access and analysis [90]. In this chapter, we have introduced the state of the art of music data mining, navigating from the basic data mining tasks to specific music data mining applications. Some popular music data mining tasks, such as music genre classification, singer identification, emotion detection, instrument recognition, and so on, provide substantial benefits for the real-world music management applications and softwares. In addition to the research discussed above, there are many other research issues in music data mining.

Data issues: These include mining various music data characteristics and information from heterogeneous music databases, and the use of culture knowledge.

- *Mining different kinds of features in music*: Music, to some extent, represents an information combination involving cultures, artists' interests, and so on. The very nature of music data may provide various unexplored features for research purposes, which denote diverse audio aspects in terms of music itself. For example, when an artist is playing a melody, the keynote (or mood) may change accompanied by the progress of the melody. By extracting related features that can detect the emotion of the artists, we can easily identify the emotion variation in this melody, which helps us to understand the artist's sentiment on the objects of reference when he/she is composing.

- *Mining information from heterogeneous music data sources*: Various types of data describing different aspects related to music emerge in the wake of the explosion of music information, such as music reviews, music tags, and so on. The procedure of music data mining tasks may be facilitated by taking into account such information. For instance, by virtue of music tags, we can somehow improve music genre classification; by incorporating music reviews, music emotion detection could be easier and more reasonable. Moreover, we can explore the inner relationship among different music data sources, for example, users, songs, artists, tags, reviews, and so on, and then deduce a high-level music social network.

- *Incorporation of culture knowledge*: A particular characteristic of music data is the culture difference. Music is an art form that can be shared by people from different cultures because it crosses the barriers of national languages and cultural backgrounds. For example, Western classical music has passionate followers in China, and many persons in Europe are keen on classical Indian music: all of them can enjoy music without the need of a translation, which is normally required for accessing foreign textual works [90]. Therefore, how to eliminate cultural differences in a reasonable and understandable way is a special research direction.

Methodology issues:

- *Interactive mining of music information*: Interactive mining allows users to focus on the search for patterns, providing and refining data mining requests based on returned results [44]. It is an important aspect of music information retrieval systems, since effective interactive mining can help users to better describe their music information needs. For example, we can utilize dynamic query form to facilitate interactions between users and MIR systems.

Performance issues: These include efficiency and scalability of music data mining algorithms, and the visualization of music data mining results.

- *Efficiency and scalability of music data mining algorithms*: Music databases consist of a huge amount of data, for example, music metadata, acoustic features, and so on. A natural question is how to effectively extract information from such databases. This requires music data mining algorithms to be efficient and scalable [44]. In other words, the running time of a music data mining algorithm must be predictable and acceptable in large databases. The efficiency and scalability of algorithms are then becoming key issues in the implementation of music data mining systems.

- *Visualization of music data mining results*: Music data mining results are usually represented as tables or simple plots, which cannot be vividly analyzed. In addition, some patterns hidden in the results cannot be easily identified by reviewing these simple representations. This is especially crucial if the music data mining system is to be interactive.

The above issues are regarded as major challenges for the further evolution of music data mining technology.

Bibliography

[1] A. Adli, Z. Nakao, and Y. Nagata. A content dependent visualization system for symbolic representation of piano stream. In *Knowledge-Based Intelligent Information and Engineering Systems*, pages 287–294, 2010.

[2] R. Agrawal, T. Imieliński, and A. Swami. Mining association rules between sets of items in large databases. *ACM SIGMOD Record*, 22(2):207–216, 1993.

[3] R. Agrawal and R. Srikant. Fast algorithms for mining association rules. In *Proceedings of the 20th International Conference on Very Large Data Bases, VLDB*, volume 1215, pages 487–499, 1994.

[4] A. Anglade, Q. Mary, R. Ramirez, and S. Dixon. Genre classification using harmony rules induced from automatic chord transcriptions. In *Proceedings of the International Conference on Music Information Retrieval*, pages 669–674, 2009.

[5] J.J. Aucouturier and F. Pachet. Music similarity measures: What's the use? In *Proceedings of the International Conference on Music Information Retrieval*, pages 157–163, 2002.

[6] B. Bakker and T. Heskes. Task clustering and gating for bayesian multitask learning. *Journal of Machine-Learning Research*, 4:83–99, 2003.

[7] A.J. Bell and T.J. Sejnowski. An information-maximization approach to blind separation and blind deconvolution. *Neural Computation*, 7(6):1129–1159, 1995.

[8] J.P. Bello. Audio-based cover song retrieval using approximate chord sequences: testing shifts, gaps, swaps and beats. In *Proceedings of the International Conference on Music Information Retrieval*, pages 239–244, 2007.

[9] E. Benetos and C. Kotropoulos. A tensor-based approach for automatic music genre classification. In *Proceedings of the 16th European Conference on Signal Processings*, 2008.

[10] A. Berenzweig, D.P.W. Ellis, and S. Lawrence. Using voice segments to improve artist classification of music. In *AES 22nd International Conference*, pages 79–86, 2002.

[11] A. Berenzweig, B. Logan, D.P.W. Ellis, and B. Whitman. A large-scale evaluation of acoustic and subjective music-similarity measures. *Computer Music Journal*, 28(2):63–76, 2004.

[12] D. Blei, T.L. Griffiths, M.I. Jordan, and J.B. Tenenbaum. Hierarchical topic models and the nested Chinese restaurant process. *Advances in Neural Information Processing Systems*, 16:17–24, 2004.

[13] E. Brazil et al. Audio information browsing with the sonic browser. In *Proceedings of International Conference on Coordinated and Multiple Views in Exploratory Visualization*, pages 26–31, 2003.

[14] A. Camacho. Detection of pitched/unpitched sound using pitch strength clustering. In *Proceedings of the International Conference on Music Information Retrieval*, pages 533–537, 2008.

[15] R. Caruana. Multitask learning. *Machine Learning*, 28(1):41–75, 1997.

[16] M.A. Casey and A. Westner. Separation of mixed audio sources by independent subspace analysis. In *Proceedings of the International Computer Music Conference*, 2000.

[17] A. Ceglar and J.F. Roddick. Association mining. *ACM Computing Surveys (CSUR)*, 38(2), 2006.

[18] C.H. Chen, M.F. Weng, S.K. Jeng, and Y.Y. Chuang. Emotion-based music visualization using photos. *Advances in Multimedia Modeling*, pages 358–368, 2008.

[19] P. Ciaccia. Multimedia Data Indexing. In L. Liu and Ö.M. Tamer, editors, *Encyclopedia of Database Systems*, pages 1804–1808. Springer-Verlag, New York, 2009.

[20] R. Cilibrasi, P. Vitányi, and R. Wolf. Algorithmic clustering of music based on string compression. *Computer Music Journal*, 28(4):49–67, 2004.

[21] L.P. Clarisse, J.P. Martens, M. Lesaffre, B. De Baets, H. De Meyer, and M. Leman. An auditory model based transcriber of singing sequences. In *Proceedings of the International Conference on Music Information Retrieval*, pages 116–123, 2002.

[22] M. Cooper and J. Foote. Automatic music summarization via similarity analysis. In *Proceedings of IRCAM*, pages 81–85, 2002.

[23] M. Cooper and J. Foote. Summarizing popular music via structural similarity analysis. In *Applications of Signal Processing to Audio and Acoustics, 2003 IEEE Workshop*, pages 127–130, 2003.

[24] M. Crampes, S. Ranwez, F. Velickovski, C. Mooney, and N. Mille. An integrated visual approach for music indexing and dynamic playlist composition. In *Proceedings of SPIE*, volume 6071, pages 24–39, 2006.

[25] R.B. Dannenberg and C. Raphael. Music score alignment and computer accompaniment. *Communications of the ACM*, 49(8):38–43, 2006.

[26] A.P. de Vries. *Content and Multimedia Database Management Systems.* Enschede: Centre for Telematics and Information Technoloy (CTIT), 1999.

[27] C. DeCoro, Z. Barutcuoglu, and R. Fiebrink. Bayesian aggregation for hierarchical genre classification. In *Proceedings of the International Conference on Music Information Retrieval*, pages 77–80, 2007.

[28] F. Deliège, B.Y. Chua, and T.B. Pedersen. High-level audio features: Distributed extraction and similarity search. In *Proceedings of the International Conference on Music Information Retrieval*, pages 565–570, 2008.

[29] T.G. Dietterich, R.H. Lathrop, and T. Lozano-Pérez. Solving the multiple instance problem with axis-parallel rectangles. *Artificial Intelligence*, 89(1-2):31–71, 1997.

[30] W.J. Dowling and D.L. Harwood. *Music Cognition.* Academic Press, New York, 1986.

[31] A. Eronen and A. Klapuri. Musical instrument recognition using cepstral coefficients and temporal features. In *IEEE International Conference on Acoustics, Speech and Signal Processing*, volume 2, 2000.

[32] S. Essid, G. Richard, and B. David. Musical instrument recognition by pair-wise classification strategies. *IEEE Transactions on Audio, Speech, and Language Processing*, 14(4):1401–1412, 2006.

[33] P.G.D. Ferreira. A survey on sequence pattern mining algorithms (unpublished).

[34] D. FitzGerald, M. Cranitch, and E. Coyle. Sound source separation using shifted non-negative tensor factorisation. In *Proceedings on the IEEE Conference on Audio and Speech Signal Processing*, 2006.

[35] J. Foote. A similarity measure for automatic audio classification. In *Proc. AAAI 1997 Spring Symposium on Intelligent Integration and Use of Text, Image, Video, and Audio Corporation*, 1997.

[36] J. Foote. Visualizing music and audio using self-similarity. In *Proceedings of the 7th ACM International Conference on Multimedia (Part 1)*, pages 77–80, ACM, Press, New York, 1999.

[37] J. Foote and M. Cooper. Visualizing musical structure and rhythm via self-similarity. In *Proceedings of the 2001 International Computer Music Conference*, pages 419–422, 2001.

[38] J. Foote, M. Cooper, and U. Nam. Audio retrieval by rhythmic similarity. In *Proceedings of the International Conference on Music Information Retrieval*, volume 3, pages 265–266, 2002.

[39] H. Fujihara, T. Kitahara, M. Goto, K. Komatani, T. Ogata, and H.G. Okuno. Singer identification based on accompaniment sound reduction and reliable frame selection. In *Proceedings of the International Conference on Music Information Retrieval*, pages 329–336, 2005.

[40] A. Gelman. *Bayesian Data Analysis*. CRC Press, Boca Raton, FL, 2004.

[41] O. Gillet and G. Richard. Supervised and unsupervised sequence modelling for drum transcription. In *Proceedings of the International Conference on Music Information Retrieval*, pages 219–224, 2007.

[42] M. Grimaldi, P. Cunningham, and A. Kokaram. A wavelet packet representation of audio signals for music genre classification using different ensemble and feature selection techniques. In *Proceedings of the 5th ACM SIGMM International Workshop on Multimedia Information Retrieval*, pages 102–108, ACM Press, New York, 2003.

[43] A. Guo and H. Siegelmann. Time-warped longest common subsequence algorithm for music retrieval. In *Proceedings of the International Conference on Music Information Retrieval*, pages 258–261, 2004.

[44] J. Han and M. Kamber. *Data Mining: Concepts and Techniques*. Morgan Kaufmann, San Francisco, 2006.

[45] P. Herrera-Boyer, G. Peeters, and S. Dubnov. Automatic classification of musical instrument sounds. *Journal of New Music Research*, 32(1):3–21, 2003.

[46] X. Hu, J.S. Downie, and A.F. Ehmann. Lyric text mining in music mood classification. In *Proceedings of the International Conference on Music Information Retrieval*, pages 411–416, 2009.

[47] X. Hu, J.S. Downie, K. West, and A. Ehmann. Mining music reviews: Promising preliminary results. In *Proceedings of the International Conference on Music Information Retrieval*, pages 536–539, 2005.

[48] E. Isaacson. What you see is what you get: On visualizing music. In *Proceedings of the International Conference on Music Information Retrieval*, pages 389–395, 2005.

[49] T. Jehan. Perceptual segment clustering for music description and time-axis redundancy cancellation. In *Proceedings of the International Conference on Music Information Retrieval*, pages 124–127, 2004.

[50] B.S. Jensen. Exploratory datamining in music. Master's thesis, Informatics and Mathematical Modelling, Technical University of Denmark, DTU, Richard Petersens Plads, Building 321, DK-2800 Kgs. Lyngby, 2006.

[51] K.S. Jones and P. Willett. *Readings in Information Retrieval*. Morgan Kaufmann, San Francisco, 1997.

[52] H. Kameoka, T. Nishimoto, and S. Sagayama. Harmonic temporal-structured clustering via deterministic annealing EM algorithm for audio feature extraction. In *Proceedings of the International Conference on Music Information Retrieval*, pages 115–122, 2005.

[53] Y.E. Kim and B. Whitman. Singer identification in popular music recordings using voice-coding features. In *Proceedings of the International Conference on Music Information Retrieval*, pages 13–17, 2002.

[54] P. Knees, E. Pampalk, and G. Widmer. Artist classification with web-based data. In *Proceedings of the International Conference on Music Information Retrieval*, pages 517–524, 2004.

[55] I. Knopke. Sound, music and textual associations on the World Wide Web. In *Proceedings of the International Symposium on Music Information Retrieval*, pages 484–488, 2004.

[56] T.G. Kolda and B.W. Bader. Tensor decompositions and applications. *SIAM Review*, 51(3):455–500, 2009.

[57] I. Kononenko. Semi-naive Bayesian classifier. In *Machine Learning-EWSL-91*, pages 206–219, 1991.

[58] B. Kostek. Musical instrument classification and duet analysis employing music information retrieval techniques. *Proceedings of the IEEE*, 92(4):712–729, 2004.

[59] R. Kraft, Q. Lu, and S. Teng. Method and apparatus for music summarization and creation of audio summaries, 2001. U.S. Patent 6,225,546.

[60] C.L. Krumhansl. Music: A link between cognition and emotion. *Current Directions in Psychological Science*, 11(2):45–50, 2002.

[61] F.F. Kuo, M.F. Chiang, M.K. Shan, and S.Y. Lee. Emotion-based music recommendation by association discovery from film music. In *Proceedings of the 13th Annual ACM International Conference on Multimedia*, pages 507–510, ACM Press, New York, 2005.

[62] P. Lamere. Social tagging and music information retrieval. *Journal of New Music Research*, 37(2):101–114, 2008.

[63] A.S. Lampropoulos, P.S. Lampropoulou, and G.A. Tsihrintzis. Musical genre classification enhanced by improved source separation techniques. In *Proceedings of the International Conference on Music Information Retrieval*, pages 576–581, 2005.

[64] M. Lenzerini. Data integration: A theoretical perspective. In *Proceedings of the 21th ACM SIGMOD-SIGACT-SIGART Symposium on Principles of Database Systems*, pages 233–246, ACM Press, New York, 2002.

[65] M. Li and R. Sleep. Genre classification via an LZ78-based string kernel. In *Proceedings of the International Conference on Music Information Retrieval*, pages 252–259, 2005.

[66] T. Li. A general model for clustering binary data. In *Proceedings of the 11th ACM SIGKDD International Conference on Knowledge Discovery in Data Mining*, pages 188–197, ACM Press, New York, 2005.

[67] T. Li, Q. Li, S. Zhu, and M. Ogihara. A survey on wavelet applications in data mining. *ACM SIGKDD Explorations Newsletter*, 4(2):49–68, 2002.

[68] T. Li and M. Ogihara. Detecting emotion in music. In *Proceedings of the International Symposium on Music Information Retrieval*, pages 239–240, 2003.

[69] T. Li and M. Ogihara. Music artist style identification by semi-supervised learning from both lyrics and content. In *Proceedings of the 12th Annual ACM International Conference on Multimedia*, pages 364–367, ACM Press, New York, 2004.

[70] T. Li and M. Ogihara. Music genre classification with taxonomy. In *Proceedings of the IEEE International Conference on Acoustics, Speech, and Signal Processing*, pages 197–200, 2005.

[71] T. Li and M. Ogihara. Content-based music similarity search and emotion detection. In *Proceedings of the IEEE International Conference on Acoustic, Speech, and Signal Processing*, pages 17–21, 2006.

[72] T. Li, M. Ogihara, and Q. Li. A comparative study on content-based music genre classification. In *Proceedings of the 26th Annual International ACM SIGIR Conference on Research and Development in Informaion Retrieval*, pages 282–289. ACM, 2003.

[73] T. Li, M. Ogihara, and S. Zhu. Integrating features from different sources for music information retrieval. In *Proceedings of the 6th International Conference on Data Mining*, pages 372–381, 2006.

[74] C. Liao, P. Wang, and Y. Zhang. Mining association patterns between music and video clips in professional MTV. *Advances in Multimedia Modeling*, pages 401–412, 2009.

[75] C.C. Liu and C.S. Huang. A singer identification technique for content-based classification of MP3 music objects. In *Proceedings of the 11th International Conference on Information and Knowledge Management*, pages 438–445, ACM Press, New York, 2002.

[76] D. Liu, L. Lu, and H.J. Zhang. Automatic mood detection from acoustic music data. In *Proceedings of the International Symposium on Music Information Retrieval*, pages 81–87, 2003.

[77] N. Liu, B. Zhang, J. Yan, Z. Chen, W. Liu, F. Bai, and L. Chien. Text representation: From vector to tensor. In *Proceedings of the 5th IEEE International Conference on Data Mining*, pages 725–728, Washington, DC, IEEE Computer Society, 2005.

[78] Y. Liu, Y. Wang, A. Shenoy, W.H. Tsai, and L. Cai. Clustering music recordings by their keys. In *Proceedings of the International Conference on Music Information Retrieval*, pages 319–324, 2008.

[79] B. Logan and S. Chu. Music summarization using key phrases. In *Proceedings of the IEEE International Conference on Acoustics, Speech, and Signal Processing*, volume 2, pages 749–752, 2000.

[80] B. Logan and A. Salomon. A content-based music similarity function. *Cambridge Research Labs-Tech Report*, 2001.

[81] D. Lübbers. Sonixplorer: Combining visualization and auralization for content-based exploration of music collections. In *Proceedings of the International Conference on Music Information Retrieval*, pages 590–593, 2005.

[82] H. Lukashevich, J. Abeßer, C. Dittmar, and H. Grossmann. From multi-labeling to multi-domain-labeling: A novel two-dimensional approach to music genre classification. In *Proceedings of the International Conference on Music Information Retrieval*, pages 459–464, 2009.

[83] M. Mandel and D. Ellis. Multiple-instance learning for music information retrieval. In *Proceedings of the International Conference on Music Information Retrieval*, 2008.

[84] C. McKay and I. Fujinaga. Automatic genre classification using large high-level musical feature sets. In *Proceedings of the International Conference on Music Information Retrieval*, pages 525–530, 2004.

[85] A. Meng, P. Ahrendt, and J. Larsen. Improving music genre classification by short time feature integration. In *Proceedings of the IEEE International Conference on Acoustics, Speech, and Signal Processing*, pages 497–500, 2005.

[86] A. Meng and J. Shawe-Taylor. An investigation of feature models for music genre classification using the support vector classifier. In *Proceedings of the International Conference on Music Information Retrieval*, pages 604–609, 2005.

[87] U. Nam and J. Berger. Addressing the "same but different-different but similar" problem in automatic music classification. In *Proceedings of the International Conference on Music Information Retrieval*, pages 21–22, 2001.

[88] K. Ni, J. Paisley, L. Carin, and D. Dunson. Multi-task learning for analyzing and sorting large databases of sequential data. *IEEE Transactions on Signal Processing*, 56(8):3918–3931, 2008.

[89] N.M. Norowi, S. Doraisamy, and R. Wirza. Factors affecting automatic genre classification: An investigation incorporating non-Western musical forms. In *Proceedings of the International Conference on Music Information Retrieval*, pages 13–20, 2005.

[90] N. Orio. Music retrieval: A tutorial and review. *Foundations and Trends Information Retrieval*, (1)1: 1–90, Now Publishers, 2006.

[91] E. Pampalk, A. Flexer, and G. Widmer. Improvements of audio-based music similarity and genre classification. In *Proceedings of the International Conference on Music Information Retrieval*, pages 628–633, 2005.

[92] E. Pampalk, A. Rauber, and D. Merkl. Content-based organization and visualization of music archives. In *Proceedings of the 10th ACM International Conference on Multimedia*, pages 570–579, ACM Press, New York, 2002.

[93] G. Peeters. Sequence representation of music structure using higher-order similarity matrix and maximum-likelihood approach. In *Proceedings of the International Conference on Music Information Retrieval*, pages 35–40, 2007.

[94] G. Peeters, A. La Burthe, and X. Rodet. Toward automatic music audio summary generation from signal analysis. In *Proceedings of the International Conference on Music Information Retrieval*, pages 94–100, 2002.

[95] W. Peng, T. Li, and M. Ogihara. Music clustering with constraints. In *Proceedings of the International Conference on Music Information Retrieval*, pages 27–32, 2007.

[96] A. Pienimäki and K. Lemström. Clustering symbolic music using paradigmatic and surface level analyses. In *Proceedings of the International Conference on Music Information Retrieval*, pages 262–265, 2004.

[97] A. Pryakhin. Similarity Search and Data Mining Techniques for Advanced Database Systems. PhD thesis, Ludwig-Maximilians-Universität München, 2006.

[98] L. Rabiner and B.H. Juang. *Fundamentals of Speech Recognition*. Prentice Hall, Upper Saddle River, NJ, 1993.

[99] A. Rauber, E. Pampalk, and D. Merkl. Content-based music indexing and organization. In *Proceedings of the 25th Annual International ACM SIGIR Conference on Research and Development in Information Retrieval*, pages 409–410, ACM Press, New York, 2002.

[100] A. Rauber, E. Pampalk, and D. Merkl. Using psycho-acoustic models and self-organizing maps to create a hierarchical structuring of music by sound similarity. In *Proceedings of the International Conference on Music Information Retrieval*, pages 71–80, 2002.

[101] J. Reed and C.H. Lee. A study on music genre classification based on universal acoustic models. In *Proceedings of the International Conference on Music Information Retrieval*, pages 89–94, 2006.

[102] J.F. Roddick and M. Spiliopoulou. A survey of temporal knowledge discovery paradigms and methods. *IEEE Transactions on Knowledge and Data Engineering*, 14:750–767, 2002.

[103] A. Rodriguez, DB Dunson, and AE Gelfand. The nested Dirichlet process (with discussion). *Journal of American Statistical Association*, 103:1131–1144, 2008.

[104] S. Sagayama, K. Takahashi, H. Kameoka, and T. Nishimoto. Specmurt analysis: A piano-roll visualization of polyphonic music signal by deconvolution of log-frequency spectrum. In *ISCA Tutorial and Research Workshop (ITRW) on Statistical and Perceptual Audio Processing*, 2004.

[105] V. Sandvold, F. Gouyon, and P. Herrera. Percussion classification in polyphonic audio recordings using localized sound models. In *Proceedings of the International Conference on Music Information Retrieval*, pages 537–540, 2004.

[106] D. Schnitzer, A. Flexer, G. Widmer, and A. Linz. A filter-and-refine indexing method for fast similarity search in millions of music tracks. In *Proceedings of the International Conference on Music Information Retrieval*, pages 537–542, 2009.

[107] R. Scholz and G. Ramalho. Cochonut: Recognizing complex chords from MIDI guitar sequences. In *Proceedings of the International Conference on Music Information Retrieval*, pages 27–32, 2008.

[108] X. Shao, C. Xu, Y. Wang, and M.S. Kankanhalli. Automatic music summarization in compressed domain. In *Proceedings of the IEEE International Conference on Acoustics, Speech, and Signal Processing*, pages 261–264, 2004.

[109] J. Shen, B. Cui, J. Shepherd, and K.L. Tan. Toward efficient automated singer identification in large music databases. In *Proceedings of the 29th Annual International ACM SIGIR Conference on Research and Development in Information Retrieval*, pages 59–66. ACM, 2006.

[110] J. Shen, J. Shepherd, and A. Ngu. InMAF: Indexing music databases via multiple acoustic features. In *Proceedings of the 2006 ACM SIGMOD International Conference on Management of Data*, pages 778–780, ACM Press, New York, 2006.

[111] M. Slaney, K. Weinberger, and W. White. Learning a metric for music similarity. In *Proceedings of the International Conference on Music Information Retrieval*, pages 313–381, 2008.

[112] J. Sun, D. Tao, and C. Faloutsos. Beyond streams and graphs: dynamic tensor analysis. In *Proceedings of the 12th ACM SIGKDD International Conference on Knowledge Discovery and Data Mining*, pages 374–383, ACM Press, New York, 2006.

[113] P.N. Tan, M. Steinbach, and V. Kumar. *Introduction to Data Mining*. Pearson Addison Wesley, Boston, 2006.

[114] Y.W. Teh, M.I. Jordan, M.J. Beal, and D.M. Blei. Hierarchical dirichlet processes. *Journal of American Statistical Association*, 101(476):1566–1581, 2006.

[115] S. Thrun and J. O'Sullivan. Discovering structure in multiple learning tasks: The TC algorithm. In *Proceedings of the International Conference on Machine Learning*, pages 489–497, 1996.

[116] M. Torrens, P. Hertzog, and J.L. Arcos. Visualizing and exploring personal music libraries. In *Proceedings of the International Conference on Music Information Retrieval*, pages 421–424, 2004.

[117] W.H. Tsai, D. Rodgers, and H.M. Wang. Blind clustering of popular music recordings based on singer voice characteristics. *Computer Music Journal*, 28(3):68–79, 2004.

[118] W.H. Tsai, H.M. Wang, and D. Rodgers. Automatic singer identification of popular music recordings via estimation and modeling of solo

vocal signal. In *Proceedings of the 18th European Conference on Speech Communication and Technology*, pages 2993–2996, 2003.

[119] K.T.G. Tsoumakas, G. Kalliris, and I. Vlahavas. Multi-label classification of music into emotions. In *Proceedings of the International Conference on Music Information Retrieval*, pages 325–330, 2008.

[120] R. Typke, F. Wiering, and R.C. Veltkamp. A survey of music information retrieval systems. In *Proceedings of the International Conference on Music Information Retrieval*, pages 153–160, 2005.

[121] G. Tzanetakis and P. Cook. Marsyas3D: A prototype audio browser-editor using a large scale immersive visual and audio display. In *Proceedings of the International Conference on Auditory Display (ICAD)*, pages 250–254, 2001.

[122] G. Tzanetakis and P. Cook. Musical genre classification of audio signals. *IEEE Transactions on Speech and Audio Processing*, 10(5):293–302, 2002.

[123] F. Wang, X. Wang, B. Shao, T. Li, and M. Ogihara. Tag integrated multi-label music style classification with hypergraph. In *Proceedings of the International Conference on Music Information Retrieval*, pages 363–368, 2009.

[124] E. Wold, T. Blum, D. Keislar, and J. Wheaten. Content-based classification, search, and retrieval of audio. *IEEE Multimedia*, 3(3):27–36, 1996.

[125] L. Xiao, A. Tian, W. Li, and J. Zhou. Using a statistic model to capture the association between timbre and perceived tempo. In *Proceedings of the International Conference on Music Information Retrieval*, pages 659–662, 2008.

[126] C. Xu, N.C. Maddage, and X. Shao. Automatic music classification and summarization. *IEEE Transactions on Speech and Audio Processing*, 13(3):441–450, 2005.

[127] C. Xu, X. Shao, N.C. Maddage, M.S. Kankanhalli, and Q. Tian. Automatically summarize musical audio using adaptive clustering. In *2004 IEEE International Conference on Multimedia and Expo, 2004. ICME'04*, pages 2063–2066, 2004.

[128] Y. Xue, X. Liao, L. Carin, and B. Krishnapuram. Multi-task learning for classification with Dirichlet process priors. *Journal of Machine-Learning Research*, 8:35–63, 2007.

[129] Y.H. Yang, C.C. Liu, and H.H. Chen. Music emotion classification: A fuzzy approach. In *Proceedings of the 14th Annual ACM International Conference on Multimedia*, pages 81–84, ACM Press, New York, 2006.

[130] W. You and R. Dannenberg. Polyphonic music note onset detection us-
      ing semi-supervised learning. In *Proceedings of the International Con-
      ference on Music Information Retrieval*, pages 279–282, 2007.

[131] T. Zhang. Automatic singer identification. In *Proceedings of the Inter-
      national Conference on Multimedia and Expo*, pages 33–36, 2003.

[132] T. Zhang and C.C.J. Kuo. Hierarchical system for content-based audio
      classification and retrieval. In *Conference on Multimedia Storage and
      Archiving Systems III, SPIE*, volume 3527, pages 398–409, 1998.

2

Audio Feature Extraction

George Tzanetakis

University of Victoria

CONTENTS

The automatic analysis of music stored as a digital audio signal requires a sophisticated process of distilling information. For example, a three-minute song stored as uncompressed digital audio is represented digitally by a sequence almost 16 million numbers (3 [minutes] * 60 [seconds] * 2 [stereo channels] * 44100 [sampling rate]). In the case of tempo induction, these 16 million numbers need to somehow be converted to a single numerical estimate of the tempo of the piece.

Audio feature extraction forms the foundation for any type of music data mining. It is the process of distilling the huge amounts of raw audio data into much more compact representations that capture higher level information about the underlying musical content. As such, it is much more specific to music than the data mining processes that typically follow it. The most basic digital audio representation is a sequence of quantized pulses in time corresponding to the discretized displacement of air pressure that occurred when the music was recorded. Humans (and many other organisms) make sense of their auditory environment by identifying periodic sounds with specific frequencies. At a very fundamental level music consists of sounds (some of which are periodic) that start and stop at different moments in time. Therefore, representations of sound that have a separate notion of time and frequency are commonly used as the first step in audio feature extraction and are the topic of the first section of this chapter. First, the short-time Fourier transform, which is the most common audio representation used for feature extraction, is reviewed followed by short descriptions of other audio representations such as wavelets and filterbanks.

A common way of grouping audio features (or *descriptors* as they are sometimes called) is based on the type of information they are attempting to capture [56]. On an abstract level, one can identify different high-level aspects of a music recording. The hierarchical organization in time is referred to as *rhythm* and the hierarchical organization in frequency or pitch is referred to as *harmony*. Timbre is the quality that distinguishes sounds of the same pitch and loudness generated by different methods of sound production. We will use the term *timbral texture* to refer to the more general quality of the mixture of sounds present in music that is independent of *rhythm* and *harmony*. For example the same piece of notated music played at the same tempo by a string quartet and a saxophone quartet would have different *timbral texture* characteristics in each configuration. In this chapter various audio feature extraction methods that attempt to capture these three basic aspects of musical information are reviewed. Some additional audio features that cover other aspects of musical information are also briefly described. The chapter ends with a short description of audio genre classification as a canonical case study of how audio feature extraction can be used as well as some references to freely available software that can be used for audio feature extraction. Audio feature extraction is a big topic and it would be impossible to cover it fully in this chapter. Although our coverage is not complete we believe we describe most of the common audio feature extraction approaches and the bibliography is representative of the "state of the art" in this area in 2010.

2.1   Audio Representations

To analyze music stored as a recorded audio signal we need to devise representations that roughly correspond to how we perceive sound through the auditory system. At a fundamental level, such audio representations will help determine when things happen in time and how fast they repeat (frequency).

Therefore, the foundation of any audio analysis algorithm is a representation that is structured around time and frequency. In audio signal processing an important property of time and frequency transform is invertibility which means that the original audio signal or a very close approximation can be reconstructed from the values of the transform. As the goal of audio feature extraction is analysis, typically a lot of information needs to be discarded and therefore perfect reconstruction is not as important.

Before discussing time and frequency representations, we briefly discuss how audio signals are represented digitally. Sound is created when air molecules are set into motion by some kind of vibration. The resulting changes in air pressure can be represented as a continuous signal over time. In order to represent the continuous process in a finite amount of memory the continuous signal is sampled at regular periodic intervals. The resulting sequence of samples which still has continuous values is then converted to a sequence of discretized samples through the process of quantization. For example CD quality audio has a sampling rate of 44,100 Hz and a dynamic range of 16 bit. This means each second of sound is represented as 44,100 samples equally spaced in time and each one of those samples is represented by 16 bits. One fundamental result in signal processing is the Nyquist-Shannon sampling theorem [59] which states that if a function $x(t)$ contains no frequencies higher than B Hertz, it can be completely reconstructed from a series of points spaced $\frac{1}{2B}$ seconds apart. What this means is that if the highest frequency we are interested in is B Hertz then we need to sample the signal at $2B$ Hertz or higher. As the data rates of audio signals are very high this has important implications. For example telephone quality speech typically has a sampling rate of 16,000 Hz whereas CD audio has 44,100 Hz.

The short-time Fourier transform (STFT) is arguably the most common time-frequency representation and has been widely used in many domains in addition to music processing. In addition, other audio feature representations such as the Mel-frequency cepstral coefficients (MFCCs) and chroma are based on the STFT. An important factor in the wide use of the STFT is the high speed with which it can be computed in certain cases when using the fast Fourier transform algorithm.

2.1.1   The Short-Time Fourier Transform

The fundamental idea behind the short-time Fourier transform (STFT) as well as many other time-frequency representations is to express a signal as a linear combination of basic elementary signals that can be more easily understood and manipulated. The resulting representation contains information about how the energy of the signal is distributed in both time and frequency. The STFT is essentially a discrete Fourier transform (DFT) adapted to provide localization in time. The DFT has its origins in the Fourier series in which any complicated continuous periodic function can be written as an infinite discrete sum of sine and cosine signals. Similarly, the DFT can be viewed as a

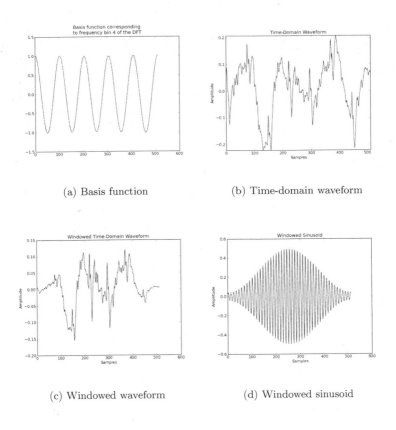

(a) Basis function   (b) Time-domain waveform

(c) Windowed waveform   (d) Windowed sinusoid

Figure 2.1
Example of discrete Fourier transform. Using the DFT, a short sequence of time-domain samples such as the one shown in (b) is expressed as a linear combination of simple sinusoidal signals such as the one shown in (a). Windowing applies a bell-shaped curve to reduce artifacts due to discontinuities when the waveform is periodically repeated (c) and (d).

similar process of representing any finite, discrete signal (properties required for processing by a computer) by a finite, discrete sum of discretized sine and cosine signals. Figure 2.1(a) shows such a simple sinusoidal signal.

It is possible to calculate the DFT of an entire audio clip and show how the energy of the signal is distributed among different frequencies. However, such an analysis would provide no information about when these frequencies start and stop in time. The idea behind the STFT is to process small segments of the audio clip at a time and the DFT of each segment. The output of the DFT is called a *spectrum*. The resulting sequence of spectrums (or spectra) contains information about time as well as frequency. The process of obtaining a small segment from a long audio signal can be viewed as a multiplication of the original audio signal with signal that has the value 1 during the time period of interest and the value 0 outside it. Such a signal is called a rectangular window.

Any type of Fourier analysis assumes infinite periodic signals so processing finite signals is done by effectively repeating them to form a periodic signal. If the finite signal analyzed has been obtained by rectangular windowing then there will be a large discontinuity at the point where the end of the signal is connected to the start of the signal in the process of periodic repetition. This discontinuity will introduce significant energy in all frequencies distorting the analysis. Another way to view this is that the sharp slope of the rectangular window causes additional frequencies in the entire spectrum. This distortion of the spectrum is called *spectral leakage*. To reduce the effects of spectral leakage, instead of using a rectangular window, a nonnegative smooth "bell-shaped" curve is used. There are several variants named after the person who proposed them with slightly different characteristics. Examples include: Hann, Hamming, and Blackman. Figures 2.1(c) and 2.1(d) show the effect of windowing on a time-domain waveform and a single sinusoid respectively. Figure 2.2 shows the spectra of a mixture of two sinusoids with frequencies of 1500 Hz and 3000 Hz sampled at 22050 Hz and weighted with rectangular, Hamming, and Hann windows. As can be seen, windowing makes the spectrum less spread and closer to the ideal theoretical result (which should be two single peaks at the corresponding frequencies).

Formally, the DFT is defined as:

$$X[k] = \sum_{t=0}^{N-1} x[t]e^{-jkt\frac{2\pi}{N}} \qquad k = 0...N-1 \tag{2.1}$$

where $X[k]$ is the complex number corresponding to frequency bin k or equivalently a frequency of $k\frac{F_s}{N}$ where F_s is the sampling frequency and N is the number of frequency bins as well as the number of the time-domain samples $x[n]$.

The notation above is based on Euler's relation:

$$e^{j\theta} = cos\theta + jsin\theta \tag{2.2}$$

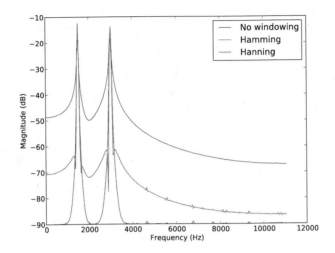

Figure 2.2
Effect of windowing to the magnitude spectrum of the mixture of two sinusoids.

Therefore, one can think of the representation $x[t]$ as a weighted sum (with weights $X[k]$) of sinusoidal signals of a particular frequency $k\frac{F_s}{N}$ with magnitudes:

$$|X[k]| = \sqrt{Re(X[k])^2 + Im(X[k])^2} \qquad (2.3)$$

and phases:

$$\phi[k] = -\arctan\left(\frac{Im(X[k])}{Re(X[k])}\right) \qquad (2.4)$$

The obvious implementation of the DFT requires N multiplications for each frequency bin resulting in a computational complexity of $O(N^2)$. It turns out that it is possible to calculate the DFT much faster with complexity $O(N \log N)$ if the size N is a power of 2. For example, for 512 points (a very common choice in audio signal processing) the FFT is approximately 56.8 times faster than the direct FFT ($512^2/512 \log_2 512$).

The identity of a sound is mostly affected by the magnitude spectrum although the phase plays an important role especially near transient parts of the signal such as percussion hits. Therefore, in the majority of audio feature extraction for analyzing music only the magnitude spectrum is considered.

The human ear has a remarkably large dynamic range in audio perception. The ratio of sound intensity of the quietest sound that the ear can hear to the loudest sound that can cause permanent damage exceeds a trillion (10^{12}). Psychophysic experiments have determined that humans perceive intensities approximately on a logarithmic scale in order to cope with this large dynamic

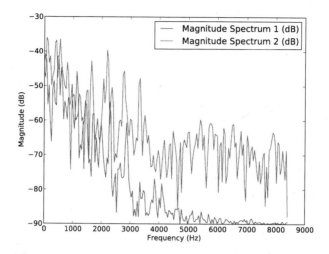

Figure 2.3
Two magnitude spectra in dB corresponding to 20-msec excerpts from music clips.

range. The decibel is commonly used in acoustics to quantify sound levels relative to a 0 dB reference (defined as the sound pressure level of the softest sound a person with average hearing can detect). The difference in decibels between two sounds playing with power P_1 and and P_2 is:

$$10 \log_{10} \frac{P_2}{P_1} \qquad (2.5)$$

For example if the second sound has twice the power the difference is:

$$10 \log_{10} \left(\frac{P_2}{P_1} \right) = 10 \log_{10} 2 = 3\text{dB} \qquad (2.6)$$

The base-10 logarithm of 1 trillion is 12 resulting in an audio level of 120 dB. Digital samples measure levels of sound pressure. The power in a sound wave is the square of pressure. Therefore, the difference between two sounds with pressure p_1 and p_2 is:

$$10 \log_{10} \frac{P_2}{P_1} = 10 \log_{10} \frac{p_2^2}{p_1^1} = 20 \log_{10} \frac{p_2}{p_1} \qquad (2.7)$$

The spectrum in dB can then be defined as:

$$|X[k]|_{dB} = 20 \log_{10}(|X[k]| + \epsilon) \qquad (2.8)$$

where ϵ is a very small number to avoid taking the logarithm of 0. Figure 2.3 shows two magnitude spectra in dB corresponding to 20-millisecond (msec) frames from two different pieces of music.

2.1.2   Filterbanks, Wavelets, and Other Time-Frequency Representations

In signal processing, a filterbank refers to any system that separates the input signal into several subbands using an array (bank) of filters. Typically, each subband corresponds to a subset of frequencies. There is no requirement that the subband-frequency ranges are not overlapping or that the transformation is invertible. The STFT, wavelets, and many other types of signal decompositions can all be viewed as filterbanks with specific constraints and characteristics.

A common use of filterbanks in audio processing is as approximations to how the human auditory system processes sound. For example from experiments in psychophysics it is known that the sensitivity of the human ear to different frequencies is not uniformly spaced along the frequency axis. For example, a common approximation is the Mel filterbank in which filters are uniformly spaced before 1 kHz and logarithmically spaced after 1 kHz. In contrast the DFT can be thought of as a filterbank with very narrow-band filters that are linearly spaced between 0 Hz and $F_s/2$ where F_s is the sampling frequency.

Music and audio signals in general change over time therefore we are interested in their frequency content locally in time. The STFT addresses this by windowing a section of the signal and then taking its DFT. The time and frequency resolution of the STFT is fixed based on the size of the window and the sampling rate. Increasing the size of the window can make the estimation of frequencies more precise (high-frequency resolution) but makes the detection of when they take place (time resolution) less accurate. This time-frequency trade-off is fundamental to any type of time-frequency analysis.

Wavelet analysis is performed by using variable time-frequency resolution so that low frequencies are more precisely detected (high-frequency resolution) but are not very accurately placed in time (low-time resolution) and high frequencies are less precisely detected (low-frequency resolution) but are more accurately placed in time (high-time resolution). The most common dyadic type of the discrete wavelet transform can be viewed as a filterbank in which each filter has half the frequency range of the filter with the closest center frequency that is higher. Essentially, this means that each filter spans an octave in frequency and has half/twice the bandwidth of the corresponding adjacent filters. As an example, the filters in a dyadic discrete wavelet transform for a sampling rate of 10,000 Hz would have the following bandwidths 2500–5000 Hz, 1250–2500, 725–1250, and so on. Each successive band (from high-center frequency to low) can be represented by half the number of samples of the previous one according to the Nyquist-Shannon sampling theorem. That way, the discrete wavelet transform is a transform that has the same number of coefficients (similar to the DFT) as the original discrete signal.

Both the DFT and the DWT are mathematical and signal-processing tools that do not take into account the characteristics of the human auditory system. There are also alternative representations that take this information into account typically termed *auditory models*. They are informed by experiments in psychoacoustics and our increasing knowledge of the biomechanics of the human ear and differ in the amount of detail and accuracy of their modeling [61]. The more efficient versions are constructed on top of the DFT but the more elaborate models do direct filtering followed by stages simulating the inner hair cell mechanism of the ear. In general, they tend to be heavier computationally and for some tasks such as automatic music genre classification so far they do not show any advantages compared to more direct approaches. However, they exhibit many of the perception characteristics of the human ear so it is likely that they will be used especially in tasks that require rich, structured representations of audio signals [50].

2.2 Timbral Texture Features

Features representing timbral information have been the most widely used audio features and ones that have so far provided the best results when used in isolation. Another factor in their popularity is their long history in the area of speech recognition. There are many variations in timbral feature extraction but most proposed systems follow a common general process. First, some form of time-frequency analysis such as the STFT is performed followed by summarization steps that result in a feature vector of significantly smaller dimensionality. A similar approach can be used with other time-frequency representations.

2.2.1 Spectral Features

Spectral features are directly computed on the magnitude spectrum and attempt to summarize information about the general characteristics of the distribution of energy across frequencies. They have been motivated by research in timbre perception of isolated notes of instruments [21]. The spectral centroid is defined as the center of gravity of the spectrum and is correlated with the perceived "brightness" of the sound. It is defined as:

$$C_n = \frac{\sum_{k=0}^{N-1} |X[k]|_n k}{\sum_{k=0}^{N-1} k} \tag{2.9}$$

where n is the frame number to be analyzed, k is the frequency bin number, and $|X(k)|_n$ is the corresponding magnitude spectrum.

The spectral rolloff is defined as the frequency R_n below which 85% of the energy distribution of the magnitude spectrum is concentrated:

$$\sum_{n=0}^{R_n-1} = 0.85 * \sum_{n=0}^{N-1} |X[k]|_n \qquad (2.10)$$

2.2.2 Mel-Frequency Cepstral Coefficients

The Mel-frequency cepstral coefficients (MFCCs) [10] are the most common representation used in automatic speech recognition systems and have been frequently used for music modeling. Their computation consists of three stages: (1) Mel-scale filterbank, (2) Log energy computation, and (3) discrete cosine transform.

A computationally inexpensive method of computing a filterbank is to perform the filtering by grouping STFT bins using triangular windows with specific center frequencies and bandwidths. The result is a single energy value per STFT frame corresponding to the output of each subband. The most common implementation of MFCC is calculated using 13 linearly spaced filters separated by 133.33 Hz between their center frequencies, followed by 27 log-spaced filters (separated by a factor of 1.0711703 in frequency) resulting in 40 filterbank values for each STFT frame.

The next step consists of computing the logarithm of the magnitude of each of the filterbank outputs. This can be viewed as a simple step of dynamic compression, making feature extraction less sensitive to variations in dynamics.

The final step consists of reducing the dimensionality of the 40 filterbank outputs by performing a discrete cosine transform (DCT) which is similar to the discrete Fourier transform but uses only real numbers. It expresses a finite set of values in terms of a sum of cosine functions oscillating at different frequencies. The DCT is used frequently in compression applications because for typical signals of interest it tends to concentrate most of the signal information in few of the lower-frequency components and therefore higher-frequency components can be discarded for compression purposes. In the "classic" MFCC implementation the lower 13 coefficients are retained after transforming the 40 logarithmically compressed Mel filterbank outputs.

2.2.3 Other Timbral Features

There have been many other features proposed to describe short-term timbral information. In this subsection we briefly mention them and provide citations to where the details of their calculation can be found. Time-domain zero-crossings can be used to measure how noisy is the signal and also somewhat correlate to high-frequency content [56, 5]. Spectral bandwidth [5, 38, 34] and octave-based spectral contrast [27, 34] are other features that attempt to summarize the magnitude spectrum. The spectral flatness measure and spectral crest factor [1] are low-level descriptors proposed in the context of the MPEG-7 standard [7].

Daubechies Wavelet Coefficient Histogram (DWCH) is a technique for audio feature extraction based on the Discrete Wavelet Transform (DWT) [32]. It is applied in 3 seconds of audio using the db_8 Daubechies wavelet filter [9] with seven levels of decomposition. After the decomposition, the histograms of the coefficients in each band are calculated. Finally, each histogram is characterized by its mean, variance, and skewness as well as the subband energy defined as the mean of the absolute coefficient values. The result is 7 (subbands) * 4 (quantities per subband) = 28 features.

One of the major innovations that enabled the explosive growth of music represented digitally is perceptual audio compression [40]. Audio compression is frequently used to reduce the high data requirements of audio and music. Audio data does not compress well using generic data compression algorithms so specialized algorithms have been developed. The majority of audio coding algorithms are lossy meaning that the original data can not be reconstructed exactly from the compressed signal. They are frequently based on some form of time-frequency representation and they achieve compression by allocating different number of bits to different parts of the signal. One of the key innovations in audio coding is the use of psychoacoustics (i.e., the scientific study of sound perception by humans) to guide this process so that any artifacts introduced by the compression are not perceptible.

There has been some interest in computing audio features directly in the compressed domain as part of the decompression process. Essentially, this takes advantage of the fact that a type of time-frequency analysis has already been performed for encoding purposes and it is not necessary to repeat it after decompression. The type of features used are very similar to the ones we have described except for the fact that they are computed directly on the compressed data. Early works mainly focused on the MPEG audio compression standard and the extraction of timbral texture features [55, 45]. More recently, the use of newer sparse overcomplete representations as the basis for audio feature extraction has been explored [47] covering timbral texture, rhythmic content, and pitch content.

2.2.4 Temporal Summarization

The features that have been described so far characterize the content of a very short segment of music audio (typically around 20–40 milliseconds) and can be viewed as different ways of summarizing frequency information. Frequently, a second step of temporal summarization is performed to characterize the evolution of the feature vectors over a longer time frame of around 0.5 to 1 seconds. We can define a "texture" window $T_M[n]$ of size M corresponding to analysis frame n as the sequence of the previous $M - 1$ computed feature vectors including the feature vector corresponding to n:

$$T_M[n] = F[n - M + 1]...F[n] \qquad (2.11)$$

Temporal integration is performed by summarizing the information contained in the texture window to a single feature vector. In other words, the sequence of M past feature vectors is "collapsed" to a single feature vector. At the one extreme, the texture window can be advanced by one analysis frame at a time in which case the resulting sequence of temporally summarized feature vectors has the same sampling rate as the original sequence of feature vectors [56]. At the other extreme, temporal integration can be performed across the entire length of the audio clip of interest resulting in a single feature vector representing the clip (sometimes such features are termed *song level* [35] or *aggregate features* [5]). Figure 2.4 shows schematically a typical feature extraction process starting with a time-frequency representation based on the STFT, followed by the calculation of MFCC (frequency) summarization and ending with summarization of the resulting feature vector sequence over the texture window.

There is no consistent terminology describing temporal summarization. Terms that have been used include: *dynamic features* [44, 36], *aggregate features* [5], *temporal statistics* [38], *temporal feature integration* [37], *fluctuation patterns* [46], and *modulation features* [30].

Statistical moments such as the mean, standard deviation, and the covariance matrix can be used to summarize the sequence of feature vectors over the texture window into a single vector. For example a common approach is to compute the means and variances (or standard deviations) of each feature over the texture window [56]. Figure 2.5 shows the original trajectory of spectral centroids over a texture window as well as the trajectory of the running mean and standard deviation of the spectral centroid for two pieces of music.

Another alternative is to use the upper triangular part of the covariance matrix as the feature vector characterizing the texture window [35]. Such statistics capture the variation of feature vectors within a texture window but do not directly represent temporal information.

Another possibility is to utilize techniques from multivariate time-series modeling to characterize the feature vector sequence that better preserve temporal information. For example, multivariate autoregressive models have been used to model temporal feature evolution [37]. The temporal evolution of the feature vectors can also be characterized by performing frequency analysis on their trajectories over time and analyzing their periodicity characteristics. Such modulation features show how the feature vectors change over time and are typically calculated at rates that correspond to rhythmic events. Any method of time-frequency analysis can be used to calculate modulation features but a common choice is the short-time Fourier transform.

As a representative example of calculating modulation frequencies we briefly describe the approach proposed by McKinney and Breebaart [36]. A standard MFCC computation is performed resulting in a sequence of 64 feature vectors each with 13 dimensions characterizing a texture window of 743

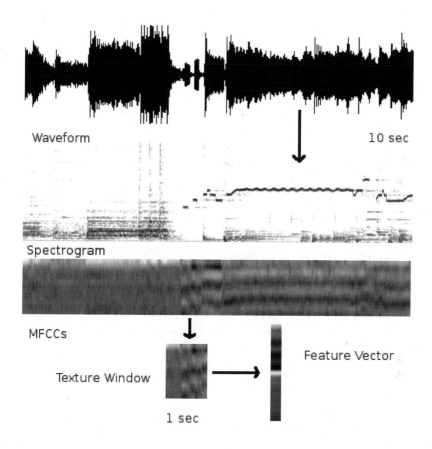

Waveform

10 sec

Spectrogram

MFCCs

Texture Window

1 sec

Feature Vector

Figure 2.4
Feature extraction showing how frequency and time summarization with a texture window can be used to extract a feature vector characterizing timbral texture.

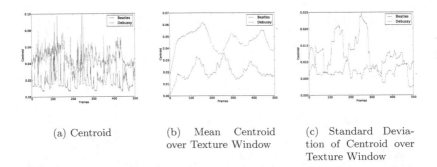

(a) Centroid   (b)   Mean   Centroid   (c) Standard Devia-
                            over Texture Window   tion of Centroid over
                                                         Texture Window

Figure 2.5
The time evolution of audio features is important in characterizing musical
content. The time evolution of the spectral centroid for two different 30-second
excerpts of music is shown in (a). The result of applying a moving mean
and standard deviation calculation over a texture window of approximately 1
second is shown in (b) and (c).

milliseconds of audio. A power spectrum of size 64 using a DFT is calculated
for each of the 13 coefficients/dimensions resulting in 13 power spectra which
are then summarized in four frequency bands ($0\ Hz$, $1 - 2\ Hz$, $3 - 15\ Hz$,
$20 - 43\ Hz$) that roughly correspond to musical beat rates, speech syllabic
rates, and modulations contributing to perceptual roughness. So, the final
representations for the 13×64 matrix of feature values of the texture window
is $4 \times 13 = 52$ dimensions.

A final consideration is the size of the texture windows and the amount of
overlap between them. Although the most common approach is to use fixed
window and hop sizes there are alternatives. Aligning texture windows to
note events can provide more consistent timbral information [60]. In many
music analysis applications such as cover song detection it is desired to obtain
an audio feature representation that is to some extent tempo invariant. One
way to achieve this is using so-called beat-synchronous features which as their
name implies are calculated using estimated beat locations as boundaries [14].
Although more commonly used with features that describe pitch content they
can also be used for timbral texture modeling.

2.2.5   Song-Level Modeling

Frequently, in music data mining the primary unit of consideration is an entire
track or excerpt of a piece of music. The simplest and most direct type of representation is a single feature vector that represents the entire piece of music under consideration. This is typically accomplished by temporal summarization

techniques such as the ones described in the previous section applied to the entire sequence of feature vectors. In some cases, two stages of summarization are performed: one at the texture window level and one across the song [56]. The effectiveness of different parameter choices and temporal summarization methods has also been explored [5].

In other music data mining problems the entire sequence of feature vectors is retained. These problems typically deal with the internal structure within a music piece rather than the relations among a collection of pieces. For example in audio segmentation [16, 54] algorithms the locations in time where the musical content changes significantly (such the transition from a singing part to an electric guitar solo) are detected. Audio structure analysis goes a step further and detects repeating sections of a song and their form such as the well-known AABA form [8]. Sequence representations are also used in cover song detection [14].

A representation that is frequently utilized in structure analysis is the self-similarity matrix. This matrix is calculated by calculating pair-wise similarities between feature vectors $\mathbf{v_i}$ and $\mathbf{v_j}$ derived from audio analysis frames i and j.

$$s(i, j) = sim(\mathbf{v_i}, \mathbf{v_j}) \qquad (2.12)$$

where sim is a function that returns a scalar value corresponding to some notion of similarity (or symmetrical distance) between the two feature vectors. Note that music is generally self-similar with regular structure and repetitions, which can be revealed through the self-similarity matrix.

Figure 2.6 shows an example of a self-similarity matrix calculated for a piece of HipHop/Jazz fusion using simply energy contours (shown to the left and bottom of the matrix). The regular structure at the beat and measure level as well as some sectioning can be observed.

The final variation in song-level representations of audio features is to model each music piece as a distribution of feature vectors. In this case, a parametric distribution form (such as a Gaussian Mixture Model [2]) is assumed and its parameters are estimated from the sequence of feature vectors. Music data mining tasks such as classification are performed by considering distance functions between distributions typically modeled as mixture models such as the KL-divergence or Earth Mover's Distance [35, 26].

2.3 Rhythm Features

Automatically extracting information related to rhythm is also an important component of audio MIR systems and has been an active area of research for over 20 years. A number of different subproblems within this area have been identified and explored. The most basic approach is finding the average tempo of the entire recording which can be defined as the frequency with which a

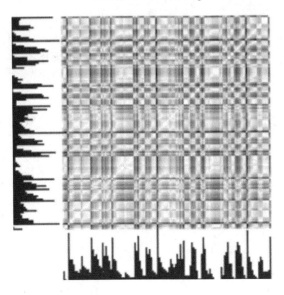

Figure 2.6
Self-similarity matrix using RMS contours.

human would tap their foot while listening to the same piece of music. The more difficult task of beat tracking consists of estimating time-varying tempo (frequency) as well as the locations in time of each beat (phase). Rhythmic information is hierarchical in nature and tempo is only one level of the hierarchy. Other levels frequently used and estimated by audio MIR systems are tatum (defined as the shortest commonly occurring time interval), beat or tactus (the preferred human tapping tempo), and bar or measure. For some MIR applications such as automatic classification it is possible to use a a representation that provides a salience value for every possible tempo, for example, the beat histograms described in [56]. Rhythm analysis approaches can be characterized in different ways. The first and most important distinction is by the type of input: most of the earlier beat tracking algorithms used a symbolic representation while audio signals have been used more recently. Symbolic algorithms can still be utilized with audio signals provided an intermediate transcription step is performed, typically audio onset detection. Another major distinction between the algorithms is the broad approach used which includes rule-based, autocorrelative, oscillating filters, histogramming, multiple agent, and probabilistic. A good overview of these approaches can be found in Chapter 4 of Klapuri and Davy [23].

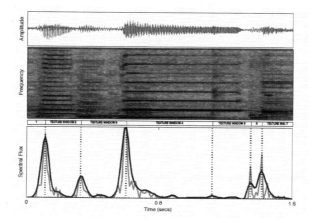

Figure 2.7

Time-domain representation and onset detection function. The top panel depicts the time-domain representation of a fragment of a polyphonic jazz recording, below which is displayed its corresponding spectrogram. The bottom panel plots both the onset detection function $SF(n)$ (gray line), as well as its filtered version (black line). The automatically identified onsets are represented as vertical dotted lines.

2.3.1 Onset Strength Signal

Frequently, the first step in rhythm feature extraction is the calculation of the onset strength signal. The goal is to calculate a signal that has high values at the onsets of musical events. By analyzing the onset strength signal to detect common recurring periodicities it is possible to perform tempo induction, beat tracking as well as other more sophisticated forms of rhythm analysis. Other names used in the literature include onset strength function and novelty curve.

The onset detection algorithm described is based on a recent tutorial article by Dixon [12], where a number of onset detection algorithms were reviewed and compared on two data sets. Dixon concluded that the use of a spectral flux detection function for onset detection resulted in the best performance versus complexity ratio.

The spectral flux as an onset detection function is defined as:

$$SF(n) = \sum_{k=0}^{N/2} H(|X(n,k)| - |X(n-1,k)|) \qquad (2.13)$$

where $H(x) = \frac{x+|x|}{2}$ is the half-wave rectifier function, $X(n,k)$ represents the k-th frequency bin of the n-th frame of the power magnitude (in dB) of the short-time Fourier transform, and N is the corresponding Hamming window size. The bottom panel of Figure 2.7 plots the values over time of the onset detection function $SF(n)$ for an jazz excerpt example.

The onsets are subsequently detected from the spectral flux values by a causal peak picking algorithm, that attempts to find local maxima as follows. A peak at time $t = \frac{nH}{fs}$ is selected as an onset if it satisfies the following conditions:

$$SF(n) \geq SF(k) \quad \forall k : n - w \leq k \leq n + w \tag{2.14}$$

$$SF(n) > \frac{\sum_{k=n-w}^{k=n+w} SF(k)}{mw + w + 1} \times thres + \delta \tag{2.15}$$

where $w = 6$ is the size of the window used to find a local maximum, $m = 4$ is a multiplier so that the mean is calculated over a larger range before the peak, $thres = 2.0$ is a threshold relative to the local mean that a peak must reach in order to be sufficiently prominent to be selected as an onset, and $\delta = 10^{-20}$ is a residual value to avoid false detections on silent regions of the signal. All these parameter values were derived from preliminary experiments using a collection of music signals with varying onset characteristics.

As a way to reduce the false detection rate, the onset detection function $SF(n)$ is smoothed (see bottom panel of Figure 2.7), using a Butterworth filter defined as:

$$H(z) = \frac{0.1173 + 0.2347z^{-1} + 0.1174z^{-2}}{1 - 0.8252z^{-1} + 0.2946z^{-2}} \tag{2.16}$$

In order to avoid phase distortion (which would shift the detected onset time away from the $SF(n)$ peak) the input data is filtered in both the forward and reverse directions. The result has precisely zero-phase distortion, the magnitude is the square of the filter's magnitude response, and the filter order is double the order of the filter specified in the equation above.

2.3.2 Tempo Induction and Beat Tracking

Many pieces of music are structured in time on top of an underlying semi-regular sequence of pulses frequently accentuated by percussion instruments especially in modern popular music. The basic tempo of a piece of music is the rate of musical beats/pulses in time, sometimes also called the *foot-tapping* rate for obvious reasons. Tempo induction refers to the process of estimating the tempo of an audio recording. Beat tracking is the additional related problem of locating the positions in time of the associated beats.

In this section we describe a typical method for tempo induction as a representative example and provide pointers to additional literature in the topic. The first step is the calculation of the onset strength signal as described above. Figure 2.8 shows an example of an onset strength signal for a piece of HipHop/Jazz fusion showing periodicities at the beat and measure level. The

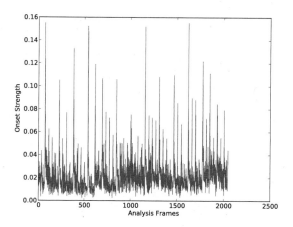

Figure 2.8
Onset strength signal.

autocorrelation of the onset strength signal will exhibit peaks at the lags corresponding to the periodicities of the signal. The autocorrelation values can be warped to form a beat histogram which is indexed by tempos in beats-per-minute (BPM) and has values proportional to the sum of autocorrelation values that map to the same tempo bin. Typically, either the highest or the second highest peak of the beat histogram corresponds to the tempo and can be selected with peak picking heuristics. Figure 2.9 shows two example beat histograms from 30-second clips of HipHop Jazz (left) and Bossa Nova (right). As can be seen in both histograms the prominent periodicities or candidate tempos are clearly visible. Once the tempo of the piece is identified the beat locations can be calculated by locally fitting tempo hypothesis with regularly spaced peaks of the onset strength signal. There has been a systematic evaluation of different tempo induction methods [20] in the context of the Music Information Retrieval Evaluation Exchange (MIREX) [13].

Frequently, a subband decomposition is performed so that periodicities at different frequency ranges can be identified. For example, the bass drum sound will mostly affect low frequency whereas a snare hit will affect all frequencies. Figure 2.10 shows a schematic diagram of the beat histogram calculation method described originally by Tzanetakis and Cook [56]. A discrete wavelet transform filterbank is used as the front-end, followed by multiple channel envelope extraction and periodicity detection.

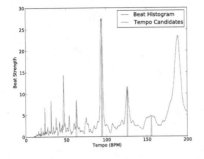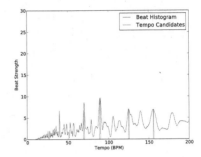

Figure 2.9
Beat histograms of HipHop/Jazz and Bossa Nova.

2.3.3   Rhythm Representations

The beat histogram described in the previous section can be viewed as a
song-level representation of rhythm. In addition to the tempo and related periodicities, the total energy and/or height of peaks represent the amount of
self-similarity that the music has. This quantity has been termed *beat strength*
and has been shown to be a perceptually salient characteristic of rhythm. For
example, a piece of rap music with a tempo of 80 BPM will have more beat
strength than a piece of country music at the same tempo. The spread of
energy around each peak indicates the amount of tempo variations and the
ratios between the tempos of the prominent peaks give hints about the time
signature of the piece. A typical approach is to further reduce the dimensionality of the beat histogram by extracting characteristic features such as the
location of the highest peak and its corresponding height [56]. A thorough
investigation of various features for characterizing rhythm has been presented
by Gouyon et al. [19].

An alternative method of computing a very similar representation to the
beat histogram is based on the self-similarity matrix and termed the *beat spectrum* [17]. Another approach models the rhythm characteristics of patterns as
a sequence of audio features over time [42]. A dynamic time warping algorithm can be used to align the time axis of the two sequences and allow their
comparison. Another more recent approach is to identify rhythmic patterns
that are characteristic of a particular genre automatically [52] and then use
their occurrence histogram as a feature vector.

One interesting question is whether rhythm representations should be
tempo invariant or variant. To some extent, the answer depends on the task.
For example if one is trying to automatically classify speed metal (a genre of
rock music) pieces then absolute tempo is a pretty good feature to include.
On the other hand, classifying something as a waltz has more to do with
the ratios of periodicities rather than absolute tempo. Representations that

BEAT HISTOGRAM CALCULATION FLOW DIAGRAM

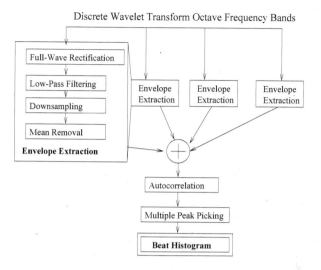

Figure 2.10
Beat histogram calculation.

are to some degree tempo invariant have also been explored. Dynamic periodicity wrapping is a dynamic programming technique used to align average periodicity spectra obtained from the onset strength signal [24]. Tempo invariance is achieved through the alignment process. The fast Melin transform (FMT) is a transform that is invariant to scale. It has been used to provide a theoretically tempo invariant (under the assumption that tempo is scaled uniformly throughout the piece) representation by taking the FMT of the autocorrelation of the onset strength function [25]. An exponential grouping of the lags of the autocorrelation function of the onset strength signal can also be used for a tempo-invariant representation [62]. Beat histograms can also be used as the basis for a tempo invariant representation by using a logarithmically spaced lag-axis [22]. The algorithm requires the estimation of a reference point. Experiments comparing four periodicity representations in the spectral or temporal domains using the autocorrelation and the discrete Fourier transform of the onset strength signal have also been conducted [43].

2.4   Pitch/Harmony Features

In other cases, for example in cover song identification or automatic chord detection, it is desired to have a representation that is related to the pitch content of the music rather than the specifics of the instruments and voices that are playing. Conceivably, a fully automatically transcribed music score could be used for this purpose. Unfortunately, current automatic transcription technology is not robust enough to be used reliably.

Instead, the most common pitch content representation is the Pitch and Pitch Class Profile (PCP) (other alternative names used in literature are pitch histograms and chroma vectors). The pitch profile measures the occurrence of specific discrete musical pitches in a music segment and the pitch class profile considers all octaves equivalent essentially folding the pitch profile into 12 pitch classes. The pitch profile and pitch class profile are strongly related to the underlying harmony of the music piece. For example, a music segment in C major is expected to have many occurrences of the discrete pitch classes C, E, and G that form the C major chord. These representations are used for a variety of tasks including automatic key identification [49, 31], chord detection [49, 31], cover song detection [14, 48, 33], and polyphonic audio-score alignment [39].

There are two major approaches to the computation of the PCP. The first approach directly calculates the PCP by appropriately mapping and folding all the magnitude bins of a fast Fourier transform. The terms *chroma* and *chromagram* are used to describe this process [4]. Each FFT bin is mapped to its closest note, according to:

$$f(p) = 440 * 2^{p-69/12} \tag{2.17}$$

where p is the note number. This is equivalent to segmenting the magnitude spectrum into note regions. The average magnitude within each region is calculated resulting in a pitch histogram representation. The pitch histogram is then folded so that all octaves map to the same pitch class resulting into a vector of size 12. The FFT-based calculation of chroma has the advantage that it is efficient to compute and has consistent behavior throughout the song. However, it is affected by nonpitched sound sources such as drums and the harmonics of pitch sound sources are mapped to pitches which reduces the potential precision of the representation.

An alternative is to utilize multiple pitch detection to calculate the pitch and pitch class profiles. If the multiple pitch detection was perfect then this approach would eliminate the problems of chroma calculation; however, in practice pitch detection especially in polyphonic audio is not particularly robust. The errors tend to average out but there is no consensus about whether the pitch-based or FFT-based approach is better. Multiple-pitch detection typically operates in an incremental fashion. Initially, the dominant pitch is detected and the frequencies corresponding to the fundamental and the associated harmonics are removed for example by using comb filtering. Then the process is repeated to detect the second most dominant pitch.

Pitch detection has been a topic of active research for a long time mainly due to its importance in speech processing. In this chapter we briefly outline two common approaches. The first approach is to utilize time-domain autocorrelation of the signal [6]:

$$R(\tau) = \frac{1}{N} \sum_{n=0}^{N-1-m} x[n]x[n+m] \quad 0 \le m < M \qquad (2.18)$$

An alternative used in the YIN pitch extraction method [11] is based on the difference function:

$$dt = \sum_{n=0}^{N-1} (x[n] - x[n+\tau])^2 \qquad (2.19)$$

The dips in the difference function correspond to periodicities. In order to reduce the occurrence of subharmonic errors, YIN employs a cumulative mean function which de-emphasizes higher period dips in the difference function.

Independently of their method of calculation, PCPs can be calculated using a fixed size analysis window. This occasionally will result in inaccuracies when a window contains information from two chords. The use of beat synchronous features [14] can help improve the results as by considering windows that are aligned with beat locations it is more likely the the pitch content information remains stable within an analysis window.

2.5 Other Audio Features

In addition to timbral texture, rhythm, and pitch content there are other facets of musical content that have been used as the basis for audio feature extraction. Another important source of information is the instrumentation of audio recordings i.e., what instruments are playing at any particular time of an audio recording. For example it is more likely that a saxophone will be part of a jazz recording than a recording of country music although of course there are exceptions. The classification of musical instrument in a polyphonic context has been explored [15] but so far has not been evaluated in the context of other music data mining tasks such as classification.

Until now, all the features described characterize the mixture of sound sources that comprise the musical signal. Another possibility is to characterize individually each sound source. Feature extraction techniques for certain types of sound sources have been proposed but they are not widely used partly due to the difficulty of separating and/or characterizing individual sound source in a complex mixture of sounds such as music. An example is automatic singer identification. The most basic approach is to use features developed for voice/speech identification directly on the mixed audio signal [63]. More sophisticated approaches first attempt to identify the parts of the signal containing vocals and in some cases even attempt to separately characterize the singing voice and reduce the effect of accompaniment [18]. Another type of sound source that has been explored for audio feature extraction is the bass line [28, 53].

In many modern pop and rock recordings each instrument is recorded separately and the final mix is created by a recording producer/engineer(s) who among other transformations add effects such as reverb or filtering and spatialize individual tracks using stereo panning cues. For example the amount of stereo panning and placement of sources remains constant in older recordings that tried to reproduce live performances compared to more recent recordings that would be almost impossible to realize in a live setting. Stereo panning features have recently been used for audio classification [57, 58].

2.6   Musical Genre Classification of Audio Signals

In the last 10 years musical genre classification of audio signals has been widely studied and can be viewed as a canonical problem in which audio feature extraction has been used. Frequently, new approaches to audio feature extraction are evaluated in the context of musical genre classification. They include sparse overcomplete representations [47] as well as bio-inspired joint acoustic and modulation-frequency representations [41].

Audio classification has a long history in the area of Speech Recognition but has only recently been applied to music. Even though there was some earlier work [29, 51] a good starting point for audio-based musical genre classification is the system that was proposed in 2002 by Tzanetakis [56].

Once the audio features are extracted they need to be used to "train" a classifier using supervised learning techniques. This is accomplished using all the labeled audio feature representations for a training collection track. If the audio feature representation has been summarized over the entire track to a single high-dimensional feature vector then this corresponds to the "classic" formulation of classification and any machine-learning classifier can be used directly. Examples of classifiers that have been used in the context of audio-based music classification include: Gaussian Mixture Models [3, 56], support vector machines [35], and AdaBoost [5]. Another alternative is to perform classification in smaller segments and then aggregate the results using majority voting. A more complicated approach (frequently called bag-of-frames) consists of modeling each track using distribution estimation methods, for example, a Gaussian Mixture Model trained using the EM-algorithm. In this case, each track is represented as a probability distribution rather than a single high-dimensional point (feature vector). The distance between the probability distributions can be estimated for example using KL-divergence or approximations of it for example using the Monte-Carlo method depending on the particular parametric form used for density estimation. By establishing a distance metric between tracks it is possible to perform retrieval and classification by simple techniques such as k-nearest neighbors [3].

Evaluation of classification is relatively straightforward and in most ways identical to any classification task. The standard approach is to compare the predicted labels with ground truth labels. Common metrics include classification accuracy as well as retrieval metrics such as precision, recall, and f-measure. When retrieval metrics are used it is assumed that for a particular query relevant documents are the tracks with the same genre label. Cross-validation is a technique frequently used in evaluating classification where the labeled data is split into training and testing sets in different ways to ensure that the metrics are not influenced by the particular choice of training and testing data. One detail that needs to be taken into account is the so-called album effect in which classification accuracy improves when tracks from the same album are included in both training and testing data. The cause is recording production effects that are common between tracks in the same album. The typical approach is to ensure that when performing cross-validation tracks from the same album or artist only go to either the training or testing data set.

Classification accuracy on the same data set and using the same cross-validation approach can be used for comparing the relative performance of different algorithms and design choices. Interpreting the classification accuracy in absolute terms is trickier because of the subjective nature of genre labeling as has already been discussed in the section on ground truth labeling. In the early years of research in audio-based musical genre classification each research group utilized different data sets, cross-validation approaches, and metrics making it hard to draw any conclusions about the merits of different approaches. Sharing data sets is harder due to copyright restrictions. The Music Information Retrieval Evaluation Exchange [13] is an annual event where different MIR algorithms are evaluated using a variety of metrics on different tasks including several audio-based classification tasks. The participants submit their algorithms and do not have access to the data which addresses both the copyright problem and the issue of overfitting to a particular data set. Table 2.1 shows representative results of of the best performing system in different audio classification tasks from MIREX 2009. With the exception of audio tag classification all the results are percentages of classification accuracy. For audio tag classification, the average f-measure is used instead.

Genre Classification	66.41
Genre Classification (Latin)	65.17
Audio Mood Classification	58.2
Artist Identification	47.65
Classical Composer Identification	53.25
Audio Tag Classification	0.28

Table 2.1
Audio-Based Classification Tasks for Music Signals (MIREX 2009)

Name	URL	Programming Language
Auditory Toolbox	tinyurl.com/3yomxwl	MATLAB
CLAM	clam-project.org/	C++
D. Ellis Code	tinyurl.com/6cvtdz	MATLAB
HTK	htk.eng.cam.ac.uk/	C++
jAudio	tinyurl.com/3ah8ox9	Java
Marsyas	marsyas.info	C++/Python
MA Toolbox	pampalk.at/ma/	MATLAB
MIR Toolbox	tinyurl.com/36500jm	MATLAB
Sphinx	cmusphinx.sourceforge.net/	C++
VAMP Plugins	www.vamp-plugins.org/	C++

Table 2.2
Software for Audio Feature Extraction

2.7 Software Resources

Although frequently researchers implement their own audio feature extraction algorithms there are several software collections that are freely available that contain many of the methods described in this chapter. They have enabled researchers more interested in the data mining and machine-learning aspects of music analysis to build systems more easily. They differ in the programming language/environment they are written, the computational efficiency of the extraction process, their ability to deal with batch processing of large collections, their facilities for visualizing feature data, and their expressiveness/flexibility in describing complex algorithms.

Table 2.2 summarizes information about some of the most commonly used software resources as of 2010. The list is by no means exhaustive but does provide reasonable coverage of what is available. Several of the figures in this chapter were created using *Marsyas* and some using custom MATLAB™ code.

2.8 Conclusion

Musical signals in the audio domain contain an enormous amount of information that needs to be distilled in order to apply music mining techniques. The most common facets of music that are modeled are timbral texture, rhythmic structure, and pitch content. Most audio features are built on top of time-frequency representations and a large variety of different feature sets have been proposed in the literature to model different aspects of music. It is relatively straightforward using freely available software to extract audio features and perform fundamental music mining tasks such as automatic musical genre classification.

Bibliography

[1] E. Allamanche, J. Herre, O. Hellmuth, B. Froba, and T. Kastner. Content-based identification of audio material using MPEG-7 low-level description. In *International Conference on Music Information Retrieval (ISMIR)*, 2001.

[2] J.J. Aucouturier and F. Pachet. Music similarity measures: What's the use? In *International Conference on Music Information Retrieval (IS-MIR)*, 2002.

[3] J.J. Aucouturier and F. Pachet. Representing musical genre: A state of the art. *Journal of New Music Research*, 32(1):1–2, 2003.

[4] M.A. Bartsch and G.H. Wakefield. To catch a chorus: using chroma-based representations for audio thumbnailing. In *Workshop on Applications of Signal Processing to Audio and Acoustics*, pages 15–18, 2001.

[5] J. Bergstra, N. Casagrande, D. Erhan, D. Eck, and B. Kegl. Aggregate features and AdaBoost for music classification. *Machine Learning*, 65(2–3):473–484, 2006.

[6] P. Boersma. Accurate short-term analysis of the fundamental frequency and the harmonics-to-noise ratio of a sampled sound. *Proceedings of the Institute of Phonetic Sciences*, 17:97–110, 1993.

[7] M. Casey. Sound classification and similarity tools. In B.S. Manjunath, P. Salembier, and T. Sikora, editors, *Introduction to MPEG-7: Multimedia Content Description Language*, pages 309–323. Wiley, Hoboken, New Jersey, 2002.

[8] R. Dannenberg and N. Hu. Pattern discovery techniques for music audio. *Journal of New Music Research*, June 2003:153–164, 2003.

[9] I. Daubechies. Orthonormal bases of compactly supported wavelets. *Communications on Pure and Applied Math*, 41:909–996, 1988.

[10] S. Davis and P. Mermelstein. Experiments in syllable-based recognition of continuous speech. *IEEE Transcactions on Acoustics, Speech and Signal Processing*, 28:357–366, August 1980.

[11] A. de Cheveigne and H. Kawahara. YIN, a fundamental frequency estimator for speech and music. *Journal of Acoustical Society of America*, 111(4):1917–1930, 2002.

[12] S. Dixon. Onset detection revisited. In *International Conference on Digital Audio Effects (DAFx)*, 2006.

[13] J.S. Downie. The music information retrieval evaluation exchange (2005-2007): A window into music information retrieval. *Acoustical Science and Technology*, 29(4):247–255, 2008.

[14] D.P.W. Ellis and G.H Poliner. Identifying cover songs with chroma features and dynamic programming beat tracking. In *International Conference on Acoustics, Speech, and Signal Processing (ICASSP)*, pages IV-1429–IV-1432, 2007.

[15] S. Essid, G. Richard, and B. David. Musical instrument recognition by pair-wise classification. *IEEE Transactions on Audio, Speech and Language Processing*, 14(4):1401–1412, 2006.

[16] J. Foote. Visualizing music and audio using self-similarity. In *ACM Multimedia*, pages 77–80, 1999.

[17] J. Foote, M. Cooper, and U. Nam. Audio retrieval by rhythmic similarity. In *International Conference on Music Information Retrieval (ISMIR)*, 2002.

[18] H. Fujihara, M. Goto, T. Kitahara, and H.G. Okuno. A modeling of singing voice robust to accompaniment sounds and its application to singer identification and vocal-timbre similarity based music information retrieval. *IEEE Transactions on Audio, Speech and Language Processing*, 18(3):638–648, 2010.

[19] F. Gouyon, S. Dixon, E. Pampalk, and G. Widmer. Evaluating rhythmic descriptors for musical genre classification. In *AES 25th International Conference on Metadata for Audio*, pages 196–204, 2002.

[20] F. Gouyon, A. Klapuri, S. Dixon, M. Alonso, G. Tzanetakis, C. Uhle, and P. Cano. An experimental comparison of audio tempo induction algorithms. *IEEE Transactions on Audio, Speech and Language Processing*, 14(5):1832–1844, 2006.

[21] J.M. Grey. Multidimensional perceptual scaling of musical timbres. *Journal of the Acoustical Society of America*, 61(5):1270–1277, 1977.

[22] M. Gruhne, C. Dittmar, and D. Gaertner. Improving rhythmic similarity computation by beat histogram transformations. In *International Conference on Music Information Retrieval*, 2009.

[23] S. Hainsworth. Beat tracking and musical metre analysis. In A. Klapuri and M. Davy, editors, *Signal Processing Methods for Music Transcription*, pages 101–129. Springer, New York, 2006.

[24] A. Holzapfel and Y. Stylianou. Rhythmic similarity of music based on dynamic periodicity warping. In *IEEE International Conference on Acoustics, Speech and Signal Processing (ICASSP)*, pages 2217–2220, 2008.

[25] A. Holzapfel and Y. Stylianou. A scale transform based method for rhythmic similarity of music. In *IEEE International Conference on Acoustics, Speech and Signal Processing (ICASSP)*, pages 317–320, 2009.

[26] J.H. Jensen, D.P.W. Ellis, M.G. Christensen, and S.H. Jensen. Evaluation of distance measures between Gaussian mixture models of MFCCs. In *International Conference on Music Information Retrieval (ISMIR)*, 2007.

[27] D.N. Jiang, L. Lu, H.J. Zhang, and J.H. Tao. Music type classification by spectral contrast feature. In *International Conference on Multimedia and Expo (ICME)*, 2002.

[28] T. Kitahara, Y. Tsuchihashi, and H. Katayose. Music genre classification and similarity calculation using bass-line features. In *10th IEEE International Symposium on Multimedia*, 2008.

[29] T. Lambrou, P. Kudumakis, R. Speller, M. Sandler, and A. Linnery. Classification of audio signals using statistical features on time and wavelet transform domains. In *International Conference on Acoustics, Speech and Signal Processing (ICASSP)*, 1998.

[30] C.H. Lee, J.L. Shih, K.M. Yu, and H.S. Lin. Automatic music genre classification based on modulation spectral analysis of spectral and cepstral features. *IEEE Transactions on Multimedia*, 11(4):670–682, 2009.

[31] K. Lee and M. Slaney. Acoustic chord transcription and key extraction from audio using key-dependent HMMS trained on synthesized audio. *IEEE Transactions on Audio, Speech and Language Processing*, 16(2):291–301, 2008.

[32] T. Li and M. Ogihara. Toward intelligent music information retrieval. *IEEE Transactions on Multimedia*, 8(3):564–574, 2006.

[33] C. Liem and A. Hanjalic. Cover song retrieval: A comparative study of system component choices. In *International Conference on Music Information Retrieval (ISMIR)*, 2009.

[34] L. Lu, D. Lie, and H.-J. Zhang. Automatic mood detection and tracking of music audio signals. *IEEE Transactions on Speech and Audio Processing*, 14(1):5–18, 2006.

[35] M. Mandel and D. Ellis. Song-level features and SVMS for music classification. In *International Conference on Music Information Retrieval (ISMIR)*, 2005.

[36] M. F. McKinney and J. Breebaart. Features for audio and music classification. In *International Conference on Music Information Retrieval (ISMIR)*, 2003.

[37] A. Meng, P. Ahrendt, and J. Larsen. Temporal feature integration for music genre classification. *IEEE Transactions on Audio, Speech and Language Processing*, 5(15):1654–1664, 2007.

[38] F. Morchen, A. Ultsch, M. Thiers, and I. Lohken. Modeling timbre distance with temporal statistics from polyphonic music. *IEEE Transactions on Audio, Speech and Language Processing*, 8(1):81–90, 2006.

[39] G. Tzanetakis, N. Hu, and R.B. Dannenberg. Polyphonic audio matching and alignment for music retrieval. In *Workshop on Applications of Signal Processing to Audio and Acoustics*, 2003.

[40] T. Painer and A. Spanias. Perceptual coding of digital audio. *Proceedings of the IEEE*, 88(4):451–515, 2000.

[41] Y. Panagakis, C. Kotropoulos, and G.C. Arce. Non-negative multilinear principal component analysis of auditory temporal modulations for music genre classification. *IEEE Transactions on Audio, Speech and Language Processing*, 18(3):576–588, 2010.

[42] J. Paulus and A. Klapuri. Measuring the similarity of rhythmic patterns. In *International Conference on Music Information Retrieval (ISMIR)*, 2002.

[43] G. Peeters. Spectral and temporal periodicity representations of rhythm for the automatic classification of music audio signal. *IEEE Transactions on Audio, Speech, and Language Processing*, in press.

[44] G. Peeters, A.L. Burthe, and X. Rodet. Toward automatic music audio summary generation from signal analysis. In *International Conference on Music Information Retrieval (ISMIR)*, 2002.

[45] D. Pye. Content-based methods for management of digital music. In *International Conference on Acoustics, Speech and Signal Processing (ICASSP)*, 2000.

[46] A. Rauber, E. Pampalk, and D. Merkl. The SOM-enhanced JukeBox: Organization and visualization of music collections based on perceptual models. *Journal of New Music Research*, 32(2):193–210, 2003.

[47] E. Ravelli, G. Richard, and L. Daudet. Audio signal representations for indexing in the transform domain. *IEEE Transactions on Audio, Speech, and Language Processing*, 18(3):434–446, 2010.

[48] J. Serr and E. Gómez. Audio cover song identification based on tonal sequence alignment. In *IEEE International Conference on Acoustics, Speech and Signal Processing (ICASSP)*, pages 61–64, 2008.

[49] A. Sheh and D.P.W. Ellis. Chord segmentation and recognition using em-trained hidden Markov models. In *International Conference on Music Information Retrieval (ISMIR)*, 2003.

[50] M. Slaney and R.F. Lyon. On the importance of time—A temporal representation of sound. In S. Beete and M. Cooke, editors, *Visual Representations of Speech Signals*, pages 95–116, J. Wiley and Sons, New York, 1993.

[51] H. Sotlau, T. Schultz, M. Westphal, and A. Waibel. Recognition of music types. In *International Conference on Acoustics, Speech and Signal Processing (ICASSP)*, 1998.

[52] E. Tsunoo. Audio genre classification using percussive pattern clustering combined with timbral features. In *International Conference on Multimedia and Expo (ICME)*, 2009.

[53] E. Tsunoo, N. Ono, and S. Sagayama. Musical bass-line pattern clustering and its application to audio genre classification. In *International Conference on Music Information Retrieval (ISMIR)*, 2009.

[54] G. Tzanetakis and P. Cook. Multi-feature audio segmentation for browsing and annotation. In *Workshop on Applications of Signal Processing to Audio and Acoustics*, 1999.

[55] G. Tzanetakis and P. Cook. Sound analysis using mpeg compressed audio. In *International Conference on Acoustics, Speech and Signal Processing (ICASSP)*, 2000.

[56] G. Tzanetakis and P. Cook. Musical genre classification of audio signals. *IEEE Transactions on Speech and Audio Processing*, 10(5):293–302, 2002.

[57] G. Tzanetakis, R. Jones, and K. McNally. Stereo panning features for classifying recording production style. In *International Conference on Music Information Retrieval (ISMIR)*, 2007.

[58] G. Tzanetakis, L.G. Martins, K. McNally, and R. Jones. Stereo panning information for music information retrieval tasks. *Journal of the Audio Engineering Society*, 58(5):409–417, 2010.

[59] M. Unser. Sampling—50 years after Shannon. *Proceedings of the IEEE*, 88(4):569–587, 2000.

[60] K. West and S. Cox. Finding an optimal segmentation for audio genre classification. In *International Conference on Music Information Retrieval (ISMIR)*, 2005.

[61] R.F. Lyon, A.C. Katsiamis, and E.M. Drakakis. History and future of auditory filter models. In *IEEE International Symposium on Circuits and Systems (ISCAS)*, pages 3809–3812, 2010.

[62] J. Jensen, M. Christensen, and S. Jensen. A temp-sensitive representation of rhythmic patterns. In *European Signal Processing Conference (EUSIPCO)*, 2009.

[63] Y.E. Kim and B. Whitman. Singer identification in popular music recordings using voile coding features. In *International Conference on Music Information Retrieval (ISMIR)*, 2002.

Part II

Classification

3

Auditory Sparse Coding

Steven R. Ness

University of Victoria

Thomas C. Walters

Google Inc.

Richard F. Lyon

Google Inc.

CONTENTS

The concept of sparsity has attracted considerable interest in the field of machine learning in the past few years. Sparse feature vectors contain mostly values of zero and one or a few nonzero values. Although these feature vectors can be classified by traditional machine learning algorithms, such as Support Vector Machines (SVMs), there are various recently developed algorithms that explicitly take advantage of the sparse nature of the data, leading to massive speedups in time, as well as improved performance. Some fields that have benefited from the use of sparse algorithms are finance, bioinformatics, text mining [1], and image classification [4]. Because of their speed, these algorithms perform well on very large collections of data [2]; large collections are becoming increasingly relevant given the huge amounts of data collected and warehoused by Internet businesses.

In this chapter, we discuss the application of sparse feature vectors in the field of audio analysis, and specifically their use in conjunction with preprocessing systems that model the human auditory system. We present early results that demonstrate the applicability of the combination of auditory-based processing and sparse coding to content-based audio analysis tasks.

We present results from two different experiments: a search task in which ranked lists of sound effects are retrieved from text queries, and a music information retrieval (MIR) task dealing with the classification of music into genres.

3.1   Introduction

Traditional approaches to audio analysis problems typically employ a short-window fast Fourier transform (FFT) as the first stage of the processing pipeline. In such systems a short, perhaps 25 ms, segment of audio is taken from the input signal and windowed in some way, then the FFT of that segment is taken. The window is then shifted a little, by perhaps 10 ms, and the process is repeated. This technique yields a two-dimensional spectrogram of the original audio, with the frequency axis of the FFT as one dimension, and time (quantized by the step-size of the window) as the other dimension.

While the spectrogram is easy to compute, and a standard engineering tool, it bears little resemblance to the early stages of the processing pipeline in the human auditory system. The mammalian cochlea can be viewed as a bank of tuned filters the output of which is a set of band-pass filtered versions of the input signal that are continuous in time. Because of this property, fine-timing information is preserved in the output of cochlea, whereas in the spectrogram described above, there is no fine-timing information available below the 10 ms hop-size of the windowing function.

This fine-timing information from the cochlea can be made use of in later stages of processing to yield a three-dimensional representation of audio, the stabilized auditory image (SAI) [11], which is a movie-like representation of sound which has a dimension of "time-interval" in addition to the standard dimensions of time and frequency in the spectrogram. The periodicity of the waveform gives rise to a vertical banding structure in this time interval dimension, which provides information about the sound which is complementary to that available in the frequency dimension. A single example frame of a stabilized auditory image is shown in Figure 3.1.

While we believe that such a representation should be useful for audio analysis tasks, it does come at a cost. The data rate of the SAI is many times that of the original input audio, and as such some form of dimensionality reduction is required in order to create features at a suitable data rate for use in a recognition system. One approach to this problem is to move from

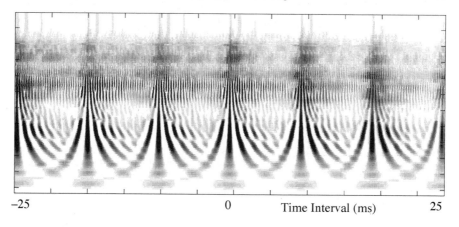

-25   0   Time Interval (ms)   25

Figure 3.1

An example of a single SAI of a sound file of a spoken vowel sound. The vertical axis is frequency with lower frequencies at the bottom of the figure and higher frequencies on the top. The horizontal axis is the autocorrelation lag. From the positions of the vertical features, one can determine the pitch of the sound.

a dense representation of the SAI to a sparse representation, in which the overall dimensionality of the features is high, but only a limit number of the dimensions are nonzero at any time.

In recent years, machine-learning algorithms that utilize the properties of sparsity have begun to attract more attention and have been shown to outperform approaches that use dense feature vectors. One such algorithm is the passive-aggressive model for image retrieval (PAMIR) [4, 6], a machine-learning algorithm that learns a ranking function from the input data, that is, it takes an input set of documents and orders them based on their relevance to a query. PAMIR was originally developed as a machine vision method and has demonstrated excellent results in this field.

There is also growing evidence that in the human nervous system sensory inputs are coded in a sparse manner; that is, only small numbers of neurons are active at a given time [10]. Therefore, when modeling the human auditory system, it may be advantageous to investigate this property of sparseness in relation to the mappings that are being developed. The nervous systems of animals have evolved over millions of years to be highly efficient in terms of energy consumption and computation. Looking into the way sound signals are handled by the auditory system could give us insights into how to make our algorithms more efficient and better model the human auditory system.

One advantage of using sparse vectors is that such coding allows very fast computation of similarity, with a trainable similarity measure [4]. The efficiency results from storing, accessing, and doing arithmetic operations on only the nonzero elements of the vectors. In one study that examined the performance of sparse representations in the field of natural language processing, a 20- to 80-fold speedup over LIBSVM was found [7]. They comment that kernel-based methods, like SVM, scale quadratically with the number of training examples and discuss how sparsity can allow algorithms to scale linearly based on the number of training examples.

In this chapter, we use the stabilized auditory image (SAI) as the basis of a sparse feature representation which is then tested in a sound ranking task and a music information retrieval task. In the sound ranking task, we generate a two-dimensional SAI for each time slice, and then sparse-code those images as input to PAMIR. We use the ability of PAMIR to learn representations of sparse data in order to learn a model which maps text terms to audio features. This PAMIR model can then be used to rank a list of unlabeled sound effects according to their relevance to some text query. We present results that show that in certain tasks our methods can outperform highly tuned FFT-based approaches. We also use similar sparse-coded SAI features as input to a music genre classification system. This system uses an SVM classifier on the sparse features, and learns text terms associated with music. The system was entered into the annual Music Information Retrieval Evaluation Exchange (MIREX 2010) evaluation.

Results from the sound-effects ranking task show that sparse auditory model-based features outperform standard MFCC features, reaching precision about 73% for the top-ranked sound, compared to about 60% for standard MFCC and 67% for the best MFCC variant. These experiments involved ranking sounds in response to text queries through a scalable online machine learning approach to ranking.

3.1.1   The Stabilized Auditory Image

In our system, we have taken inspiration from the human auditory system in order to come up with a rich set of audio features that are intended to more closely model the audio features that we use to listen and process music.

Such fine-timing relations are discarded by traditional spectral techniques. A motivation for using auditory models is that the auditory system is very effective at identifying many sounds. This capability may be partially attributed to acoustic features that are extracted at the early stages of auditory processing. We feel that there is a need to develop a representation of sounds that captures the full range of auditory features that humans use to discriminate and identify different sounds, so that machines have a chance to do so as well.

This SAI representation generates a 2-D image from each section of waveform from an audio file. We then reduce each image in several steps: first cutting the image into overlapping boxes converted to fixed resolution per box; second, finding row and column sums of these boxes and concatenating those into a vector; and finally vector quantizing the resulting medium-dimensionality vector, using a separate codebook for each box position. The Vector Quantization (VQ) codeword index is a representation of a 1-of-N sparse code for each box, and the concatenation of all of those sparse vectors, for all the box positions, makes the sparse code for the SAI image. The resulting sparse code is accumulated across the audio file, and this histogram (count of number of occurrences of each codeword) is then used as input to an SVM [5] classifier[3]. This approach is similar to that of the "bag of words" concept, originally from natural language processing, but used heavily in computer vision applications as "bag of visual words"; here we have a "bag of auditory words," each "word" being an abstract feature corresponding to a VQ codeword. The bag representation is a list of occurrence counts, usually sparse.

3.2 Algorithm

In our experiments, we generate a stream of SAIs using a series of modules that process an incoming audio stream through the various stages of the auditory model. The first module filters the audio using the pole–zero filter cascade (PZFC) [9], then subsequent modules find strobe points in this audio, and generate a stream of SAIs at a rate of 50 per second. The SAIs are then cut into boxes and are transformed into a high-dimensional dense feature vector [12] which is vector quantized to give a high-dimensional sparse feature vector. This sparse vector is then used as input to a machine learning system which performs either ranking or classification. This whole process is shown in diagrammatic form in Figure 3.2.

3.2.1 Pole–Zero Filter Cascade

We first process the audio with the pole–zero filter cascade (PZFC) [9], a model inspired by the dynamics of the human cochlea as shown in Figure 3.3. The PZFC is a cascade of a large number of simple filters with an output tap after each stage. The effect of this filter cascade is to transform an incoming audio signal into a set of band-pass filtered versions of the signal. In our case we used a cascade with 95 stages, leading to 95 output channels. Each output channel is half-wave rectified to simulate the output of the inner hair cells along the length of the cochlea. The PZFC also includes an automatic gain control (AGC) system that mimics the effect of the dynamic compression mechanisms seen in the cochlea. A smoothing network, fed from the output of each channel, dynamically modifies the characteristics of the individual filter stages. The AGC can respond to changes in the output on the timescale of milliseconds, leading to very fast-acting compression. One way of viewing this filter cascade is that its outputs are an approximation of the instantaneous neuronal firing rate as a function of cochlear place, modeling both the frequency filtering and the automatic gain control characteristics of the human cochlea [8]. The PZFC

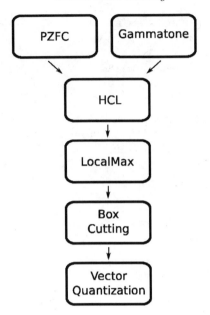

Figure 3.2

A flowchart describing the flow of data in our system. First, either the PZFC or gamma tone filterbank filters the input audio signal. Filtered signals then pass through a half-wave rectification module (HCL), and trigger points in the signal are then calculated by the local-max module. The output of this stage is the SAI, the image in which each signal is shifted to align the trigger time to the zero lag point in the image. The SAI is then cut into boxes with the box-cutting module, and the resulting boxes are then turned into a codebook with the vector-quantization module.

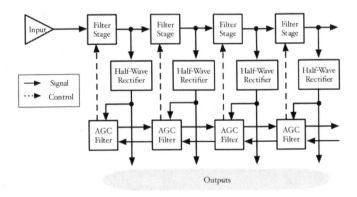

Figure 3.3

The cochlear model, a filter cascade with half-wave rectifiers at the output taps, and an automatic gain control (AGC) filter network that controls the tuning and gain parameters in response to the sound.

parameters used for the sound-effects ranking task are described by Lyon et al. [9]. We did not do any further tuning of this system to the problems of genre, mood, or song classification; this would be a fruitful area for further research.

3.2.2 Image Stabilization

The output of the PZFC filterbank is then subjected to a process of strobe finding where large peaks in the PZFC signal are found. The temporal locations of these peaks are then used to initiate a process of temporal integration whereby the stabilized auditory image is generated. These strobe points "stabilize" the signal in a manner analogous to the trigger mechanism in an oscilloscope. When these strobe points are found, a modified form of autocorrelation, known as strobed temporal integration is performed. The strobed temporal integration is like a sparse version of autocorrelations where only the strobe points are correlated against the signal. Strobed temporal integration has the advantage of being considerably less computationally expensive than full autocorrelation.

3.2.3 Box Cutting

We then divide each image into a number of overlapping boxes using the same process described by Lyon et al. [9]. We start with rectangles of size 16 lags by 32 frequency channels, and cover the SAI with these rectangles, with overlap. Each of these rectangles is added to the set of rectangles to be used for vector quantization. We then successively double the height of the rectangle up to the largest size that fits in an SAI frame, but always reducing the contents of each box back to 16 by 32 values. Each of these doublings is added to the set of rectangles. We then double the width of each rectangle up to the width of the SAI frame and add these rectangles to the SAI frame. The output of this step is a set of 44 overlapping rectangles. The process of box cutting is shown in Figure 3.4.

To reduce the dimensionality of these rectangles, we then take their row and column marginals and join them together into a single vector.

3.2.4 Vector Quantization

The resulting dense vectors from all the boxes of a frame are then converted to a sparse representation by vector quantization.

We first preprocessed a collection of 1,000 music files from 10 genres using a PZFC filterbank followed by strobed temporal integration to yield a set of SAI frames for each file. We then take this set of SAI and apply the box-cutting technique described above followed by the calculation of row and column marginals. These vectors are then used to train dictionaries of 200 entries, representing abstract "auditory words," for each box position, using a k-means algorithm.

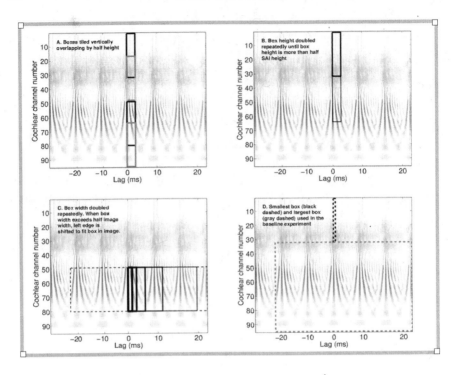

Figure 3.4
Stabilized auditory images. The boxes, or multiscale regions, used to analyze the stabilized auditory images are generated in a variety of heights, widths, and positions.

This process requires the processing of large amounts of data, just to train the VQ codebooks on a training corpus.

The resulting dictionaries for all boxes are then used in the MIREX experiment to convert the dense features from the box-cutting step on the test corpus songs into a set of sparse features where each box was represented by a vector of 200 elements with only one element being nonzero. The sparse vectors for each box were then concatenated, and these long spare vectors are histogrammed over the entire audio file to produce a sparse feature vector for each song or sound effect. This operation of constructing a sparse bag of auditory words was done for both the training and testing corpora.

3.2.5   Machine Learning

For this system, we used the support vector machine-learning system from LIBSVM which is included in the Marsyas [13] framework. Standard Marsyas SVM parameters were used in order to classify the sparse bag of auditory words representation of each song. It should be noted that SVM is not the ideal algorithm for doing classification on such a sparse representation, and if time permitted, we would have instead used the PAMIR machine learning algorithm as described by Lyon et al. [9]. This algorithm has been shown to outperform SVM on ranking tasks, both in terms of execution speed and quality of results.

3.3   Experiments

3.3.1   Sound Ranking

We performed an experiment in which we examined a quantitative ranking task over a diverse set of audio files using tags associated with the audio files.

For this experiment, we collected a data set of 8,638 sound effects, which came from multiple places. Of the sound files, 3,855 were from commercially available sound effect libraries, of these 1,455 were from the BBC sound effects library. The other 4,783 audio files were collected from a variety of sources on the Internet, including findsounds.com, partnersinrhyme.com, acoustica.com, ilovewaves.com, simplythebest.net, wav-sounds.com, wav-source.com, and wavlist.com.

We then manually annotated this data set of sound effects with a small number of tags for each file. Some of the files were already assigned tags and for these, we combined our tags with this previously existing tag information. In addition, we added higher level tags to each file, for example, files with the tags "cat," "dog," and "monkey" were also given the tags "mammal" and "animal." We found that the addition of these higher level tags assist retrieval by inducing structure over the label space. All the terms in our database were stemmed, and we used the Porter stemmer for English, which left a total of 3,268 unique tags for an average of 3.2 tags per sound file.

To estimate the performance of the learned ranker, we used a standard three-fold cross-validation experimental setup. In this scheme, two thirds of the data is used for training and one third is used for testing; this process is then repeated for all three splits of the data and results of the three are averaged. We removed any queries that had fewer than five documents in either the training set or the test set, and if the corresponding documents had no other tags, these documents were removed as well.

To determine the values of the hyperparameters for PAMIR we performed a second level of cross-validation where we iterated over values for the aggressiveness parameter C and the number of training iterations. We found that in general system performance was good for moderate values of C and that lower values of C required a longer training time. For the agressiveness parameter, we selected a value of C = 0.1, a value which was also found to be optimal in other research [6]. For the number of iterations, we chose 10 M, and found that in our experience, the system was not very sensitive to the exact value of these parameters.

We evaluated our learned model by looking at the precision within the top k audio files from the test set as ranked by each query. Precision at top k is a commonly used measure in retrieval tasks such as these and measures the fraction of positive results within the top k results from a query.

The stabilized auditory image generation process has a number of parameters which can be adjusted including the parameters of the PZFC filter and the size of rectangles that the SAI is cut into for subsequent vector quantization. We created a default set of parameters and then varied these parameters in our experiments. The default SAI box cutting was performed with 16 lags and 32 channels, which gave a total of 49 rectangles. These rectangles were then reduced to their marginal values which gives a 48 dimension vector, and a codebook of size 256 was used for each box, giving a total of $49 \times 256 = 12,544$ feature dimensions. Starting from these, we then made systematic variations to a number of different parameters and measured their effect on precision of retrieval. For the box-cutting step, we adjusted various parameters including the smallest sized rectangle, and the maximum number of rectangles used for segmentation. We also varied the codebook sizes that we used in the sparse coding step.

In order to evaluate our method, we compared it with results obtained using a very common feature extraction method for audio analysis, MFCCs (Mel-frequency cepstral coefficients). In order to compare this type of feature extraction with our own, we turned these MFCC coefficients into a sparse code. These MFCC coefficients were calculated with a Hamming window with initial parameters based on a setting optimized for speech. We then changed various parameters of the MFCC algorithm, including the number of cepstral coefficients (13 for speech), the length of each frame (25 ms for speech), and the number of codebooks that were used to sparsify the dense MFCC features for each frame. We obtained the best performance with 40 cepstral coefficients, a window size of 40 ms and codebooks of size 5000. The performance comparison is shown in Table 3.1.

We investigated the effect of various parameters of the SAI feature extraction process on test-set precision, these results are displayed graphically in Figure 3.5 where the precision of the top ranked sound file is plotted against the number of features used. As one can see from this graph, performance saturates when the number of features approaches $\approx 10^5$ which results from the use of 4,000 codewords per codebook, with a total of 49 codebooks. This particular set of parameters led to a performance of 73%, significantly better than the best MFCC result which achieved a performance of 67%, which represents a smaller error of 18% (from 33% to 27% error). It is also notable that SAI can achieve better precision-at-top-k consistently for all values of k, albeit with a smaller improvement in relative precision.

Table 3.2 shows results of three queries along with the top five sound files

Top-k	SAI	MFCC	Percent Error Reduction
1	27	33	18%
2	39	44	12%
5	60	62	4%
10	72	74	3%
20	81	84	4%

Table 3.1

A Comparison of the Best SAI and MFCC Configurations (This table shows the percent error at top-k, where error is defined as 1 − precision.)

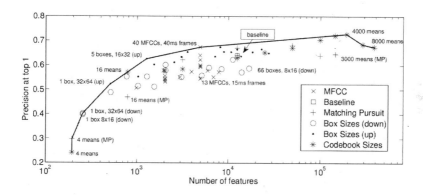

Figure 3.5

Ranking at top-1 retrieved result for all the experimental runs described in this section. A few selected experiment names are plotted next to each point, and different experiments are shown by different icons. The convex hull that connects the best-performing experiments is plotted as a solid line.

that were returned by the best SAI-based and MFCC-based systems. From this table, one can see that the two systems perform in different ways; this can be expected when one considers the basic audio features that these two systems extract. For example, for the query "gulp," the SAI system returns "pouring" and "water-dripping," all three of these share the similarity of involving the movement of water or liquids.

When we calculated performance, it was based on textual tags, which are often noisy and incomplete. Due to the nature of human language and perception, people often use different words to describe sounds that are very similar, for example, a Chopin Mazurka could be described with the words "piano," "soft," "classical," "Romantic," and "mazurka." To compound this difficulty, a song that had a female vocalist singing could be labelled as "woman," "women," "female," "female vocal," or "vocal." This type of multilabel

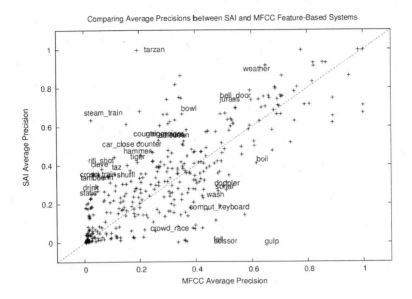

Figure 3.6

A comparison of the average precision of the SAI- and MFCC-based systems. Each point represents a single query, with the horizontal position being the MFCC average precision and the vertical position being the SAI average precision. More of the points appear above the y = x line, which indicates that the SAI based system achieved a higher mean average precision.

problem is common in the field of content-based retrieval. It can be alleviated by a number of techniques, including the stemming of words, but due to the varying nature of human language and perception, will continue to remain an issue.

In Figure 3.6, the performance of the SAI- and MFCC-based systems are compared to each other with respect to their average precision. A few select full tag names are placed on this diagram; for the rest, only a plus is shown. This is required because otherwise the text would overlap to such a great degree that it would be impossible to read.

In this diagram we plot the average precision of the SAI-based system against that of the MFCC-based system, with the SAI precision shown along the vertical axis and the MFCC precision shown along the horizontal axis. If the performance of the two systems was identical, all points would lie on the line y = x. Because more points lie above the line than below the line, the performance of the SAI-based system is better than that of the MFCC-based system.

Query	SAI File (labels)	MFCC File (labels)
tarzan	Tarzan-2 (tarzan, yell) (tarzan, yell)	TARZAN (tarzan, yell)
	tarzan2 (tarzan, yell)	175 orgs (steam, whistle)
	203 (tarzan)	mosquito-2 2 (mosquito)
	wolf (mammal, wolves, wolf, ...)	evil-witch-laugh (witch, evil, laugh)
	morse (morse, code)	Man-Screams (horror, scream, man)
applause audience	27-Applause-from-audience 30-Applause-from-audience	26-Applause-from-audience phase1 (trek, phaser, star)
	golf50 (golf)	fanfare2 (fanfare, trumpet)
	firecracker	45-Crowd-Applause (crowd, applause)
	53-ApplauseLargeAudienceSFX	golf50
gulp	tite-flamn (hit, drum, roll)	GULPS (gulp, drink)
	water-dripping (water, drip)	drink (gulp, drink)
	Monster-growling (horror, monster, growl)	california-myotis-search (blip)
	Pouring (pour,soda)	jaguar-1 (bigcat, jaguar, mammal, ...)

Table 3.2
The Top Documents That Were Obtained for Queries Which Performed Significantly Differently between the SAI and MFCC Feature-Based Systems.

3.3.2 MIREX 2010

All of these algorithms were then ported to the Marsyas music information retrieval framework from AIM-C, and extensive tests were written as described above. These algorithms were submitted to the MIREX 2010 competition as C++ code, which was then run by the organizers on blind data. As of this date, only results for two of the four train/test tasks have been released. One of these is for the task of classifying classical composers and the other is for classifying the mood of a piece of music. There were 40 groups participating in this evaluation, the most ever for MIREX, which gives some indication about how this classification task is increasingly important in the real world. Below we present the results for the best entry, the average of all entries, our entry, and the other entry for the Marsyas system. It is instructive to compare our

Algorithm	Classification Accuracy
SAI/VQ	0.4987
Marsyas MFCC	0.4430
Best	0.6526
Average	0.455

Table 3.3
Classical Composer Train/Test Classification Task

Algorithm	Classification Accuracy
SAI/VQ	0.4861
Marsyas MFCC	0.5750
Best	0.6417
Average	0.49

Table 3.4
Music Mood Train/Test Classification Task

results to that of the standard Marsyas system because in large part we would like to compare the SAI audio feature to the standard MFCC features, and since both of these systems use the SVM classifier, we partially negate the influence of the machine learning part of the problem.

For the classical composer task, the results are shown in Table 3.3 and for the mood classification task, results are shown in Table 3.4.

From these results we can see that in the classical composer task we outperformed the traditional Marsyas system which has been tuned over the course of a number of years to perform well. This gives us the indication that the use of these SAI features has promise. However, we underperform the best algorithm, which means that there is work to be done in terms of testing different machine learning algorithms that would be better suited to this type of data. However, in a more detailed analysis of the results, which is shown in Figure 3.7, it is evident that each of the algorithms has a wide range of performance on different classes. This graph shows that the most well predicted in our SAI/VQ classifier overlap significantly with those from the highest scoring classification engines.

In the mood task, we underperform both Marsyas and the leading algorithm. This is interesting and might speak to the fact that we did not tune the parameters of this algorithm for the task of music classification, but instead used the parameters that worked best for the classification of sound effects. Music mood might be a feature that has spectral aspects that evolve over longer time periods than other features. For this reason, it would be important to search for other parameters in the SAI algorithm that would perform well for other tasks in music information retrieval.

For these results, due to time constraints, we only used the SVM classifier on the SAI histograms. This has been shown by Lyon et al. [9] to be

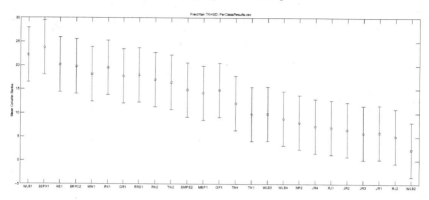

Figure 3.7
Per class results for classical composer.

an inferior classifier for this type of sparse, high-dimensional data than the PAMIR algorithm. In the future, we would like to add the PAMIR algorithm to Marsyas and to try these experiments using this new classifier. It was observed that the MIR community is increasingly becoming focused on advanced machine learning techniques, and it is clear that it will be critical to both try different machine learning algorithms on these audio features as well as to perform wider sweeps of parameters for these classifiers. Both of these will be important in increasing the performance of these novel audio features.

3.4 Conclusion

The use of physiologically plausible acoustic models combined with a sparsification approach has shown promising results in both the sound effects ranking and MIREX 2010 experiments. These features are novel and hold great promise in the field of MIR for the classification of music as well as other tasks. Some of the results obtained were better than that of a highly tuned MIR system on blind data. In this task we were able to expose the MIR community to these new audio features. These new audio features have been shown to outperform MFCC features in a sound-effects ranking task, and by evaluating these features with machine learning algorithms more suited for these high dimensional, sparse features, we have great hope that we will obtain even better results in future MIREX evaluations.

Bibliography

[1] S. Balakrishnan and D. Madigan. Algorithms for sparse linear classifiers in the massive data setting. *Journal Machine Learning Research*, 9:313–337, 2008.

[2] L. Bottou, O. Chapelle, D. DeCoste, and J. Weston. *Large-Scale Kernel Machines (Neural Information Processing)*. MIT Press, Cambridge, Massachusetts, 2007.

[3] O. Chappelle, B. Scholkopf, and A. Zien. *Semi-Supervised Learning*. MIT Press, Cambridge, Massachusetts, 2006.

[4] G. Chechik, V. Sharma, U. Shalit, and S. Bengio. Large-scale online learning of image similarity through ranking. *Journal Machine Learning Research*, 11:1109–1135, 2010.

[5] Y. EL-Manzalawy and V. Honavar. *WLSVM: Integrating LibSVM into Weka Environment*, 2005. (Software available at: http://www.cs.iastate.edu/~yasser/wilson.)

[6] D. Grangier and S. Bengio. A discriminative kernel-based approach to rank images from text queries. *IEEE Transactions Pattern Anal. Machine Intelligence*, 30(8):1371–1384, 2008.

[7] P. Haffner. Fast transpose methods for kernel learning on sparse data. In *Proceedings of the 23rd International Conference on Machine Learning*, pages 385–392, ACM Press, New York, 2006.

[8] R.F. Lyon. Automatic gain control in cochlear mechanics. In P. Dallos et al., editor. *The Mechanics and Biophysics of Hearing*, pages 395–420, Springer-Verlag, New York, 1990.

[9] R.F. Lyon, M. Rehn, S. Bengio, T.C. Walters, and G. Chechik. Sound retrieval and ranking using auditory sparse-code representations. *Neural Computation*, 22, 2010.

[10] B.A. Olshausen and D.J. Field. Sparse coding of sensory inputs. *Current Opinion in Neurobiology*, 14(4):481–487, 2004.

[11] R.D. Patterson. Auditory images: How complex sounds are represented in the auditory system. *Journal of the Acoustical Society of America*, 21:183–190, 2000.

[12] M. Rehn, R.F. Lyon, S. Bengio, T.C. Walters, and G. Chechik. Sound ranking using auditory sparse-code representations. *ICML 2009 Workshop on Sparse Methods for Music Audio*, 2009.

[13] G. Tzanetakis. *Marsyas-0.2: A case study in implementing music information retrieval systems*, chapter 2, pages 31–49. Intelligent Music Information Systems: Tools and Methodologies. Information Science Reference, 2008.

4

Instrument Recognition

Jayme Garcia Arnal Barbedo

State University of Campinas, Brazil

CONTENTS

The task of identifying instruments is performed with relative ease by human listeners with some training. Even untrained listeners can succeed in discriminating different classes of instruments. The emulation of those human abilities has been one of the main fields of research in digital audio, and many techniques have been proposed in the last decade. This chapter presents an overview of this important area of research, focusing on the main problems posed by the subject and describing some of the solutions proposed in the literature.

4.1 Introduction

For many years, the research on music information retrieval and data mining was limited by the lack of computational resources to process the great amount of information present in a musical signal. In the last decade, with the available computational power finally matching the demand, this area of research has seen the number of publications increase greatly. Among the new technologies emerging from this new reality is the automatic recognition of musical instruments.

One interesting aspect of the music processing area is the interdependence of different subareas of research. In this context, advances on instrument recognition can, for instance, benefit the research on sound source separation, as the knowledge about the instruments may allow the introduction of instrument-specific strategies that can potentially improve the quality of the separation. Conversely, a good algorithm of sound source separation can split a polyphonic signal into several individual instrument streams, whose monophonic nature makes it much easier to identify the instrument. In another example, the identification of the instruments can greatly benefit the area of music genre classification—as each genre usually has a well-defined set of possible instruments, knowing which instruments are present can narrow down the set of genres to be searched. Conversely, knowing the genre of a given song may narrow down the set of probable instruments. Other areas that have such a symbiotic relation with instrument recognition are the multiple fundamental frequency estimation, the automatic music transcription, and the music clustering, among others.

As can be seen, the area of music processing as a whole can benefit from advances in instrument recognition. As a response, several new techniques have been proposed in the last years. However, the challenges posed by real musical signals are such that the techniques proposed so far are either limited to a tightly defined set of signals, or have low accuracy, or both. The main objective of this article is to present the main challenges involved in the endeavor of designing an algorithm for instrument recognition, to present some of the most successful solutions proposed so far, and to discuss some future research directions and emerging trends.

The chapter is organized as follows. Section 4.2 shows how the scope of an instrument recognition algorithm is determined. Section 4.3 presents the main stages involved in the development of an instrument recognition algorithm. Section 4.4 presents some of the most successful methods proposed in the literature, focusing on their basic ideas, strengths and limitations. Finally, Section 4.5 presents some possible directions for future research and potential solutions for the problems still lingering.

4.2 Scope Delimitation

Ideally, an instrument recognition algorithm should be able to identify any kind of instrument and should work properly for any kind of musical signal. However, this is not a mature technology, hence proposals with limited scope are still ubiquitous. This section describes some of the most important factors that delineate the scope of instrument recognition algorithms.

4.2.1 Pitched and Unpitched Instruments

Musical instruments can be divided into categories according to many criteria. One of the most basic forms of categorization is dividing the instruments as pitched and non-pitched. Pitched instruments are those that have a well-defined pitch or fundamental frequency—in this case, normally the pitch (a subjective frequency measure) matches the fundamental frequency (an objective frequency measure). For that reason, it is common that both terms be used interchangeably. Most pitched instruments are also harmonic, meaning that their partials (spectral peaks associated to F0) are close to being harmonics, that is, the frequencies of the partials are almost exact multiples of the F0. However, some percussion instruments have a defined pitch, but their partials tend to be nonharmonic. Nonpitched instruments, on the other hand, do not have a discernible pitch, and their spectral content as a whole is usually noise-like. Most nonpitched instruments are percussive. Figure 4.1 shows typical spectra associated to each kind of instrument.

Virtually all instrument recognition algorithms are able to identify pitched harmonic instruments, and most of them can also deal with pitched non-harmonic instruments. On the other hand, the noise-like characteristics of non-pitched instruments make their identification very challenging. Some methods specifically designed to identify this kind of instrument have been proposed, as discussed in Section 4.4.

4.2.2 Signal Complexity

The signal complexity is arguably the most important factor that defines the scope of an instrument recognition algorithm. Musical signals can roughly be divided into four groups according to their complexity [32]:

(1) Isolated sounds: Those are the simplest signals, composed by only an isolated sound generated by an instrument playing a single note only once. In this case, there is no interference from sources other than eventual background noise, the characteristics of the signal are relatively homogeneous throughout its entire duration, and there are no note transitions to be taken into account. Those characteristics made this kind of signal the ideal target for most early instrument recognition algorithms, as will be seen in Section 4.4. However, it is

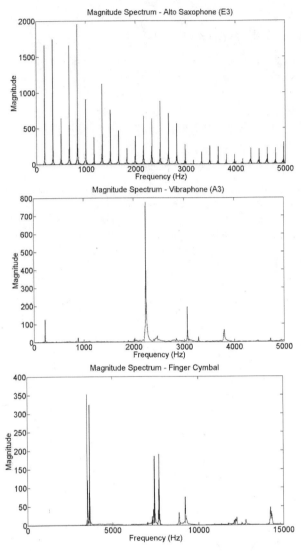

Figure 4.1

Examples of typical magnitude spectra for a pitched harmonic instrument (top), a pitched nonharmonic instrument (middle), and a nonpitched instrument (bottom). The magnitude spectrum for the pitched harmonic instrument (alto saxophone) has the partial peaks well-defined and equally spaced. The spectrum for the pitched nonharmonic instrument (vibraphone) has a relatively small—yet clearly defined—fundamental frequency at 220 Hz, and some peaks loosely related to the F0 that are spread throughout the spectrum. Finally, the spectrum for the nonpitched instrument (finger cymbal) has several unrelated high-frequency peaks along the entire spectrum. The unpredictable behavior of nonpitched instruments makes their identification difficult.

important to point out that even when the simplest signals are considered, the task of identifying instruments is not trivial. As described in Section 4.3.2, the characteristics of a type of instrument may vary greatly depending on factors like musician, manufacturer, temperature, and so forth. This makes it hard to find strong commonalities between different samples of a same instrument, no matter how well behaved is the signal.

(2) Solo phrases: The nature of this kind of signal is still monophonic, as there is at most one tone being played at any time. Therefore, the characteristics for this kind of signal are the same as those presented for isolated tones, except that here there are note transitions. As will be seen in Sections 4.3.1 and 4.3.2, the signal under analysis is characterized through variables called features. The features should be extracted from homogeneous excerpts of the signal, meaning that the position of note transitions should be correctly identified, and the signal should be segmented accordingly. However, that identification is not an easy task even in signals with only one instrument, especially if such an instrument has smooth note transitions, like violin. On the other hand, since there are several notes available, there is also more information to be explored and more cues to be collected, increasing the probability of a correct identification.

(3) Duets: Identifying instruments in signals of this group is a much more challenging task. The signals no longer have a monophonic nature, as the instruments in the duet can play simultaneously—they overlap in time. More than that, instruments in a duet are usually played in a consonant way, meaning that their fundamental frequencies will have some kind of harmonic relationship, causing them to significantly overlap in the frequency domain. Identifying instruments with that kind of interference is challenging, and currently there are three main strategies to solve this problem. The first one is applying some kind of source separation so the polyphonic signal is split into monophonic instrument streams. This would be the ideal solution if the problem of sound source separation was solved, but unfortunately such a technology is still quite crude. Nevertheless, even a defective separation can preserve some of the characteristics of individual instruments, improving the algorithm's recognition ability. The second possible strategy is to design the features in such a way the impact of inter-instrument interference is reduced. Finally, the third strategy consists in identifying regions on the time-frequency plane where a single instrument strongly dominates. In the time domain, this means searching for excerpts in which only one instrument is playing (monophonic excerpt); in the frequency domain, this means identifying partials that either do not collide with any partial from the other instrument, or is much stronger than the interfering partial.

(4) Higher polyphonies: All observations made for duets are also valid in the case of higher polyphonies (more than two instruments). The only difference is that in this case the signal can have any number of instruments. As a result, the interference in time is more severe, the spectrum becomes populated with a large number of partial peaks that often collide with each other,

and a new difficult problem arises as the number of instruments is usually unknown, and its estimation is a very difficult task by itself [3]. Despite the difficulties involved in identifying instruments under such difficult conditions, most real signals have several instruments. Because of that, the number of proposals able to deal with this kind of signal has increased in the last years. Most of those proposals still have some kind of limitation—the number of simultaneous instruments may be limited, the set of instruments that can be classified may be limited, the possible combinations of instruments may be set *a priori*, and so forth. This last limitation is particularly important, because it may be seen as a trick that artificially transforms a polyphonic problem into a monophonic one, as the mixture is treated as a single entity, just like an individual instrument—a polytimbral one, but still a single instrument. As an example, an algorithm could be trained to identify the combinations flute, piano+violin, and oboe+saxophone+viola. The developers may claim that the algorithm can recognize six instruments in polyphonic signals, but if a combination like piano+flute is to be dealt with, the algorithm will probably fail, since this entity was not present in the training. Thus, it makes a big difference if the single entities treated by an algorithm are individual instruments or combinations of instruments. This does not mean that proposals using the combination-as-an-entity approach are worthless, but the scope restrictions caused by such an approach must be taken into consideration. It is important to remark, however, that those restrictions can be greatly reduced if all, or at least the most common, possible combinations of instruments are included in the training, which may be impractical if many instruments are considered. More details about the effects of the constraints faced by algorithms that deal with higher polyphonies can be found in Section 4.4.

In short, the factor that most influences the difficulty involved in identifying the instruments is the number of simultaneous sounds in an excerpt and the interference they cause on each other. Figure 4.2 illustrates the time and frequency differences between monophonic isolated sounds (first two rows) and a polyphonic signal resulting from mixing the two isolated sounds (third row). The example of Figure 4.3 shows a strong time overlap and a relatively mild frequency overlap. In practice, it is common that instruments be played in unison (same note), and the presence of more than two simultaneous instruments is also a common occurrence. Further complications can arise from the presence of nonharmonic instruments, particularly percussive ones, whose spectral contents can be distributed throughout the entire spectrum.

4.2.3   Number of Instruments

A factor that may also have a strong influence over the scope of an instrument recognition algorithm is the number and type of instruments used for training it. There is a large number of different instruments in the world, and training an algorithm to be able to identify all of them is close to impossible, as most available databases only include the most common instruments in Western

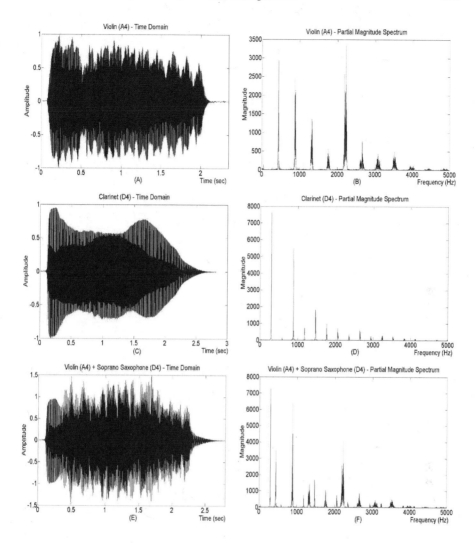

Figure 4.2

Example of the differences among monophonic and polyphonic signals. The first two rows of the figure show the time and frequency representations of two monophonic signals, each consisting of an instrument (violin and clarinet) playing a single note. The third row shows the result of mixing both signals. As can be seen, they overlap almost entirely in the time domain (not entirely because the clarinet note is slightly longer), and since the notes D4 and A4 form a perfect fifth interval (3:2), every second harmonic of the violin and every third harmonic of the clarinet will coincide in the frequency domain – this can be seen clearly in plot (F), where the partials at 880 Hz of both instruments completely merge. Therefore, there is a significant cross-interference both in time and frequency.

music—the Real-World Computing (RWC) database [28] is an exception, as it also includes Eastern instruments and other so-called exotic instruments, but it still does not encompass all instruments. Yet, some algorithms are trained with really narrow sets of instruments, and those instruments normally belong to different families. For example, a method may reach a high accuracy in identifying instruments in polyphonic signals, but if the set of possible instruments includes only violin, clarinet, piano, and guitar, the scope of the algorithm is clearly strongly limited. Moreover, the algorithm's generalization abilities cannot be properly inferred with such a small set of instruments. This does not mean that such a proposal would be worthless, as the ideas brought to light might be interesting and useful. However, any algorithm intended to be used in practice should be able to identify a large number of instruments— there is no hard lower limit for such a number, but any amount above 20 instruments seems to be reasonable. If the algorithm is intended to deal only with a specific musical genre (e.g., jazz), the number of possible instruments is limited and, in that case, a narrower set of instruments is acceptable.

4.3   Problem Basics

4.3.1   Signal Segmentation

A digital audio signal is a sequence of numbers generated from the sampling of the original analog signal. A major step in the recognition of an instrument is the extraction of features from such a sequence of numbers in order to characterize the signal under analysis (see Section 4.3.2). Although the features can be extracted for the signal as a whole, this is rarely done in practice, because a musical signal normally has several instruments playing several notes at different instants. In this context, a feature will not be able to effectively fulfill its purpose of characterizing the signal, because there are too many variations to take into account. To avoid this problem, the features are normally extracted from smaller segments and then integrated over the whole signal. Those segments must be as homogeneous as possible, in such a way the information provided by the features is consistent.

The most straightforward way to segment the signal is simply dividing it into frames of fixed length (normally between 20 ms and 100 ms); consecutive frames usually overlap by 50%, and a window function (usually Hamming) is applied to each frame in order to avoid spectral distortions caused by an abrupt end at the edges of the frame. The main problem with this kind of segmentation is that events—which consist of a new note or instrument— may occur in the middle of a frame, causing the features calculated for that frame to be inconsistent. Also, the signal may be segmented more times than it would be necessary, and considering that the frames usually overlap, the computational effort is significantly increased. A better solution is segmenting the signal according to the events, in which case the boundaries of a frame are located exactly where the events occur. In this case, the frames may have different lengths. Although clearly more efficient in a computational effort point of view, this strategy depends on the accurate estimation of the events locations. This is a difficult problem by itself, and the few strategies proposed so far (for example, see Klapuri [48], Rodet and Jaillet [73], Bello et al. [6]) still have limitations. Because of that, almost all instrument recognition algorithms use fixed frame lengths.

4.3.2 Feature Extraction

There is a large number of features proposed in the literature (see Chapter 2), and choosing the most appropriate set is a crucial step in the development of the algorithm [13]. The highly varying nature of sound production by means of musical instruments makes this task far from trivial. Factors like musician, manufacturer, temperature, room acoustics, among others, can greatly affect the characteristics of the sound produced by a given instrument. Even if most variables that influence the sound production are repeated as exactly as possible, the resulting sound will certainly be different each time it is produced due to the natural variations of a human execution. Figure 4.3 shows how different the characteristics of an instrument can be.

Ideally, a feature should assume a small range of values for each instrument, and the resulting ranges should never overlap, in which case the problem of instrument recognition would be easily solved. However, as a result of those behavior variations, a feature usually can assume wide value ranges for the instruments, increasing the probability of overlap between them. When several features are considered together—which is usually the case—they generate a feature space with as many dimensions as the number of features. The samples of an instrument will populate a part of that space and delineate the neighborhood associated to that instrument. The features should be chosen so the overlap between the neighborhoods is minimized. Also, the number of features must be limited in order to avoid excessive redundancy and to avoid the well-known curse of dimensionality [5], which is the problem caused by the exponential increase of the training set associated with adding new dimensions to the feature space. In many cases, the adopted solution is to extract a large set of features, and then some method to reduce dimensionality, like linear discriminant analysis (LDA) and Principal Component Analysis (PCA), is applied. Some strategies to properly select features have been performed by Deng et al. [13], Guyon and Elisseeff [31], and Yu and Liu [85].

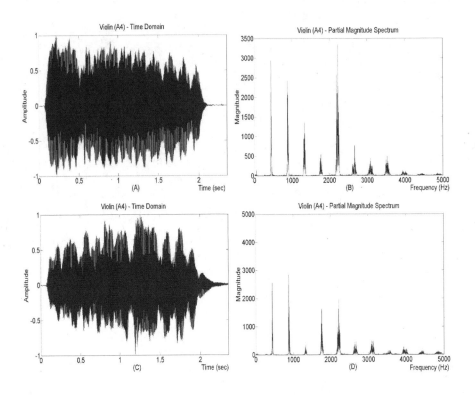

Figure 4.3
Example of variations between different samples of an instrument. The plots
in the first and second rows were generated by two violins from different man-
ufacturers and played by different musicians, but playing the same note (A4).
As can be seen comparing plots (A) and (C), the vibrato characteristics are
quite different. The magnitude spectra have also significant differences—for
instance, the third harmonic is almost absent in plot (D), but it is the fourth
most prominent harmonic in plot (B). This example shows the kind of differences that occur in most cases—extreme cases in which different samples are
almost identical or completely different are less common, but occur relatively
often.

4.3.3   Classification Procedure

4.3.3.1   Classification Systems

Once the features have been extracted, they are usually used to feed some kind of classification system. Those classification systems have the role of suitably combining all the information contained in the features so the correct instrument is returned as output. They are all very effective if the instruments are highly disjoint in the feature space. Since this is not the case, the most appropriate classifier is the one able to mitigate the problems caused by the lack of disjointness between the instruments. Since the steps and strategies used prior to the classifier vary greatly from proposal to proposal, the chosen classifiers tend to be diverse, as will be seen in Section 4.4. Five of the main classification systems are briefly described in the following.

(a) k-Nearest Neighbors (k-NN): The first step in the implementation of this kind of classifier is the construction of a dictionary composed by vectors of feature values. Those vectors are labeled according to the instrument they best represent. Thus, each instrument has a number of vectors associated. The classification is obtained by comparing the vector of features extracted from the signal to be classified with the vectors in the dictionary. In that comparison, some distance measure (usually Euclidean) is used to select the k vectors in the dictionary that are closer to the extracted vector. Finally, the classification is given by the instrument that appears the greatest number of times among the k selected vectors. This is a largely used classifier due to its easy implementation. On the other hand, it demands that large amounts of data be stored and a large number of computations be performed.

(b) Gaussian Mixture Models (GMM): This technique takes the feature vectors associated to a given instrument and uses them to infer a probability density function (PDF) that models that instrument. Such a PDF results from a weighted combination of a number of Gaussian PDFs, hence the name Gaussian Mixture Model. The parameters of a GMM are usually adjusted by means of an iterative process called Expectation Maximization (EM). Once trained, each GMM is used to estimate the probability that a feature vector was generated by the instrument associated to that GMM. The final classification is given by the instrument with greatest probability. This kind of classifier has been used with success in many audio classification problems.

(c) Support Vector Machines (SVM): The objective of this kind of classifier is to find the hyperplane that best separates observations (feature values) pertaining to two different classes in a multidimensional space. The theory does not guarantee that the best hyperplane can always be found but, in practice, a heuristic solution can always be obtained. Since the instrument classification is a multiclass problem, and the SVM is a binary classifier, some kind of adjustment must be made. The most common approach is to reduce the single multiclass problem into multiple binary problems. The implementation of this classifier is not simple, but it has several desirable characteristics, like the ability to control the complexity of the learning process, no matter the dimension of the problem.

(d) Artificial Neural Networks (ANN): An ANN is a processing structure (network) composed by a number of interconnected units (artificial neurons). Each unit presents a specific input/output behavior (local computation), determined by its transference, by the interconnections with other units, and possibly by external outputs. The neurons are divided into layers—one input layer, a number of hidden layers, and an output layer. Each neuron has three parts: the synapses, characterized by the associated weights; the summing node, which combines the weighted input signals; the activation function, which limits the neuron output and introduces nonlinearities to the model. In its training, the neural network is fed with a number of feature vectors representing each instrument, and the synaptic weights are adjusted so the neuron corresponding to the correct instrument is activated at its output. Once trained, the ANN is fed with the feature vector extracted from the signal to be classified, and the label of the neuron that is activated at its output reveals the estimated instrument. There are several types of ANNs, but in music processing the most used is the Multilayer Perceptron (MLP). An ANN is a good choice when the function to be learned is nonlinear. On the other hand, the training can be time-consuming, and the risk of data overfitting may incur in loss of generalization capability.

(e) Linear Discriminant Analysis (LDA): This classification strategy takes the feature data and projects it in such a way inter-feature correlations are reduced (thus reducing redundancy) and the variance of each variable is standardized. After that, Euclidean distances are calculated in this modified space to define the most likely class. As in the case of SVM, the LDA is inherently binary. In the case of a multiclass classification, a commonly used solution is successively partitioning the classes in such a way a single class is always put in one group, and all the other classes are put in the other one, and then using LDA to perform the classification. This results in as many classifiers as classes. Another common solution is the pair-wise classification, in which each pair of classes is considered separately, resulting in $N(N-1)/2$ classifiers, where N is the number of classes. In both solutions, the results obtained for all classifiers are combined to determine the final classification. In cases for which a linear boundary between the pairs of classes is not adequate, the quadratic discriminant analysis (QDA) can be applied instead.

There are many other classification strategies that are eventually used in the context of audio classification, like Decision Tree, Quadratic Gaussian Classifier, Fuzzy Classifier, Simple Binary Discriminator, among others.

4.3.3.2   Hierarchical and Flat Classifications

As in most classification systems, musical instruments can be organized in a taxonomic structure with a number of classification levels. In this kind of structure, the upper levels normally have a low number of classes, and they

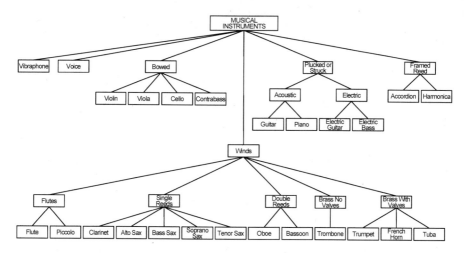

Figure 4.4

Example of musical instruments taxonomy. This example partially follows
both the Hornbostel-Sachs system [80] and objective characteristics presented
by the instruments. The actual instruments are always at the bottom of the
taxonomy, but the number of intermediate classes may vary from none to
several, depending on the criteria used to build the taxonomy—in the example,
vibraphone and voice have no parent classes, while most instruments have two
levels of parent classes.

usually are very distinct among them. On the other hand, the lower levels
normally include a large number of classes that are more tightly defined, and
the differences among them are often weak. The hierarchical structure of an
instrument taxonomy may vary according to the selected criteria. A widely
accepted taxonomy is the Hornbostel-Sachs system [80]. This system is very
popular among musicologists and organologists, but may not be the most
appropriate in the context of automatic music recognition. Essid et al. [24]
performed a study in which they compared a natural taxonomy inspired by
instrument families with a taxonomy inferred automatically by means of hierarchical clustering. They came to the conclusion that both have limitations,
and that the taxonomies should be especially designed to deal properly with
instruments that are often confused. An example of taxonomy is shown in
Figure 4.4. Other taxonomies can be found by Agostini et al. [2] and Eggink
and Klapuri [17].

The main advantage of using a hierarchical classification scheme is the
possibility of placing decision nodes at each point where the taxonomy ram-
ifies. The first decisions to be made are expected to be more accurate due
to the more marked differences between classes at the top of the taxonomic
structure, and each time a decision is made, the set of possible instruments
is reduced. As a result, when the algorithm reaches the point where the most

difficult decisions are to be made, there is a good chance that previous decisions were correct, and the number of possible instruments is greatly reduced. Another advantage of using a hierarchical approach is that each node can have a decision scheme especially tuned according to the characteristics of that particular node, increasing the chance of a correct decision. Since those decisions are nothing more than intermediate classifications, classical classification algorithms like SVM, k-NN, and GMM are often employed, and there is even the possibility of using different classifiers for different nodes. A last advantage of using this kind of approach is the modularity—if new instruments are to be included in the recognition system, it is enough to include a new branch in the taxonomic tree, and a localized new training is performed to tune the classification procedure to be used in the new node.

The flat classification can be seen as a particular case of the hierarchical classification in which there is only one decision node and no parent classes. If properly designed, the flat approach can be as effective as the hierarchical one, and it also has advantages. First, the algorithm design has fewer steps—there is no taxonomic structure to be developed, and since there is only one decision node, only one classification system is designed. Also, the training phase tends to be faster and simpler, as there is only one classifier to adjust.

As a matter of fact, the hierarchical and direct approaches have more similarities than differences—both use a number of features that feed some kind of classification scheme. The only real difference is that, in the hierarchical approach, the whole problem is split into smaller ones. The use of one or another is linked more to the personal preferences of the developers than to their effectiveness. As will be seen in the next section, both approaches are present in the group of the most effective methods proposed so far.

4.3.4   Analysis and Presentation of Results

The results obtained by a given instrument recognition system are usually presented in the form of a confusion matrix, which is a compact and practical way of showing how the system performs for each instrument and, more importantly, it clearly shows the error trends, which can potentially guide future improvements. Figure 4.5 shows an example of confusion matrix, extracted from the work by Barbedo and Tzanetakis [4].

One factor that may have a strong impact on the quality of the results obtained is the database used to calibrate and test the method. Normally, the available instrument samples are divided into two sets, one for training (used to calibrate the algorithm), and one for test (used to assess the performance of the algorithm). Under any condition a given sample can be part of both the training and test sets, because this would cause the results to be biased and unreliable. However, all the samples in both sets can come from the same database. It is important to note, however, that using the same database to train and test the algorithm may, in some cases, lead to deceptively good results [57]. Because of that, it is desirable that samples from other database be

	vi	va	ce	cb	gu	pn	eg	eb	fl	pc	cl	as	bs	ss	ts	ob	bn	tb	tp	hn	tu	ac	ha	vp	vo
vi	90	0	8	0	0	0	0	0	0	0	0	0	0	0	0	0	0	0	0	0	2	0	0	0	0
va	0	93	0	0	0	0	0	0	0	0	0	0	0	0	0	0	0	0	5	0	2	0	0	0	0
ce	10	0	90	0	0	0	0	0	0	0	0	0	0	0	0	0	0	0	0	0	0	0	0	0	0
cb	0	0	0	93	0	0	7	0	0	0	0	0	0	0	0	0	0	0	0	0	0	0	0	0	0
gu	0	0	0	0	57	18	14	11	0	0	0	0	0	0	0	0	0	0	0	0	0	0	0	0	0
pn	0	0	0	0	8	77	0	0	0	0	0	0	0	0	0	0	0	0	0	0	0	0	0	15	0
eg	0	0	0	0	14	7	72	7	0	0	0	0	0	0	0	0	0	0	0	0	0	0	0	0	0
eb	0	0	0	12	0	0	0	88	0	0	0	0	0	0	0	0	0	0	0	0	0	0	0	0	0
fl	0	0	0	3	0	0	0	0	76	3	3	0	0	0	0	2	0	0	5	0	0	0	8	0	0
pc	0	3	0	0	0	0	0	0	3	94	0	0	0	0	0	0	0	0	0	0	0	0	0	0	0
cl	0	0	0	0	0	0	0	0	0	0	95	0	0	0	0	0	0	0	0	0	0	5	0	0	0
as	0	0	0	0	0	0	0	0	0	0	0	67	6	3	0	0	9	0	15	0	0	0	0	0	0
bs	0	0	0	0	0	0	0	0	0	0	0	14	76	0	0	0	5	0	5	0	0	0	0	0	0
ss	0	0	0	0	0	0	0	0	0	0	0	24	6	58	0	6	0	6	0	0	0	0	0	0	0
ts	0	0	0	0	0	0	0	0	0	0	0	0	0	14	79	4	0	3	0	0	0	0	0	0	0
ob	0	0	0	0	0	0	0	0	6	6	0	3	0	0	0	68	12	0	3	0	0	0	2	0	0
bn	0	0	0	0	0	0	0	0	4	0	0	7	0	0	0	0	71	7	7	4	0	0	0	0	0
tb	0	0	0	0	0	0	0	0	14	3	3	3	0	0	0	0	0	66	6	5	0	0	0	0	0
tp	0	0	0	0	0	0	0	0	6	6	3	8	6	6	0	3	0	51	8	0	3	0	0	0	0
hn	0	3	0	0	0	0	0	0	0	0	0	0	4	4	0	0	0	0	14	75	0	0	0	0	0
tu	0	0	0	0	0	0	0	0	0	0	0	0	0	0	0	0	0	0	0	0	100	0	0	0	0
ac	0	0	0	0	0	0	0	0	0	0	6	0	0	0	0	6	0	0	3	0	0	82	3	0	0
ha	0	0	0	0	0	11	0	0	0	0	0	5	0	0	0	0	5	0	0	0	0	0	79	0	0
vp	0	0	0	0	0	0	0	0	0	0	0	0	0	0	0	0	0	0	0	0	0	0	0	100	0
vo	2	1	1	0	0	0	0	0	0	2	0	0	0	0	0	0	0	0	0	0	0	0	0	0	94

ac - accordion   as - alto sax   bn - bassoon   bs - bass sax   cb - contrabass   ce - cello   cl - clarinet   eb - electric bass   eg - electric guitar
fl - flute   gu - guitar   ha - harmonica   hn - French horn   ob - oboe   pc - piccolo   pn - piano   ss - soprano sax   tb - trombone
tp - trumpet   ts - tenor sax   tu - tuba   va - viola   vi - violin   vo - voice   vp - vibraphone

Figure 4.5

Example of a confusion matrix [4]. The instruments in the first column of the matrix represent the actual classifications, and the instruments in the first row represent the classification at the output of the algorithm. As a result, the main diagonal represents the correct classifications. Taking the first row as an example, the violin samples were correctly classified by the algorithm in 90% of the cases, and the violin samples were misidentified as cello and tuba in 8% and 2% of the cases, respectively. The shades of gray in the figure indicate the degree of similarity between the instruments according to the example of taxonomy shown in Figure 4.4—the darker is the shade, the more related are the instruments (lower levels in the hierarchical tree). Errors that occur inside darker areas are less severe, as at least some of the parent classes are correctly classified. In the example of the first two rows, the cases in which the violin was misidentified as cello are less severe than those in which tuba is the wrong instrument.

used in the tests to provide a cross-database validation—the results obtained in this way are usually considered stronger and more meaningful.

There is a small number of instrument sample databases available—three of the most used are the McGill University Master Samples [68], the University of Iowa musical instrument samples database [1], and the RWC database [28]—and none of them is adopted as the standard reference. Hence, the sets of samples used to test each algorithm proposed in the literature are likely to be distinct, making a direct comparison between the algorithms very difficult. As a result, in general it is not possible to draw any definite conclusions about how the algorithms compare to each other. A feasible solution to this problem would be a comparison through competitions like those organized by the Music Information Retrieval Evaluation eXchange (MIREX) community [14], which provides a framework for the formal evaluation of Music Information Retrieval (MIR) systems and algorithms. In those competitions, all proposals are tested under the same conditions by using an independent database. Until 2010, no competition of this kind has been carried out for instrument recognition algorithms.

The use of different databases is not the only factor that complicates the comparison between algorithms. As commented before, most algorithms proposed so far have some constraints regarding type of instrument, number of simultaneous instruments, etc. The degree of difficulty of the problem tackled by the algorithm is directly dependent on those constraints. If those constraints are different—they usually are—a direct comparison between algorithms becomes impossible.

When consdiering the difficulties involved in comparing algorithms it is important to take into acount the fact that instrument recognition research is still at an early stage where coming up with new ideas may be more important than figuring out which algorithm works best. As the technology matures, the number of constraints and limitations will diminish, and a direct comparison between algorithms will become more feasible.

Sometimes it is useful to compare the results achieved by an algorithm with human performances. A number of studies on the human ability to recognize instruments have been carried out [18, 41, 61, 65, 76]. The results vary greatly as the number of instruments and test conditions used in each study is different. However, a baseline can be inferred from the study carried out by Srinivasan et al. [76], since they used a large set of 27 instruments to assess the ability of conservatory students in identifying isolated tones—the recognition rate under such difficult conditions was 55.7%. Better recognition rates are expected as fewer instruments are considered. Also, better recognition rates would probably be achieved if, instead of isolated tones, complete sequences of notes were used, as the transitions between notes can provide important cues for the instrument identification. As can be seen, a rigorous comparison between machine and human performances is hardly possible, but a relative comparison taking into account the differences between the tests can provide useful information.

4.4 Proposed Solutions

This section presents a brief description of several instrument recognition algorithms proposed in the literature. This survey was designed to be as comprehensive as possible, but in an area with such a large number of publications, inevitably some of them will not be included. There are some theses and dissertations dedicated to this subject [19, 42, 61, 69], however, they were not included here—the articles that summarize them are presented instead. In the description of the algorithms presented in the following, the accuracies achieved by them are not presented because, due to the reasons stated in Section 4.3.4, a direct comparison among the methods is very difficult. Furthermore, the accuracies only make sense together with a complete understanding of the characteristics of each method, which cannot be achieved without reading the original works.

Instrument recognition methods can be grouped according to different criteria. Here, the only criterion used is if the method deals only with monophonic signals, or if it is able to deal with polyphonic signals. Other characteristics of the algorithms are presented in two tables presented later in this section.

4.4.1 Monophonic Case

This subsection provides a brief description of algorithms proposed to deal with monophonic signals, which include both isolated tones and solo phrases. Although the monophonic case tends to be simpler than the polyphonic one, the problem is still far from being solved, and many articles on this subject are still being published, as will be seen in the following. The algorithms presented next are ordered chronologically.

The early work by Kaminskyj and Materka [37] tacked the problem of identifying isolated tones of four instruments from very different families. The features consist of 80 short-term root mean square energy. The number of features is then reduced to three by means of Principal Component Analysis (PCA). The authors tested two classification schemes, an Artificial Neural Network (ANN) and a Nearest Neighbor Classifier (k-NN with $k = 1$). They remark that the results were surprisingly good as they only used temporal features.

Kaminskyj and Voumard [38] proposed a so-called multistage intelligent hybrid classification system. The algorithm extracts seven temporal and spectral features and use an ANN and a k-NN with $k = 1$ as classifiers. The article only proposes the system, thus no results are presented.

Martin and Kim [62] proposed one of the first algorithms to use a hierarchical structure with multiple decision nodes. The authors justified this choice by arguing that human listeners recognize objects and stimuli taxonomically. In total, 31 features describing temporal and spectral characteristics of the signals were extracted, and then the Fisher multiple discriminant analysis was used at each decision node to reduce the number of features, keeping only the most relevant ones. The authors tested both a Maximum a Posteriori (MAP) classifier and a k-NN classifier. The experiments were performed using isolated notes sampled from 14 instruments.

The early work proposed by Brown [8] focused on the discrimination of two very similar instruments (oboe and saxophone). The feature set is composed by 18 cepstral coefficients. It employs a probabilistic classification scheme based on k-means clustering and on Gaussian Mixture Models (GMM) used to calculate the probability density functions that describe the data. An interesting aspect of this work is that, instead of using samples from standardized databases, the authors used instrument excerpts from real recordings.

Kashino and Murase [40] proposed an algorithm that does not rely on the extraction of features to perform the instrument recognition. Instead, it uses an adaptive method for template matching that can cope with variability in musical sounds. The algorithm also includes a musical context integration step, which improves the accuracy in more than 20%. The tests were performed using real musical signals containing three instruments (violin, flute, and piano), but each part of the signal contains only one instrument, thus the algorithm works in a monophonic context.

Eggink and Klapuri [17] proposed a method based on the extraction of several temporal and cepstral features. The algorithm employs a hierarchical classification approach and, for each node, it uses either a Gaussian or a k-NN classifier depending on the characteristics of the decision to be made. The algorithm was tested using isolated tones from 30 instruments. The authors obtained better results using a flat classification approach, but they remark that the hierarchical approach can be advantageous in the classification of larger data sets with more instruments.

The work by Brown et al. [9] is very similar to that published by Brown [8]. Here, four very similar wind instruments are considered, and the feature set is composed by cepstral coefficients, bin-to-bin differences of the constant-Q coefficients, autocorrelation coefficients, and moments of time wave. The classification scheme is similar to that used by Brown [8] and briefly described above. A very thorough study about the feature dependence of the results was carried out. This work also used instrument excerpts from real recordings.

The main motivation of the algorithm proposed by Eronen [20] was the assessment of the effectiveness of several features in the task of instrument recognition. The classifier used in the tests is a k-NN, and a total of 29 instruments extracted from several databases were taken into account. The author concluded that features based on warped linear prediction (WLP) and on cepstral coefficients are effective, and that the best results were achieved by augmenting the cepstral coefficients with features describing additional characteristics of the tones.

The method proposed by Kostek and Czyzewski [51] extracts 37 features— 14 based on Fourier analysis and 23 based on wavelet analysis—and feeds them to an artificial neural network, which performs the classification. Tests were performed using a database specially built for this method. Although 21 instruments were considered in total, the tests were performed separately considering only groups of four instruments.

Agostini et al. [2] used a total of 27 instruments to test the discrimination capabilities of a number of spectral features found in the literature, and to test the effectiveness of four classifiers—Support Vector Machines (SVM), quadratic discriminant analysis (QDA), canonical discriminant analysis (CDA), and k-NN. They concluded that the most informative features are the mean of the inharmonicity, the mean and standard deviation of the spectral centroid, and the mean of the energy contained in the first partial. They also concluded that SVM and QDA are the best classifiers, but they remark that the closeness of performances among all classifiers indicate that a properly feature selection is more critical than the choice of a classification system.

In the work of Costantini et al. [12], a number of features are extracted from preprocessed versions of the signals. Preprocessing strategies based on fast Fourier transform (FFT), Constant-Q Frequency Transform, and cepstral coefficients are tested. The method uses Min-Max Neuro-Fuzzy Networks as classification model, which is synthesized using adaptive resolution training techniques. The algorithm was tested against samples from six different instruments.

Eronen [21] uses Mel-frequency cepstral coefficients and their derivatives as features. Those features are transformed to a base with maximal statistical independence using independent component analysis (ICA). Continuous Density Hidden Markov Models (HMMs) discriminatively trained were used as classification system. The algorithm was tested using two groups of data, one containing isolated tones from 27 harmonic instruments, and one containing samples from five percussive instruments.

The algorithm proposed by Piccoli et al. [71] uses the first 18 MFCC as features. Two different artificial neural networks, a MLP and a Time-Delay Neural Network (TDNN), were tested, with a slight advantage for the second one. The experiments were performed using isolated tones from nine instruments.

Essid et al. [23] proposed an algorithm focused on the instrument recognition on real solo phrases. The authors chose to use only features known to be robust, resulting in a feature set containing only Mel-frequency cepstral coefficients (MFCCs), their derivatives, and some audio spectrum flatness (ASF) features. Different features were chosen for each possible pair of instruments. The algorithm uses a SVM as classifier, for which different kernels were tested. The method was tested with solo phrases of 10 instruments, all extracted from real recordings.

A study presented by the same authors [22] focuses on the use of simple features to perform the instrument recognition. The proposed algorithm extracts 47 features, and SVM is used for classification. Experiments were carried out using solo phrases performed by amateur musicians and sound samples from 10 instruments, all extracted from commercial recordings. The authors conclude that the combination of cepstral coefficients with features describing the audio signal spectral shape is very effective in the recognition of instruments belonging to different classes.

The main objective of Kitahara et al. [46] was to develop a method capable of identifying the category (family) of an instrument that was not present in the training data (unregistered). First, the method tries to determine if a given instrument is registered; if so, the name of the instrument is identified, if not the category of the instrument is estimated. The method uses 18 features selected from a larger set of 129 elements, and uses a musical instrument hierarchy (MIH) for the category-level identification.

This proposal by Krishna and Sreenivas [53] aims to identify instruments in isolated notes and solo phrases. The features the method uses are linear predictive coefficients called line spectral frequencies (LSF) [11], which can be seen as characteristic short-term spectral envelopes, but MFCC and linear prediction cepstral coefficients (LPCCs) are also used for comparison. This choice of LSF as features was motivated by one of the major objectives of the authors, which is keeping the scalability of their method. The performances of GMM and k-NN classifiers were tested, with a slight advantage for the GMM. The experiments used isolated tones from 14 instruments, and also some short segments of solo phrases.

Livshin and Rodet [58] proposed a method to identify seven instruments in solo recordings. They initially extracted 62 features, and then applied the Gradual Descriptor Elimination (GDE) algorithm to reduce such a set to 20. Using the reduced feature set resulted in an accuracy 3% worse than using the whole set, but the authors argue that such a reduction made it possible for the algorithm to be used in real-time applications. The classification scheme consists in a combination of LDA and a k-NN classifier. All tests were performed using excerpts extracted from real recordings, and they also perform some tests with duets to show that their method can be useful in the polyphonic case.

An article by Tindale et al. [78] presents one of the few methods that deal specifically with the recognition of drum sounds. A number of temporal features and the energies of four subbands feed an artificial neural network responsible for the final classification. Several experiments were performed using drum samples generated by the authors.

Kaminskyj and Czaszejko [39] proposed an algorithm that uses 710 features selected from a set of 2,804 elements by means of PCA. They tested three types of classification architectures (hierarchical, hybrid, and flat), with k-NN classifiers being used at the decision nodes. The tests were performed using isolated tones from 19 instruments. It was concluded that, although the hierarchical and hybrid structures perform better than the flat one, such a gain is too small in comparison with the added computational effort to justify their use.

This algorithm proposed by Kitahara et al. [47] takes into consideration the pitch dependency of timbre of musical instruments. The method extracts 129 features from an instrument sound and then reduces the dimensionality of the feature space into 18 dimensions. After that, an F0-dependent mean function and an F0-normalized covariance are calculated. The key idea underlying those two parameters is to represent the features as a function of the fundamental frequency of the instruments. The final classification is given by the Bayes decision rule (BDR). The algorithm was tested using isolated tones of 19 musical instruments.

An article by Pruysers et al. [72] uses Morlet Wavelet Analysis and Wavelet Packet Analysis to generate features that, combined with six other features from a previous work, are submitted to a single stage classifier. In this classifier, each feature has its own k-NN classifier, and the individual results are combined by a so-called k-NN result combiner. The experiments used samples from 19 instruments. The authors came to the conclusion that both the proposed wavelet-based features are useful and, additionally, they complement each other in an effective way.

Benetos et al. [7] used a branch-and-bound search to select a subset of relevant features from a full set, which is composed by 41 features. They also present four classifiers based on the non-negative matrix factorization (NMF), the best of which achieving a performance only slightly worse than that achieved using GMM and HMM classifiers. The authors remark that their experiments employed unsupervised classification, in contrast to the supervised GMM and HMM classifiers. The experiments were carried out using samples from six instruments.

Chetry and Sandler [11] proposed an instrument recognition method whose features consist of line spectral frequencies. Two classification procedures, k-Means and SVM, were investigated using solo phrases of six instruments extracted from commercial recordings. The authors conclude that the SVM performs slightly better, and that better results are achieved if the models are trained using solo phrases that have been recorded in various acoustic conditions.

As in earlier works by the same authors, the recognition in this algorithm proposed by Essid et al. [26] is performed over solo phrases from real recordings. The algorithm employs two feature selection techniques, the inertia ratio maximization with feature space projection and the genetic algorithms. A selection of the most relevant features is performed separately for each possible pair of instruments. Hence, the algorithm uses a one versus one classification strategy: a winner instrument is taken for each possible pair by means of either a SVM or a GMM, and the final classification is determined according to a majority vote rule. The authors performed a thorough study about several aspects of the algorithm using solo phrases extracted from real recordings.

In another algorithm proposed by Essid et al. [24], a large set of 540 features was considered, and an automatic feature selection was used to fetch the most useful ones. Since this algorithm uses a hierarchical approach, recognition decisions are performed throughout a number of nodes. The authors tested two different hierarchical structures, one following standard instrument families, and other generated automatically by means of an agglomerative hierarchical clustering. They concluded that spreading related instruments over distant nodes may actually improve the recognition accuracy.

Fragoulis et al. [27] tackle the very specific problem of discriminating between piano and guitar notes. Although very dissimilar in terms of construction and way of playing, the timbre generated by those instruments are actually quite similar, making this pair one of the most difficult to be discerned by an instrument recognition algorithm. The authors created three discriminative features and inferred three different empirical classification criteria to perform the classification. They remark that a successful discrimination between piano and guitar is strictly related to the non-tonal spectral content of each note.

Mazarakis et al. [64] proposed an algorithm that uses a Time-Encoded Signal Processing (TESP) method to produce simple matrices from complex sound waveforms. Those matrices are submitted to a so-called Fast Artificial Neural Network (FANN), which performs the instrument recognition. The experiments were carried out using signals generated by five different synthesizers (19 instruments), and also extracted from the Iowa database (20 instruments).

As in other studies, Simmermacher et al. [74] first extract a large set of features, which is reduced using feature selection techniques. Three classification schemes were tested: k-NN, Multilayer Perceptron (MLP) ANN, and SVM. The best results were achieved using the MLP. The experiments included tests using isolated tones from 19 instruments, and tests with solo phrases representing four instruments.

Tan and Sen [77] present a study on the use of the attack transient envelope in the recognition of musical instruments. The classification is based on a pattern matching algorithm called Dynamic Time Warping (DTW). Several experiments were performed with samples from two instruments (cello and violin) and, according to the authors, the results indicate that the attack transient can indeed be useful in instrument recognition.

This method proposed by Ihara et al. [34] is based on the extraction of a large number of features (1,102), on the reduction of that number by applying two-dimension reduction techniques (PCA and LDA), and on SVM to perform the classification. The most important claim made by the authors is that the log-power spectrum suffices to represent characteristics that are essential in instrument recognition. The method was tested using samples from eight instruments collected from commercial recordings.

Deng et al. [13] focus on the feature selection instead of presenting a new complete classifier. The authors use machine learning techniques to select and evaluate features extracted using a number of different schemes. The tests used individual note samples extracted from 20 instruments, and also solo phrases of four instruments. The authors found out that the best features are the log attack time (LAT), the harmonic deviation, and the standard deviation of the flux. They remarked that there is significant redundancy between and within the features extraction schemes commonly used, and further studies will be necessary in order to improve this crucial stage in instrument recognition.

The strategy proposed by Loughran et al. [60] extracts features based on temporal and spectral envelopes, and also on the evolution of the centroid. A MLP artificial neural network is then applied. Only isolated tones of three instruments were considered in the experiments.

The purpose of an article by Joder et al. [36] was to show that midterm temporal properties of the signal, which are usually ignored, can actually carry relevant information that may be useful in several tasks of music information retrieval and data mining. The proposed algorithm has the following steps: a preprocessing stage; a feature extraction stage in which 30 features chosen from the original 162-element set are calculated; an early temporal integration stage in which the information carried by the features are summed up according to a higher time scale; a sonic unit segmentation aiming to obtain semantically meaningful segments, which are used as time frame for the early temporal integration; a normalization step; and a classification/late temporal integration stage, in which the decisions made by the classifier are combined (integrated) in some effective way. The article presents extensive tests using solo phrases from eight instruments. The authors conclude that the best results are obtained combining early and late integration over sonic units, and using a SVM with dynamic alignment kernels as classifier.

An article by Kramer and Hein [52] is one of the few studies that deal exclusively with the identification of percussive instruments. The algorithm extracts 100 conventional features, and an evolutionary model is applied in order to derive optimal subsets of different sizes. The final instrument identification is performed by a SVM. The experiments were performed using real percussion excerpts contaminated with noise.

Table 4.1 summarizes all methods described in this subsection. The first column contains the first author and the year of the publication; the second shows the classifier; the third shows the database from which the training and test signals were extracted (*self* means that the authors used their own database, and *CR* indicates commercial recordings); the fourth indicates if the method is able to recognize nonharmonic instruments; the fifth indicates if a hierarchical taxonomic structure was adopted; the sixth contains the number of instruments for which the algorithm was tested; and the seventh reveals the number of features—if there are two values, the first one indicates the final number of features, and the second indicates the total number of features extracted. N/A means that the information is either not available or not applicable. The accuracies were not included because, as stated before, they are not very useful without a deeper understanding about the method and respective tests.

Reference	Class System	Database	Non-harm.	Flat/Hier.	No. Inst.	No. Feat.
Agostini03 [2]	Several	McGill	No	Flat	27	18
Benetos06 [7]	NMF	Iowa	No	Flat	6	6 (41)
Brown99 [9]	GMM	CR	No	Flat	2	18
Brown01 [10]	GMM	CR	No	Flat	4	Variable
Chetry06 [12]	SVM, k-means	CR	No	Flat	6	16
Costantini03 [13]	ANN	N/A	No	Flat	6	N/A
Deng08 [14]	Several	Iowa	No	Flat	20	44
Eggink04 [19]	GMM	McGill, CR, Ircam, Iowa	No	Flat	5	90
Eronen00 [17]	k-NN, Gaussian	McGill	No	Both	30	43
Eronen01 [22]	k-NN	McGill, Iowa, Ircam, Self	No	Flat	29	23
Eronen03 [23]	HMM	McGill, Iowa Ircam, Self	Yes	Flat	27[1]	24
Essid04a [25]	SVM	CR	No	Flat	10	43
Essid04b [24]	SVM	CR,Self	No	Flat	10	47
Essid06a [28]	GMM, SVM	CR	No	Flat	10	160
Essid06b [26]	SVM	CR	Yes	Hier.	20	540
Fragoulis06 [29]	Empirical	Self	No	Flat	2	3
Ihara07 [37]	SVM	CR	No	Flat	8	10 (1102)
Joder09 [41]	SVM, GMM, HMM	RWC, CR	No	Flat	8	30 (162)
Kaminskyj95 [43]	ANN, k-NN[2]	Self	No	Flat	4	3 (80)
Kaminskyj96 [44]	ANN, k-NN[2]	McGill	No	Flat	19	7
Kaminskyj05 [45]	k-NN	McGill, Iowa, Ircam, Self	No	Both	19	710 (2804)
Kashino99 [46]	TA	CR	No	Flat	3	N/A
Kitahara04 [52]	MIH	RWC	No	Hier.	19	18 (129)
Kitahara05 [53]	BDR	RWC	No	Flat	19	18 (129)
Kostek01 [57]	ANN	Self	No	Flat	21	37
Kramer09 [58]	SVM	CR	Yes	Flat	7	100
Krishna04 [59]	GMM, k-NN	Iowa, RWC	No	Flat	14	N/A
Livshin04 [64]	k-NN	CR	No	Flat	7	20 (62)
Loughran08 [66]	ANN	RWC	No	Flat	3	N/A
Martin98 [71]	MAP, k-NN	McGill	No	Hier.	15	31
Mazarakis06 [73]	ANN	Iowa, Synth	No	Flat	20	N/A
Piccoli03 [80]	ANN	Iowa	No	Flat	9	128
Pruysers05 [81]	k-NN	N/A	No	Flat	19	8
Simmerm.06 [83]	k-NN, ANN, SVM	Iowa, CR	No	Flat	20	19 (44)
Tan06 [86]	DTW	N/A	No	Flat	2	N/A
Tindale04 [87]	ANN	Self	Yes	Flat	5	8

[1] The recognition was performed at an intermediate level with five classes.
[2] k-NN with $k = 1$.

Table 4.1
Algorithms of the Monophonic Group

4.4.2 Polyphonic Case

We briefly describe algorithms for dealing with polyphonic signals, including duets and higher polyphonies. It is only recently that researchers have studied these problems that there are fewer articles that have been published and so covered in this section. As in the monophonic case, the articles are presented chronologically.

An article by Eggink and Brown [15], which was one of the first to tackle the polyphonic problem, incorporates the ideas from missing feature into a GMM classifier in order to improve instrument recognition in polyphonic signals (only duets were tested). The features that feed the GMM classifier were computed by summing the energy within 60 Hz frequency bands, with overlap of 10 Hz. They were extracted for bands between 50 Hz and 6 kHz. The duets were generated both artificially by combining individual samples from five instruments, and from a real duet for flute and clarinet.

An article by the same authors [16] describes an algorithm able to identify solo instruments in the presence of an accompanying keyboard instrument or orchestra. The features are based on the 15 first harmonics of the dominant F0, generating a 90-element vector. Additional features include frame-to-frame differences (deltas) and differences of differences (delta-deltas) of both frequency and power within individual tones. A GMM is used to perform the classification. The algorithm is evaluated using isolated tones from five instruments, realistic monophonic phrases, and solo instruments with accompaniment. In this last case, the background is polyphonic, which means that the signal as a whole may be considered polyphonic. However, since in this case one instrument is strongly dominant, the challenge may be closer to the monophonic case. Therefore, depending on the criterion, the work of this article could be classified both as monophonic and as polyphonic.

Jincahitra [35] proposed a method that was trained to identify five instruments in signals containing one or two simultaneous instruments. The features are derived from the independent subspace analysis (ISA), whose objective is to decompose individual sources into statistically independent components. The author remarks that this strategy requires that the concurrent instruments be sufficiently non-overlapping, both in time and frequency domains, to work properly. Two classification schemes, k-NN and GMM, were tested. Mediocre performance led to the conclusion that a better learning and decomposition algorithm may be necessary to overcome some of the weaknesses.

The algorithm proposed by Kostek [49] is divided into two parts, the first dealing with the recognition of isolated tones, and the second dealing with duets. The first part follows the conventional steps of most instrument recognition algorithms—preprocessing, feature extraction (MPEG7-based and wavelet-based), and classification, in this case performed by an ANN. The second part begins with the decomposition of the signal into individual instruments, and then the harmonic detection step is applied to remove coincident harmonics from both resulting streams. Then, the residual signals are submitted to a modified frequency envelope distribution (FED) algorithm, which generates a new signal that tries to match the harmonic parts of each residual. Those new signals, called *envelope modulated oscillations* (EMO), are then used in the rest of the classification procedure, which also employs an ANN. Tests were performed using isolated notes from 12 instruments and duets encompassing four instruments.

Vincent and Rodet [79] investigated the use of independent subspace analysis (ISA) for instrument identification in musical recordings. The short-term log-power spectra of the signals are represented as weighted nonlinear combinations of typical note spectra plus background noise. The models for the instruments are learned using isolated tones. Those models are then used as references in the identification of the instruments. Most experiments were performed using solo phrases of five instruments, but the authors also present some preliminary results using a single polyphonic excerpt extracted from a commercial recording. Tests revealed that the method works well under noisy conditions. This strategy is also able to perform polyphonic transcription.

Kitahara et al. [43] proposed a method for instrument identification in polyphonic music. To solve this problem, the authors proposed the use of a feature vector extracted directly from polyphonic sound mixtures, the use of a pitch-dependent timbre model, and the use of musical context-based *a priori* probabilities. The instrument that maximizes the a posteriori probability (PP) of temporarily neighboring notes is taken as the final result for the note under analysis. The tests were performed using mixtures of one, two, and three instruments, generated from samples of four different instruments.

Essid et al. [25] proposed an algorithm able to recognize instruments in polyphonic music, from solos to quartets, without the need of performing multipitch estimation or source separation. The study focuses on Jazz music and respective instruments (10 in total). The possible ensembles of instruments are determined *a priori*, in such a way the algorithm tries to identify the ensemble present in the signal, and not individual instruments. Such a classification is performed by means of a SVM, which is applied to the nodes of a taxonomic structure that is obtained with hierarchical clustering. For every possible pair of classes in each node, the algorithm employs a SVM fed with 50 features selected from a larger set of 355 by means of a feature selection algorithm (FSA). The best features were found to be those based on signal-to-mask ratios (SMR). The algorithm was tested with Jazz musical excerpts extracted from commercial recordings and from the RWC Jazz music database [29].

Kitahara et al. [44] present a technique for instrument identification in polyphonic signals. The focus of the authors was to develop a system that does not rely on side information, particularly the onset detection and F0 estimation. The algorithm calculates the temporal trajectory of instrument existence probabilities for every possible F0, and the results are summarized in a spectrogram-like graphical representation called Instrogram, which is used as classifier. The probabilities are based on the information provided by 28 features. The experiments used artificially generated mixtures of three instruments—nine combinations were used in the tests. The authors claim that the Instrograms can be used not only in instrument recognition, but also in a variety of other applications.

In this article by the same authors [45], the algorithm copes with the effects of time and frequency overlap that occur in polyphonic signals by weighting the features based on how much they are affected by the overlapping. The recognition accuracy is improved by the inclusion of a musical context. Many of the steps, including the classification scheme, are very similar to those described by Kitahara et al. [43, 47]. Several experiments were carried out using mixtures of two, three, and four instruments. The authors recognize that, despite the good results, their evaluation has two limitations, namely the manual F0 feeding and the use of artificially generated mixtures. They remark, however, that most algorithms are tested that way due to the difficulties involved in automatically estimating multiple F0s, and the pervasive use of unrealistic test signals is motivated by the need for correctly labeled references.

This method proposed by Martins et al. [63] recognizes instruments in polyphonic musical signals by first performing a sound source separation that is inspired by the ideas of Computational Auditory Scene Analysis (CASA) and formulated as a graph partitioning problem. The first stage of the separation is a sinusoidal modeling of the signal, and the second part performs a spectral clustering according to a similarity space. The final classification is obtained by a comparison between those clusters and timbre models (TM) from six instruments. The algorithm was tested using mixtures of up to four simultaneous instruments.

Yoshii et al. [84] proposed an interesting original algorithm that tries to recognize three drum sounds in polyphonic signals. The system is based on a template-matching method that uses power spectrograms of drum sounds as templates. The system begins inferring an initial template of each drum sound. This template is then adapted to the actual drum-sound spectrograms. The interference of harmonic sounds is reduced by suppressing harmonic components in the song spectrogram before the adaptation and matching. The experiments were performed using songs present in the RWC Popular Music Database [29].

The vast majority of the work on instrument recognition deals with Western instruments. This proposal by Gunasekaran and Revathy [30] is an exception as its objective is the identification of 11 Indian instruments present both in solo recordings and in duets. Prior to the extraction of the 85 features, the signal is segmented by means of a fractal dimension analysis. Two classification schemes, k-NN and MLP, were tested, with the former one providing the best results.

This proposal by Lampropoulou et al. [55] tries to identify the families of the instruments that are present in artificially generated mixtures. Three instrument families (string, brass, and wind) were considered. A wavelet-based sound source separation is applied prior to the extraction of 30 features. A GMM is used to perform the final classification.

The objective of Little and Pardo [56] was to create a system able to learn how to recognize individual sound sources in a polyphonic context. The authors performed the training of the algorithm using weakly labeled mixtures, in which only the presence or absence of the target instrument is indicated. Accordingly, the objective of the algorithm is to indicate if a given target instrument is present or absent. The authors used 22 features to characterize the signals, and tested three classifiers—Extra Trees (ET), k-NN, and SVM. The mixtures were created by chaining a number of notes for each one of the four instruments considered, and then summing the resulting signals. Tests indicated that learning from weakly labeled mixtures works significantly better than learning from isolated examples when the task is identification of an instrument in novel mixtures.

This algorithm proposed by Somerville and Uitdenbogerd [75] follows the conventional approach of extracting a number of features and using some kind of classifier to recognize the instrument. The authors tested four classification schemes: k-NN, Decision Trees (DT), NaiveBayes (NB), and BayesNet (BN). The best results were obtained using the k-NN. The experiments were performed using excerpts extracted from real recordings. It is important to remark that this algorithm works for polyphonic signals, but it does not recognize individual instruments, but combinations of instruments. Therefore, it is not able to properly identify instruments in combination that were not present in the training, even if all the instruments in this new combination were individually present in some of the training signals.

McKay et al. [66] describe an algorithm to identify five instruments in polyphonic signals. The method first performs a sound source separation by means of the ADRess (Azimuth Discrimination and Resynthesis) algorithm. After that, 44 conventional features are extracted. Finally, a GMM is used for classification. All audio mixtures used by the authors are stereo (in contrast with monaural signals used in most studies) and were created using a MIDI synthesizer. According to the authors, there is still much room for improvement, as the recognition accuracy for some instruments is low.

Pei and Hsu [70] proposed a method for instrument recognition in polyphonic signals that starts by extracting 33 features (MFCC and MPEG7-based). Those features are integrated using a beat-synchronous approach. After that, a fuzzy c-means clustering (FCM) algorithm is applied to separate the information that characterizes each instrument. The resulting clusters feed a SVM, which is responsible to perform the final identification. The training of the method was performed using solo excerpts, but the experiments were carried out over samples extracted from real recordings.

Wieczorkowska and Kubera [81] focused on the identification of the dominant instrument in signals containing simultaneous sounds with the same pitch. The training was performed using two groups of signals, one composed by isolated tones, and the other composed by the same isolated tones mixed with artificial harmonic and noise sounds of lower amplitude. An SVM is used to perform the classification based on the information gathered by 219 features. The test set was generated by taking each sample of each instrument, and adding an interfering sound of same frequency generated by averaging the respective samples of all the other instruments. Before the mixing, such an interfering sound is attenuated to one of eight possible levels. The authors performed thorough experiments and came to the conclusion that the results are satisfactory given the difficulty involved in the task.

Wieczorkowska and Kubik-Komar [82] proposed an algorithm to recognize musical instruments in sound mixes for various levels of accompanying sounds. The algorithm extracts a large number of features, mostly based on MPEG7 descriptors. The information contained in the features is explored by means of Principal Component Analysis and discriminant analysis (DA), which are responsible for the classification. The experiments were performed using samples from 14 instruments contaminated by contents coming from other instruments at different levels, in a procedure similar to that described in Wieczorkowska and kubera [81]. As expected, the recognition rate became worse the stronger the disturbing sounds were.

Burred et al. [10] proposed a new computational model of musical instruments that tries to capture the dynamic behavior of the spectral envelope. They use sinusoidal modeling, frequency interpolation, PCA, and nonstationary Gaussian process modeling to generate a set of prototype curves that characterize the notes of the instruments. The performance of those prototype curves in the identification of instruments was assessed using both monophonic and polyphonic signals, which were generated from samples of five instruments.

Kubera and Ras [54] proposed a technique to identify instruments in mixtures of two instruments. They used 219 features of various types, mostly derived from the MPEG7 standard, but they also proposed some new ones. A SVM was used for classification. The test set consists of mixtures produced by two instruments playing the same note at the same time. The possible instrument combinations are set *a priori*, so the classification system does not try to identify individual instruments, but one of the possible 15 pairs.

The strategy proposed by Barbedo and Tzanetakis [4] explores the spectral disjointness among instruments by identifying isolated partials, from which a number of features are extracted. The information contained in those features is explored by means of a simple binary linear discriminator (BLD), which is used to infer which instrument is more likely to have generated that partial. Hence, the only condition for the method to work is that at least one isolated partial exists for each instrument somewhere in the signal. If several isolated partials are available, the results are summarized into a single, more accurate classification. Experimental results using 25 instruments demonstrate the good discrimination capabilities of the method.

Table 4.2 summarizes all methods described in this subsection. The columns are the same as Table 4.1.

Reference	Class System	Database	Non-harm.	Flat/ Hier.	No. Inst.	No. Features
Barbedo10 [4]	BLD	RWC, Iowa	No	Flat	25	34
Burred10 [11]	Prototype Curves	RWC	No	Flat	5	N/A
Eggink03 [18]	GMM	CR, McGill	No	Flat	5	120
Essid06 [27]	SVM	CR, RWC	Yes	Hier.	10	50 (355)
Gunasek.08 [32]	k-NN, ANN	N/A	Yes	Flat	11	85
Jincahitra04 [40]	k-NN, GMM	Iowa, McGill	No	Flat	5	N/A
Kubera10 [60]	SVM	McGill	No	Flat	8	219
Lamprop.08 [61]	GMM	Iowa	No	Flat[1]	19	30
Little08 [62]	ET, SVM, k-NN	Iowa	No	Flat	4	22
Kitahara05 [49]	PP	RWC	No	Flat	4	3 (43)
Kitahara06 [50]	Instrogram	RWC	No	Flat	4	28
Kitahara07 [51]	PP	RWC	No	Flat	5	43
Kostek04 [55]	ANN	Self, McGill	No	Flat	12	N/A
Martins07 [72]	TM	RWC	No	Flat	6	N/A
McKay09 [75]	GMM	MIDI	No	Flat	5	44
Pei09 [79]	SVM	CR	No	Flat	5	33
Somerville08 [84]	k-NN, DT, NB, BT	CR	No	Flat	8	18
Vincent04 [88]	ICA	CR, RWC	No	Flat	5	N/A
Wieczork.09a [90]	SVM	Iowa, McGill	No	Flat	12	219
Wieczork.09b [91]	DA	McGill	No	Flat	14	217
Yoshii07 [94]	TM	RWC	Yes	Flat	3	N/A

[1] There is a hierarchy, but only parent classes are identified.

Table 4.2
Algorithms of the Polyphonic Group

4.4.3   Other Relevant Work

The work on instrument recognition is not limited to the proposal of new algorithms. There are several publications that tackle problems that have direct relation with the problem of instrument recognition. This subsection lists some of them, briefly introducing their main contribution.

Although this is a relatively new area, there are already a number of good reviews and surveys available, like the two led by Herrera-Boyer [33, 32] and the one written by Kostek [50].

Livshin and Rodet investigated some interesting side aspects of the instrument recognition. They studied the importance of cross-database evaluation in sound classification, concluding that even if a given database has good variability of instruments and conditions, and even if the experiments are carefully designed so no data used in the training is used in the tests, there are still some biasing factors that may affect the results, thus making the use of a different database for validation highly advisable [57]. They tackled the problem of compiling a suitable musical instrument database, suggesting some criteria to remove bad samples from the final set [59].

Finally, there are some works that study specifically which features are better to characterize the signals to be classified. This is the case of the article written by Wieczorkowska et al. [83], which deals specifically with temporal descriptors, and of the work by Nielsen et al. [67], which presents studies on the relevance of spectral features.

4.5   Future Directions

The problem of instrument recognition is still open. Even the identification of instruments in isolated tones is far from being a mature technology. Hence, the research is still in a prospective phase, which means that there are many directions to be taken and many solutions to be tested.

The task of feature extraction has been intensely studied for a long time due to the importance it has in characterizing audio signals. Because of that, there is not much room for improvement. However, especially in the case of polyphonic signals, it may still be possible to create new features capable of extracting some kind of information that no other feature can capture.

The classifiers used by most algorithms are also well established, making significant advances even more difficult than in the case of the feature extraction. Moreover, many studies that compared a number of classifiers under the same conditions reveal that the classifiers actually have similar performances, which indicates that this may not be among the most important factors that influence the accuracy of the methods.

A better candidate for improvement may be the preprocessing of the signal. This stage of the algorithm aims to modify the signal in such a way it becomes more prone to the subsequent processing. Since musical signals are not well behaved, in the sense that they do not follow clear instrument-related rules, a novel preprocessing stage will have to incorporate some mechanism to make the signals more predictable and, more importantly, to make the characteristics of each instrument stand out. Clearly this is not an easy task, but given the ability that human listeners have to perceive even the slightest particular characteristics of a given instrument, such an objective is not infeasible.

It is important to notice that the feature/classifier combination is not the only possible way to recognize instruments. An interesting approach that has been already explored in some studies is the template matching. The main goal of this kind of strategy is to find, for each possible instrument, one or more representations (templates) that are consistently valid despite all the variability between instrument samples. Those templates have also to deal properly with the entire frequency range of each instrument. If the templates are really representative, they will match well with any representation extracted from the signal to be classified. Again, this is not an easy task, but studies performed so far [40, 84] indicate that this option has good potential.

In the specific case of polyphonic signals, the great difficulty lies in the cross-interference caused by simultaneous instruments. In this case, there are two main possible directions for future research. The first one consists in breaking the polyphonic problem into a number of monophonic ones. The problem with this option is that it depends on advances in sound source separation, which is a very difficult problem by itself, especially if there are more instruments than channels. To make things harder, to be useful in this context the source separation needs to be close to perfect, because while the instrument recognition is difficult with its temporal and spectral contents intact, it is nearly impossible if the contents are too distorted. The second option is to explore the temporal and spectral disjointness among instruments that usually occur in any signal. The idea here is identify the regions in time and/or frequency where a given signal is isolated, and then use only those clean parts to perform the identification. In the time domain, this implies in finding isolated notes, and in the frequency domain, the objective is identify partials that do not collide with any other one, and filter the signal to eliminate the remaining mixed partials. Some studies have already been carried out [4, 49], with promising results.

As can be seen, there are many possible options to be investigated and developed. Thare are many aspects of the problem that are still not well known, including the maximum information that can be extracted from a musical signal and what is the best way to explore such an information. Future studies will have the responsibility to bring those important questions closer to an answer.

Bibliography

[1] University of Iowa musical instrument samples database.

[2] G. Agostini, M. Longari, and E. Pollastri. Musical instrument timbres classification with spectral features. *EURASIP Journal Applied Signal Processing*, 2003:5–14, 2003.

[3] J.G.A. Barbedo, A. Lopes, and P.J. Wolfe. Empirical methods to determine the number of sources in single-channel musical signals. *IEEE Transactions on Audio, Speech and Language Processing*, 17:1435–1444, 2009.

[4] J.G.A. Barbedo and G. Tzanetakis. Musical instrument classification using individual partials. *IEEE Transactions on Audio, Speech and Language Processing*, 19:111–122, accepted for publication.

[5] J.P. Bello, L. Daudet, S. Abdallah, C. Duxbury, M. Davies, and M.B. Sandler. *Adaptive Control Processes: Guided Tour*. Princeton University Press, New Jersey, 1961.

[6] J.P. Bello, L. Daudet, S. Abdallah, C. Duxbury, M. Davies, and M.B. Sandler. A tutorial on onset detection in music signals. *IEEE Transactions Speech and Audio Processing*, 13:1035–1047, 2005.

[7] E. Benetos, M. Kotti, and C. Kotropoulos. Musical instrument classification using non-negative matrix factorization algorithms and subset feature selection. In *Proceedings of the IEEE International Conference on Acoustics, Speech, and Signal Processing*, pages 221–224, 2006.

[8] J.C. Brown. Computer identification of musical instruments using pattern recognition with cepstral coefficients as features. *Journal of the Acoustical Society of America*, 105:1933–1941, 1999.

[9] J.C. Brown, O. Houix, and S. McAdams. Feature dependence in the automatic identification of musical woodwind instruments. *Journal of the Acoustical Society of America*, 109:1064–1072, 2001.

[10] J.J. Burred, A. Robel, and T. Sikora. Dynamic spectral envelope modeling for timbre analysis of musical instrument sounds. *IEEE Transactions on Audio, Speech and Language Processing*, 18:663–674, 2010.

[11] N. Chetry and M. Sandler. Linear predictive models for musical instrument identification. In *Proceedings of the IEEE International Conference on Acoustics, Speech and Signal Processing*, pages 173–176, 2009.

[12] G. Costantini, A. Rizzi, and D. Casali. Recognition of musical instruments by generalized min-max classifiers. In *Proceedings of the IEEE Workshop on Neural Networks for Signal Processing*, pages 555–564, 2003.

[13] J.D. Deng, C. Simmermacher, and S. Cranefield. A study on feature analysis for musical instrument classification. *IEEE Transaction Systems Man Cybernetics—Part B: Cybernetics*, 38:429–438, 2008.

[14] J.S. Downie. The music information retrieval evaluation exchange (2005–2007): A window into music information retrieval research. *Acoustical Science and Technology*, 29:247–255, 2008.

[15] J. Eggink and G.J. Brown. A missing feature approach to instrument identification in polyphonic music. In *Proceedings of the IEEE International Conference on Acoustics, Speech and Signal Processing*, pages 553–556, 2003.

[16] J. Eggink and G.J. Brown. Instrument recognition in accompanied sonatas and concertos. In *Proceedings of the IEEE International Conference on Acoustics, Speech and Signal Processing*, pages 217–220, 2004.

[17] J. Eggink and A. Klapuri. Musical instrument recognition using cepstral coefficients and temporal features. In *Proceedings of the IEEE International Conference on Acoustics, Speech and Signal Processing*, pages 753–756, 2000.

[18] C. Elliott. Attacks and releases as factors in instrument identification. *Journal Research in Music Education*, 23:35–40, 1975.

[19] A. Eronen. Automatic musical instrument recognition. Master's thesis, Tampere University of Technology, Finland, 2001.

[20] A. Eronen. Comparison of features for music instrument recognition. In *Proceedings of the IEEE Workshop on Application Signal Processing to Audio and Acoustics*, pages 19–22, 2001.

[21] A. Eronen. Musical instrument recognition using ICA-based transform of features and discriminatively trained HMMs. In *Proceedings of the International Symposium on Signal Processing Application*, 2003.

[22] S. Essid, G. Richard, and B. David. Efficient musical instrument recognition on solo performance music using basic features. In *Proceedings of the AES International Conference*, 2004.

[23] S. Essid, G. Richard, and B. David. Musical instrument recognition based on class pair-wise feature selection. In *Proceedings of the International Conference on Music Information Ret.*, 2004.

[24] S. Essid, G. Richard, and B. David. Hierarchical classification of musical instruments on solo recordings. In *Proceedings of the IEEE International Conference on Acoustics, Speech and Signal Processing*, pages 817–820, 2006.

[25] S. Essid, G. Richard, and B. David. Instrument recognition in polyphonic music based on automatic taxonomies. *IEEE Transactions on Audio, Speech and Language Processing*, 14:68–80, 2006.

[26] S. Essid, G. Richard, and B. David. Musical instrument recognition by pair-wise classification strategies. *IEEE Transactions on Audio, Speech and Language Processing*, 14:1401–1412, 2006.

[27] D. Fragoulis, C. Papaodysseus, M. Exarhos, G. Roussopoulos, T. Panagopoulos, and D. Kamarotos. Automated classification of pianoguitar notes. *IEEE Transactions on Audio, Speech and Language Processing*, 14:1040–1050, 2006.

[28] M. Goto. Development of the RWC music database. In *Proceedings of the International Congress on Acoustics*, pages 553–556, 2004.

[29] M. Goto, H. Hashigushi, T. Nishimura, and R. Oka. RWC music database: Popular, classical, and jazz music databases. In *Proceedings of the International Conference on Music Information Retrieval*, pages 287–288, 2002.

[30] S. Gunasekaran and K. Revathy. Fractal dimension analysis of audio signals for Indian musical instrument recognition. In *Proceedings of the International Conference on Audio, Language and Image Processing*, pages 257–261, 2008.

[31] I. Guyon and A. Elisseeff. An introduction to variable and feature selection. *Journal Machine Learning Research*, 3:1157–1182, 2003.

[32] P. Herrera-Boyer, A. Klapuri, and M. Davy. Automatic classification of pitched musical instrument sounds. In *Signal Processing Methods for Music Transcription*, pages 166–200. Springer, New York, 2006.

[33] P. Herrera-Boyer, G. Peeters, and S. Dubnov. Automatic classification of musical instrument sounds. *Journal of New Music Research*, 32:3–21, 2003.

[34] M. Ihara, S.-I. Maeda, and S. Ishii. Instrument identification in monophonic music using spectral information. In *Proceedings of the IEEE International Symposium on Signal Processing and Information Technology*, pages 595–599, 2007.

[35] P. Jincahitra. Polyphonic instrument identification using independent subspace analysis. In *Proceedings of the IEEE International Conference on Multimedia and Expo*, pages 1211–1214, 2004.

[36] C. Joder, S. Essid, and G. Richard. Temporal integration for audio classification with application to musical instrument classification. *IEEE Transactions on Audio, Speech and Language Processing*, 17:174–186, 2009.

[37] I. Kaminskyj and A. Materka. Automatic source identification of monophonic musical instrument sounds. In *Proceedings of the IEEE International Conference on Neural Networks*, pages 189–194, 1995.

[38] I. Kaminskyj and P. Voumard. Enhanced automatic source identification of monophonic musical instrument sounds. In *Proceedings of the Australian, New Zealand Conference on Intelligent Information Systems*, pages 189–194, 1995.

[39] I. Kaminskyj and T. Czaszejko. Automatic recognition of isolated monophonic musical instrument sounds using *k*-NNC. *Journal Intelligent Information Systems*, 24:199–221, 2005.

[40] K. Kashino and H. Murase. A sound source identification system for ensemble music based on template adaptation and music stream extraction. *Speech Communication*, 27:337–349, 1999.

[41] R.A. Kendall. The role of acoustic signal partitions in listener categorization of musical phrases. *Music Perception*, 4:185–214, 1986.

[42] T. Kitahara. Computational musical instrument recognition and its application to content-based music information retrieval. PhD thesis, Kyoto University, Japan, 2007.

[43] T. Kitahara, M. Goto, K. Komatani, T. Ogata, and H.G. Okuno. Instrument identification in polyphonic music: Feature weighting with mixed sounds, pitch-dependent timbre modeling, and use of musical context. In *Proceedings of the International Conference on Music Information Retrieval*, pages 558–563, 2005.

[44] T. Kitahara, M. Goto, K. Komatani, T. Ogata, and H.G. Okuno. Instrogram: A new musical instrument recognition technique without using onset detection nor F0 estimation. In *Proceedings of the IEEE International Conference on Acoustics, Speech and Signal Processing*, pages 229–232, 2006.

[45] T. Kitahara, M. Goto, K. Komatani, T. Ogata, and H.G. Okuno. Instrument identification in polyphonic music: Feature weighting to minimize influence of sound overlaps. *EURASIP Journal Applied Signal Processing*, 2007.

[46] T. Kitahara, M. Goto, and H.G. Okuno. Category-level identification of non-registered musical instrument sounds. In *Proceedings of the IEEE International Conference on Acoustics, Speech and Signal Processing*, pages 253–256, 2004.

[47] T. Kitahara, M. Goto, and H.G. Okuno. Pitch-dependent identification of musical instrument sounds. *Applied Intelligence*, 23:267–275, 2005.

[48] A. Klapuri. Sound onset detection by applying psychoacoustic knowledge. In *Proceedings of the IEEE International Conference on Acoustics, Speech and Signal Processing*, pages 3089–3092, 1999.

[49] B. Kostek. Musical instrument classification and duet analysis employing music information retrieval techniques. *Proceedings of the IEEE*, 92:712–729, 2004.

[50] B. Kostek. Intelligent musical instrument sound classification. In *Perception-Based Data Processing in Acoustics*, pages 39–186. Springer-Verlag, New York, 2005.

[51] B. Kostek and A. Czyzewski. Representing musical instrument sounds for their automatic classification. *Journal Audio Eng. Society*, 49:768–785, 2001.

[52] O. Kramer and T. Hein. Stochastic feature selection in support vector machine based instrument recognition. In *KI 2009: Advances in Artificial Intelligences*, number 5803 in Lecture Notes in Computer Science, pages 727–734. Springer-Verlag, New York, 2009.

[53] A.G. Krishna and T. V. Sreenivas. Music instrument recognition: From isolated notes to solo phrases. In *Proceedings of the IEEE International Conference on Acoustics, Speech and Signal Processing*, pages 265–268, 2004.

[54] E. Kubera and Z.W. Ras. Identification of musical instruments by features describing sound changes in time. In *Advances in Intelligent Information Systems*, pages 357–366. Springer-Verlag, New York, 2010.

[55] P.S. Lampropoulou, A.S. Lampropoulos, and G.A. Tsihrintzis. Musical instrument category discrimination using wavelet-based source separation. In *New Directions in Intelligent Interactive Multimedia*, volume 142, pages 127–136. Springer-Verlag, New York, 2008.

[56] D. Little and B. Pardo. Learning musical instruments from mixtures of audio with weak labels. In *Proceedings of the International Conference on Music Information Retrieval*, pages 127–132, 2003.

[57] A. Livshin and X. Rodet. The importance of cross database evaluation in sound classification. In *Proceedings of the International Conference on Music Information Retrieval*, 2003.

[58] A. Livshin and X. Rodet. Instrument recognition beyond separate notes—Indexing continues recordings. In *Proceedings of the International Computing Music Conference*, 2004.

[59] A. Livshin and X. Rodet. Purging musical instrument sample databases using automatic musical instrument recognition methods. *IEEE Transactions on Audio, Speech and Language Processing*, 17:1046–1051, 2009.

[60] R. Loughran, J. Walker, M. ONeill, and M. OFarrell. Musical instrument identification using principal component analysis and multi-layered perceptrons. In *Proceedings of the International Conference on Audio, Language and Image Processing*, pages 643–648, 2008.

[61] K.D. Martin. Sound-source recognition: A theory and computational model. PhD thesis, Massachusetts Institute of Technology, 1999.

[62] K.D. Martin and Y.E. Kim. Musical instrument identification: A pattern-recognition approach. In *Proceedings of the Meeting of the Acoustics Society America*, 1998.

[63] L.G. Martins, J.J. Burred, G. Tzanetakis, and M. Lagrange. Polyphonic instrument recognition using spectral clustering. In *Proceedings of the International Conference on Music Information Retrieval*, 2007.

[64] G. Mazarakis, P. Tzevelekos, and G. Kouroupetroglou. Musical instrument recognition and classification using time encoded signal processing and fast artificial neural networks. In *Advances in Artificial Intelligence*, number 3955 in Lecture Notes in Artificial Intelligence, pages 246–255. Springer-Verlag, New York, 2006.

[65] S. McAdams, J. W. Beauchamp, and S. Meneguzzi. Discrimination of musical instrument sounds resynthesized with simplified spectrotemporal parameters. *Journal Acoustics Society America*, 105:882–897, 1999.

[66] J. McKay, M. Gainza, and D. Barry. Evaluating ground truth for ADRess as a preprocess for automatic musical instrument identification. In *Pres. AES Convention*, page 7816, 2009.

[67] A.B. Nielsen, S. Sigurdsson, L.K. Hansen, and J. Arenas-Garca. On the relevance of spectral features for instrument classification. In *Proceedings of the IEEE International Conference on Acoustics, Speech and Signal Processing*, pages 485–488, 2007.

[68] F. Opolko and J. Wapnick. *McGill-Queen's University Master Samples.* McGill University, Press, Canada, 1987.

[69] T.H. Park. Towards automatic musical instrument timbre recognition. PhD thesis, Princeton University, New Jersey, 2004.

[70] S.-C. Pei and N.-T. Hsu. Instrumentation analysis and identification of polyphonic music using beat-synchronous feature integration and fuzzy clustering. In *Proceedings of the IEEE International Conference on Acoustics, Speech and Signal Processing*, pages 169–172, 2009.

[71] D. Piccoli, M. Abernethy, S. Rai, and S. Khan. Applications of soft computing for musical instrument classification. In *AI 2003: Advances in Artificial Intelligence, 16th Austrian Conference on Artificial Intelligence*, number 2903 in Lecture Notes in Computer Science, pages 878–889. Springer-Verlag, New York, 2003.

[72] C. Pruysers, J. Schnapp, and I. Kaminskyj. Wavelet analysis in musical instrument sound classification. In *Proceedings of the International Symposium on Signal Processing Application*, 1998.

[73] X. Rodet and F. Jaillet. Detection and modeling of fast attack transients. *Proceedings of the International Computing Music Conference*, 2001.

[74] C. Simmermacher, D. Deng, and S. Cranefield. Feature analysis and classification of classical musical instruments: An empirical study. In *Advances of Data Mining*, volume 4065, pages 444–458. Springer-Verlag, New York, 2006.

[75] P. Somerville and A.L. Uitdenbogerd. Multitimbral musical instrument classification. In *Proceedings of the International Symposium on Computing Science Application*, pages 269–274, 2008.

[76] A. Srinivasan, D. Sullivan, and I. Fujinaga. Recognition of isolated instrument tones by conservatory students. In *Proceedings of the International Conference on Music Perception and Cognition*, pages 17–21, 2002.

[77] B. Tan and D. Sen. The use of the attack transient envelope in instrument recognition. In *Proceedings of the Australian International Conference on Speech and Science Technology*, pages 489–493, 2006.

[78] A. Tindale, A. Kapur, and I. Fujinaga. Towards timbre recognition of percussive sounds. In *Proceedings of the International Computing Music Conference*, pages 592–595, 2004.

[79] E. Vincent and X. Rodet. Instrument identification in solo and ensemble music using independent subspace analysis. In *Proceedings of the International Conference on Music Information Retrieval*, pages 576–581, 2004.

[80] E. M. von Hornbostel and C. Sachs. Systematik der musikinstrumente. *Zeitschrift*, 46:553–590, 1914.

[81] A.A. Wieczorkowska and E. Kubera. Identification of a dominating instrument in polytimbral same-pitch mixes using SVM classifiers with nonlinear kernel. *Journal Intelligent Information System*, published OnLine, 2009.

[82] A.A. Wieczorkowska and A. Kubik-Komar. Application of discriminant analysis to distinction of musical instruments on the basis of selected sound parameters. In *Man-Machine Interactions*, number 59 in Advances in Intelligent and Soft Computing, pages 407–416. Springer-Verlag, Berlin, 2009.

[83] A.A. Wieczorkowska, J. Wroblewski, P. Synak, and D. Slezak. Application of temporal descriptors to musical instrument sound recognition. *Journal Intelligent Information System*, 21:71–93, 2003.

[84] K. Yoshii, M. Goto, and H.G. Okuno. Drum-sound recognition for polyphonic audio signals by adaptation and matching of spectrogram templates with harmonic structure suppression. *IEEE Transactions on Audio, Speech and Language Processing*, 15:333–345, 2007.

[85] L. Yu and H. Liu. Efficient feature selection via analysis of relevance and redundancy. *Journal Machine Learning Research*, 5:1205–1224, 2004.

5

Mood and Emotional Classification

Mitsunori Ogihara

University of Miami

Youngmoo Kim

Drexel University

CONTENTS

Researchers have studied emotional aspects of music for a long time. A major aspect of the emotional studies of music is the following pair of questions about felt or perceived emotions and moods:

- What types of emotions and moods can be identified in a piece of music?

- How emotions and moods are communicated? How do underlying musical structures, articulation in performance, and preparation and disposition of listener contribute to the communication process?

Another aspect of the studies is the same pair of questions with respect to induced emotion:

- What types of emotions and moods can be induced by music listening?

- How emotions and moods are induced?

Recent advancements in techniques for extracting relevant information from audio recordings, metadata, and texts have encouraged researchers to explore such questions through computational analysis of music data. In particular, the very first question of identifying emotions and moods represented in music, is drawing much attention from music information retrieval researchers.

5.1   Using Emotions and Moods for Music Retrieval

The type of music a listener prefers to listen is not fixed. Rather, it is dependent on a number of factors, including her/his music training, her/his entire and recent music listening, and her/his present disposition to music listening (see, e.g., the work of Juslin and Laukka [33]).

For example, consider a scenario in which a person that has been listening to Avant Garde Jazz music by Albert Ayler for the past few days and has been finding to be enjoyable suddenly feels like listening to "something tender, happy, and loving," and so picks "Close To You" sung by The Carpenters.

In such a situation, if the person is only sure about listening to "something tender, happy, and loving," how can a music retrieval system assist the listener in finding a tune? To solve this problem, we think of endowing the retrieval system with the ability to accept terms like "Tender," "Happy," and "Love," which we call *emotion and mood labels* to mean that they refer to emotions and moods represented in a given piece of music. To make emotion and mood label queries possible, we have to go through three steps:

1. Determining which labels will be used.

2. Assigning the labels to the pieces in the collection.

3. Designing the query system so as to handle emotion and mood query terms.

All three steps are nontrivial. For the first, the existence of various organizations of emotion makes it difficult to decide which labels should be used. For the second, since emotion and mood labels are not part of metadata, those labels have to be calculated from other available information. For the last, the distance from a candidate piece of music to a given set of emotion and mood query terms has to be quantified in a meaningful manner.

Although each of the three steps deserves serious investigation, this chapter is concerned with the first two steps only. The next section, Section 5.2, discusses the first step and provides an overview of the research in taxonomies of emotion and mood and their relations to music. The subsequent two sections jointly cover the second step. Section 5.3 presents efforts in obtaining ground-truth labels for emotion and mood. Section 5.4 presents efforts in developing computational methods for calculating emotion and mood labels of music. Section 5.5 discusses future issues.

5.2 Emotion and Mood: Taxonomies, Communication, and Induction

5.2.1 What Is Emotion, What Is Mood?

What lies beneath the mood and emotional classification of music is the desire to

(i) understand music in terms of what listeners, beyond cognition of musical structures, identify in music or feel about it, and then

(ii) use that understanding for improving retrieval of music.

An important initial step toward achieving these goals is to clarify what emotions and moods are. The *New Oxford American Dictionary* [3] defines *emotion* as "a natural instinctive state of mind deriving from one's circumstances, mood, or relationships with others," "any of the particular feelings that characterize such a state of mind," and "instinctive or intuitive feeling as distinguished from reasoning or knowledge."

While an emotion is a state of mind that is instinctive and peculiar, mood is something more persistent and obscure. Again, the New Oxford American Dictionary defines *mood* as "a temporary state of mind or feeling" and, in particular, "the atmosphere or pervading tone of something." Thayer [75] defines moods to be "background feelings that last for a time and that often have no particular cause" and often the moods are just "good or bad."

Type	Duration	Awareness	Target
Emotion	Instinctive	Full	Often present
Mood	Persistent	Partial, or even absent	Absent

Table 5.1
Comparison between Emotion and Mood

These definitions enable us to draw some line between mood and emotion. While emotions are instinctive and sometimes attached to particular objects, mood may drift from one to another, people may not recognize their mood until they pay attention to them, people may not be able to figure out why they are in a specific mood, and while being in a certain mood people may experience various emotions. Table 5.1 summarizes the difference between emotion and mood.

5.2.2 A Hierarchical Model of Emotions

The literature in developmental psychology states that humans learn basic emotions in their early developmental stage, which is through simple stimuli, and more complex emotions in later stages (see, for example, Bower [7]). This developmental distinction between basic ones and more specific ones suggests a two-level hierarchical organization of emotion. By adding to it the division between positive and negative emotions, one obtains the hierarchical diagram of Schaver et al. [64]. In this hierarchy, at the top level are super-ordinate categories (positive and negative), at the middle level are the basic emotions, which Schaver et al. think of as universally understood emotions, and at the bottom level are the most specific 135 emotions, which may exhibit individual variability in understanding. Figure 5.1 shows the hierarchy in a simplified form with just a few bottom-level emotions present.

5.2.3 Labeling Emotion and Mood with Words and Its Issues

For description of emotion and mood we usually use words that represent emotion (or affect), for example, those appearing in the emotion hierarchy. The use of such words is very convenient and natural, but has peculiar characteristics that we must keep in mind.

The first is their multiplexity; i.e., all at the same time people can be in more than one mood and experience multiple emotions. As Michael Franks sang, people may feel like: "I don't know why I'm so happy and sad."[1]

The second is individual variability in emotion assessment. This phenomenon is more prominent when emotions are assessed at the subordinate

---

[1] Michael Franks, "I Don't Know Why I'm So Happy and Sad," from the album *The Art of Tea*, Reprise Records, 1976.

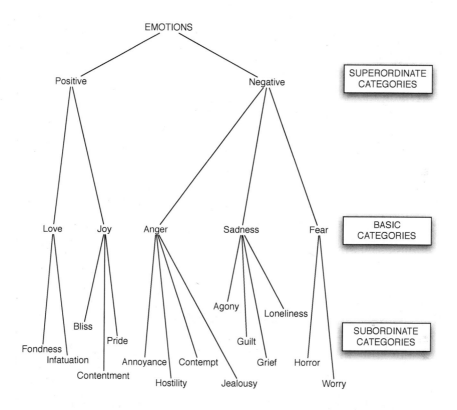

Figure 5.1
The hierarchy of emotions according to Schaver et al. [64]. At the subordinate level (the bottom level) only a small number of selected emotion categories are shown.

level than at the basic level or at the top level. On hearing a man telling the
story about a very unreasonable customer representative, a friend may think
that the man is angry but the man himself thinks that he is just frustrated.
This issue is prominent when humans provide emotion and mood labels to
music, in particular, instrumental music.

The last one, which has much to do with the second, is overlap among
emotions and among moods. Emotion and mood labels words, which are usually adjectives or nouns, have synonyms—"words having exactly or nearly the
same meaning as another word" according to the *New Oxford English Dictionary*. If the size of overlap between an emotion word, A, and another, B,
is significant but not 100%, treating the two as different makes assessment of
emotion, both by human and by machine, very difficult with respect to these
two, since the distinction among "both A and B," "A but not B," and "B but
not A" is obscure. The Schaver hierarchy contains as many as 135 emotions,
and large numbers of emotion words have been identified, for example, in
ANEW (Affective Norms for English Words) [8]. Thus, an attempt to classify
music into presence/nonpresence with such a large number of labels appears
to be extremely difficult, if not utterly impossible.

5.2.4   Adjective Grouping and the Hevner Diagram

The aforementioned characteristics suggest that to study emotions and moods
in music we will perhaps have to use groups that combine various subordinate-
level emotions and will perhaps have to treat the classification problem as a
multilabel problem.

Hevner [25] is the first to experimentally found adjective groups through
subject experiments. In this study, 450 subjects were asked to listen to 26
pieces of classical music and for each piece select from a list of 66 adjectives
any number of words that seemed "appropriate to the music." Through co-
occurrence analysis of the response Hevner divided the 66 adjectives into eight
groups and laid out the groups into a circle in such a way that the collection
of emotions represented by group members gradually changes as we go around
the circle (see Figure 5.2).

Hevner studied the scores of the pieces and attempted to correlate their
characteristics with the adjective groups chosen by the subjects. Correlations
were found in melody contour (ascending versus descending), harmony (simple
versus complex), mode (major versus minor), and rhythm (firm versus flow-
ing), with by far the clearest correlation between mode and "Happy" versus
"Graceful" as well as between mode and "Sad" versus "Dreamy."

5.2.5   Multidimensional Organizations of Emotion

Psychologists have been using two-dimensional plots for organizing emotions.
In such a representation each axis corresponds to a pair of adjectives having
opposite meanings (for example, happy and sad). A point in the space thus

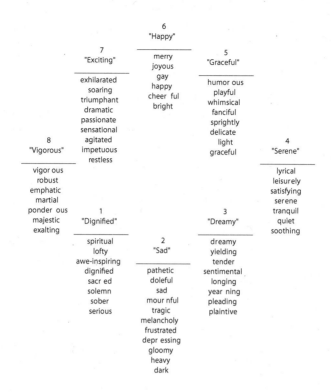

Figure 5.2
The eight adjective groups discovered by Hevner.

Figure 5.3
The facial expression categories by Schlosberg.

specifies in each axis, which of the two opposite feelings is more strongly felt. The use of two independent factors in representing emotion was first suggested by Wundt [89]. Psychological research in the late 20th century expanded the idea of two-dimensional emotion plots and produced a number of plausible representations, perhaps beginning in the seminal work of Schlosberg [67] to classify facial expressions. These ideas were further extended in the work of Russell [63], Thayer [75], and Watson and Tellegen [85].

The diagram of Schlosberg [67] is, like the one by Hevner, circular. In it the whole area of facial expressions is a circle and the area is divided into six sectors: "Fear and Suffering," "Anger and Determination," "Disgust," "Contempt," "Love, Mirth, and Happy," and "Surprise" (see Figure 5.3). This idea of circular emotional changes is further investigated by Russell [63], in which 28 emotion words are organized in a circumplex. The two components that span emotion are the Positiveness and the Arousal (i.e., excitement). The Thayer model [75] is inspired by the Russell model. Like the Russell model, the vertical axis represents the arousal level, but the horizontal axis represents the level of tension felt. The two diagrams are shown in Figure 5.4. The emotion model of Russell has been simplified in the Barrett-Russell model, which is shown in Figure 5.5. Two-dimensional emotion models assume that a human emotion can be identified with a location on the two-dimensional diagram. In all the three models, Russell, Thayer, and Barrett-Russell, there is an axis that represents positive and negative affects. Watson and Tellegen propose a model using an axis representing Positiveness and another representing Negativeness (see Figure 5.6) and view the two diagonal axes as Engagement and Pleasantness. For example, the combination of high positive and low negative represents Arousal. From this model Watson, Clark, and Tellegen developed a method called PANAS (Positive Affect–Negative Affect Schedule) for self-appraisal of affect using 20 emotional terms (provide a five-point scale answer with respect to each term) [84]. This has been further extended to PANAS-X, a schedule involving 60 labels [83].

5.2.5.1   Three and Higher Dimensional Diagrams

Some research suggests that the space of perceived emotion has at least three dimensions. Wedin [86] examined rank correlations of words in an adjective checklist on 50+ pieces of music and identifies three pairs of adjective sets. Wedin suggests that these three pairs correspond to:

- Intensity versus Softness

- Pleasantness versus Unpleasantness or Gaiety versus Gloom,

- Solemnity versus Triviality

Also, Leman et al. [44] study correlations among 15 adjective pairs by having 100 students label 60 pieces of music that cover many genres using the pairs. Through principal component analysis of the covariance matrix of the pairs,

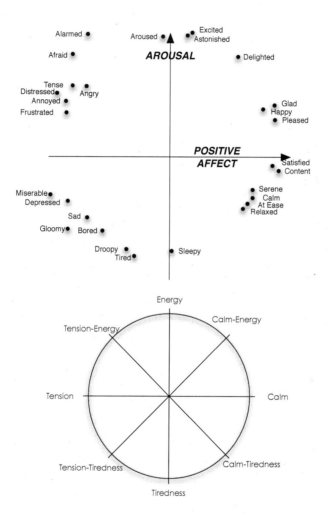

Figure 5.4
Top: The two-dimensional circumplex emotion word mapping by Russell.
Bottom: The two-dimensional emotion map suggested by Thayer.

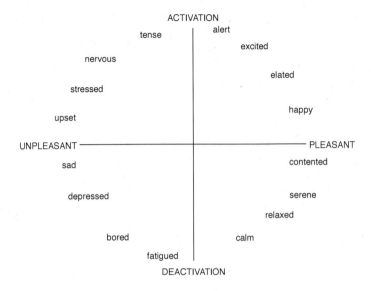

Figure 5.5
The Barrett-Russell model.

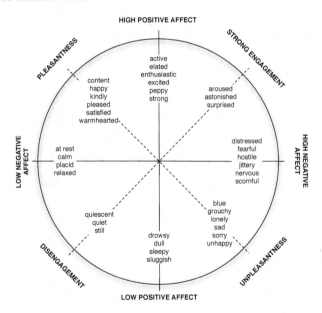

Figure 5.6
The emotion diagram by Watson and Tellegen, rotated by 45 degrees clockwise from its often used diagram so as clarify the two axes.

they find that the first three components cover 56.5% of covariance and suggest that those three together capture most of the perceived emotion in music. Each of the three components is a linear combination of six fundamental dimensions (that is, six pairs of adjectives with opposite meanings). Those dimensions were identified as Valence, Activity, and Interest.

5.2.6 Communication and Induction of Emotion and Mood

Whether music can communicate any emotion and any mood is perhaps an unsolved question. For any emotion, one can write a lyric that expresses it and attach a melody to the lyric thereby producing a song that communicates the emotion. This rather deceptive argument clearly does not apply to instrumental music.

We can tell, however, from the aforementioned work of Hevner [25], that a piece of instrumental music can communicate a certain set of emotions and moods and that such emotions and moods may have to do with the musical structures of the pieces listened to. Hampton [21] shows that there is substantial consistency in emotions recognized by listeners. The emotional expressiveness of music structures, as suggested by Hevner, has been further investigated. Gabrielsson and Lindström [20] offer a summary of the state of knowledge in this area.

In addition to the emotions and moods represented by structures of music, researchers have been studying how the music is performed has a role in emotion and mood communication. Researchers have found that performers can communicate specific emotions and to do so they use a set of musical cues [12, 19, 31, 56, 76]. Juslin [32] and Gabrielsson [18] are excellent surveys of this area. It is known at least that the basic emotions of Thayer (see Figure 5.4) can be well communicated using instrumental performances using cue. It is unclear, however, whether subordinate emotions can be well communicated.

Another important topic related to emotion and mood in music is the induction of emotion by music listening. This topic is first extensively investigated by Meyer [53]. As mentioned early the main questions are the types of emotions induced by music and their relations with musical structures and personality. Such studies have possible applications in music therapy and work performance improvement [35, 45, 58]. The extensive literature in this area [14, 23, 33, 38, 57, 65, 66, 71, 72] informs three important points we need to be careful with when studying emotion and mood classification in music. First and foremost, we should not get confused between perceived emotions and induced emotions. Second, emotions perceived in music are not necessarily induced in the listeners. Third, whether music listening successfully induces intended emotions depends on the personality and disposition of the listener.

5.3   Obtaining Emotion and Mood Labels

5.3.1   A Small Number of Human Labelers

An important step in music information retrieval research involving emotion and mood is to build a data set in which the pieces have emotion or mood labels. Given a data set without emotion and mood labels, one can use a human for manual labeling. This straightforward approach is taken in the work of Li and Ogihara [46]. In this work a single person labeled 500 pieces of instrumental music in four genres (Ambient, Classical, Jazz, and Fusion). The labels used are the 10 labels proposed by Farnsworth [15], a variation of the eight class organization by Hevner that divided Group 5 into the "Fanciful" and "Light" subgroup and the "Delicate" and "Graceful" subgroup and adds a new group of "Frustrated." The labeler answered for each piece and for each of the 10 emotion categories whether the emotion is represented in the piece, with no restriction on the number of labels to be assigned. This resulted in a total of 5,000 yes/no binary labels provided by the labeler.

The task was carried over several days. The labeler sometimes returned to the data to confirm labels to find that earlier labels were not accurate, though they seemed correct in the first round. This resulted in revisions of the labels.

This episode illustrates an important issue in human-labeled emotion and mood labels—inconsistency of human labels. The classification accuracy on this data set was not very high. The overall precision was 0.3247 and the overall recall was 0.5411. In terms of F-measure

$$\frac{2 \cdot \text{Precision} \cdot \text{Recall}}{\text{Precision} + \text{Recall}}, \tag{5.1}$$

the score was 0.4058. With respect to microaveraging (the unweighted average of class-wise performance values), the precision was 0.3621 and the recall was 0.5893, which resulted in the F-measure of 0.4487. The low accuracy can be partially attributed to the inconsistency of labels.

Another issue, one that is prevalent when dealing with music that contains structural changes, is that represented emotions and moods may change during performances. The aforementioned work by Li and Ogihara used a 30-second excerpt from each piece of music and the labels were assigned purely based on the music represented in the 30 seconds. To deal with emotions and moods of an entire piece of music one may need to track emotion/mood appraisals over entire duration of music time, as suggested by Korhonen [36] and Korhonen et al. [37], and experimented with by Liu et al. [51].

Li and Ogihara [47] followed up on their preliminary work using all the Jazz tracks from the previous experiment (235 tracks) and two labelers using a consolidated set of labels: "Cheerful" versus "Depressing," "Relaxing" versus "Exciting," and "Comforting" versus "Disturbing." The first two of these correspond to the "Pleasant"–"Unpleasant" axis and the "Arousal"–

"Nonarousal" axis of the Russell diagram discussed in Section 5.2. With this modification, classification accuracy substantially improved from the previous work. For each binary category and for each labeler, the accuracy was between almost 70% to 83%. Li and Ogihara notice that the label assignments do not agree much between the two labelers. This suggests that many labelers must be used to obtain coherent labels.

Also, Yang and Lee [90] used labels given by a single person. In the first experiment, the listener evaluated the total intensity of the two affects in the Watson–Tellegen diagram, which can be viewed as

$$\text{Positive Affect Value} + \text{Negative Affect Value},$$

that is, the sum of the x and y coordinate values. They extracted acoustic features from 500 pieces of Rock music for 20 seconds of music beginning at the end of the first third. The extracted features were: beat per minute estimation (BPM), the low-level descriptors (LLD) of MPEG-7 (12 features), and the high-level descriptors obtained by the use of Sony Extractor Discovery System [59] (12 features). They found that the intensity had a very strong linear correlation with BPM as well as with other features.

In the second experiment, from the previous pool of pieces, 145 songs were chosen. The words appearing in the lyrics were looked up in the General Inquirer (GI) database [73]. GI is a database that contains more than 8,000 words in which each word is annotated with 182 binary labels describing its word sense. Emotion-related terms (such as pleasure and pain) are among the labels, but there are many other labels that do not appear to be directly related to emotion (such as academic and economy). The labeler gives labels from PANAS-X and attempts were made to distinguish between songs that were assigned labels with a High Negative Affect and those that were not using the C4.5 decision tree algorithm [62]. The same attempt was made with respect to High Positive Affect. In both cases the accuracy was around 80% with or without using acoustic features.

5.3.2 A Large Number of Labelers

Eerola, Lartillot, and Toiviainen [13] used a clever two-stage labeling process that involved human labelers. In the first stage, a dozen experts (music students) chose 360 movie sound tracks that represent five basic emotions ("Happy," "Sad," "Tender," "Scary," and "Angry") of Schaver. On this stage, the experts also categorized the emotions into a three-dimensional affect space spanned by three dimensions of Activity, Valence, and Tension, where each dimension is discretized in seven levels (the Likert scale). The experts rated each track with respect to these scales as well. For each basic emotion they chose five tracks that represent the emotion at the strongest level and five at the moderate level. This constituted a data set of 50 tracks.

With respect to the three-dimensional representation, for each axis they sampled tracks at the top 4 percentile (20 tracks). This resulted in a data set of 60 tracks. In the second stage, they provided these data sets to as many as 112 music students to label them with respect to the five emotions as well as with respect to the three-dimensional map.

5.3.3   Mood Labels Obtained from Community Tags

The creation of music listening communities over the Internet, such as the aforementioned Last.fm [2], has opened up the possibility of inferring emotion and mood categories from the labels/tags assigned by members of the communities.

Also, there have been recent efforts using "Games With a Purpose," that is, collaborative Internet games for collecting tags and labels from the broad public [82]. While labeling tasks can be time consuming, tedious, and expensive, carefully designed games can produce large quantities of high-quality annotations. Several such games have been been developed for the collection of music tags, such as *MajorMiner* [52], *ListenGame* [78], and *TagATune* [43] (based on InputAgreement [42]), which have produced tag corpora that include many emotion- and mood-related terms. *MoodSwings* is a different style of collaborative game for time-varying annotation of emotions based on the Barrett-Russell parametric affect-arousal model, producing coordinate labels at each second of music [34, 55]. *Herd It* combines multiple types of music annotation games, including affect-arousal annotation of clips, descriptive labeling, and music trivia [5].

One thing we need to be careful of when using labels/tags from online games is that they are noisy because the people proving labels/tags are anonymous, their levels of music education are unknown, and their sincerity is unclear. The noisiness can be generally overcome by collecting inputs from many (hundreds, thousands) community members and looking for consistency within the community.

5.3.3.1   MIREX Mood Classification Data

Hu, Downie, and Ehmann [28] implemented a strategy for obtaining mood label groups and selecting representative pieces of music for them. Resulting from this effort is a mood classification benchmark data set called MIREX Mood Classification Data, part of a popular benchmark MIREX (Music Information Retrieval Exchange) data set collection at University of Illinois [29].

Their development of mood labels began by collecting from Last.fm [2] social tags assigned to candidate pieces. There were 8,000 candidate pieces. The collected social tags were analyzed in terms of POS (part-of-speech) tags to collect single-word adjectives. Next from the single-word adjectives nonemotion words as well as those appearing in social tags for only a small number of pieces were eliminated.

The remaining words were then clustered using K-Means in terms of their co-occurrences. The resulting clusters of adjectives are shown in Table 5.2.

1	2	3	4	5
Rowdy	Amiable/	Literate	Witty	Volatile
Rousing	Good-natured	Wistful	Humorous	Fiery
Confident	Sweet	Bittersweet	Whimsical	Visceral
Boisterous	Fun	Autumnal	Wry	Aggressive
Passionate	Rollicking	Brooding	Campy	Tense/Anxious
	Cheerful	Poignant	Quirky	Intense
			Silly	

Table 5.2
MIREX Mood Adjective Clusters

Then, from the pool of candidate tracks, 250 exemplar pieces were chosen for each cluster. These 250 pieces were evaluated by more than 20 people. Based on the evaluation, 120 pieces were selected such that the human evaluation showed strong agreement. This created a data set of 600 tracks total. A diagram of the construction steps is shown in Figure 5.7.

5.3.3.2 Latent Semantic Analysis on Mood Tags

The aforementioned work of Hu, Downie, and Ehmann demonstrates that the emotion and mood words appearing in listening community tags can be used to develop mood categories. How are these mood categories related to each other?

Laurier et al. [41] studied this question by analyzing co-occurrences of those emotion and mood labels in the Last.fm tags.

After elimination of infrequent words, they applied latent semantic analysis (LSA) and then K-Means clustering. The top three words of each of the four clusters are shown in Table 5.3. These clusters appear to correspond to the four emotions in the Thayer map, respectively to Tension-Energy, Calm-Energy, Calm-Tenderness, and Tension-Tenderness.

Clusters			
1	2	3	4
Angry	Sad	Tender	Happy
Aggressive	Bittersweet	Soothing	Joyous
Visceral	Sentimental	Sleepy	Bright

Table 5.3
The Top Three Words of the Four Clusters Obtained by Laurier et al. [41]

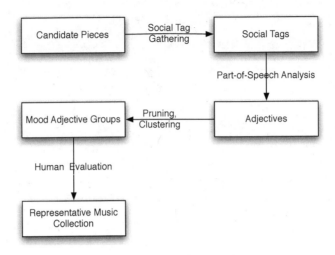

Figure 5.7
The diagram for constructing the MIREX mood data.

5.3.3.3   Screening by Professional Musicians

To achieve high level of consistency or coherence in the labels is to use experts (musicians, music students, and music teachers). The aforementioned work of Eerola, Lartillot, and Toiviainen [13] and Hu, Downie, and Ehmann [28] involved experts in labeling process.

Liu, Lu, and Zhang [50, 51] used the Thayer model and divided labels into four separate classes: "Exuberance" (Calm-Energy), "Anxious/Frantic" (Tension-Energy), "Contentment" (Calm-Tiredness), and "Depression" (Tension-Tiredness). They selected a number of styles and structures in classical music that seemed to be associated with each of the four groups, selected a number of such pieces, and then extracted four 20-second representative segments from each selected piece. Then they had musicians listen to the chosen segments so as to eliminate mislabeled tracks. The resulting database consisted of approximately 800 tracks extracted from 250 pieces.

5.4   Examples of Music Mood and Emotion Classification

5.4.1   Mood Classfication Using Acoustic Data Analysis

Liu, Lu, and Zhang [50, 51] decomposed their four-class classification problem (see Section 5.3.3.3) into a two-layer binary classification problem, in which the first level separated between "Tiredness" and "Energy" and the second level separates, within each of the two classes generated by the first level classifier between "Calm" and "Tenses." The extracted features are subband-divided spectral features, short-term Fourier transform features, and rhythmic features. They applied a transform to make these signals orthogonal and then fed the transformed feature vector into Gaussian Mixture Models consisting of four Gaussian distributions. The accuracy was around 85%.

Li and Ogihara [47] used MFCC and short-term Fourier transform by way of Marsyas [81, 80], as well as statistics (the first three moments plus the mean of absolute value) calculated over subband coefficients obtained by applying the Daubechies Db8 wavelet filter [11] on the monaural signals (see Li and Ogihara [49]). On each axis, the accuracy of binary classification was in the 70% to 80% range for both labelers.

In 2007, MIREX first included a task on audio music mood classification, using the five mood clusters mentioned previously, and performance has increased each subsequent year [29]. In 2007, Tzanetakis achieved the highest correct classification (61.5%), using MFCC, spectral shape, centroid, and rolloff features with an SVM classifier [79]. The highest performing system in 2008 by Peeters demonstrated some improvement (63.7%) by introducing a much larger feature corpus including, MFCCs, Spectral Crest/Spectral Flatness, as well as a variety of chroma based measurements [61]. The system uses a Gaussian Mixture Model (GMM) approach to classification, but first employs Inertia Ratio Maximization with Feature Space Projection (IRMFSP) to select the most informative 40 features for each task (in this case mood), and performs linear discriminant analysis for dimensionality reduction. In 2009, Cao and Li submitted a system that was a top performer in several categories, including mood classification (65.7%) [10]. Their system employs a "super vector" of low-level acoustic features, and employs a Gaussian Super Vector followed by Support Vector Machine (GSV-SVM).

5.4.2 Mood Classification Based on Lyrics

Hu, Chen, and Yang [30] designed a method for dividing sentences in a lyric into emotion groups and then extracting emotion labels from each group, using a large collection of emotion words (see Figure 5.8). The technique was applied to popular music sung in Chinese but the method should be applicable to other languages.

The first step of the method is to identify a large set of Chinese affect words and then map each collected word to a point in the two-dimensional real space, $\mathbf{R}^2$. For this purpose, they created a two-level database of approximately 3,000 emotion words. The first level was created by translating into the Chinese ANEW database [8], a large (1,031 words) collection of emotion words as follows: (a) First, 10 native speakers of Chinese independently translated these words into Chinese. During this process, if a translator found that there was no appropriate Chinese translation (mainly because of cultural difference),

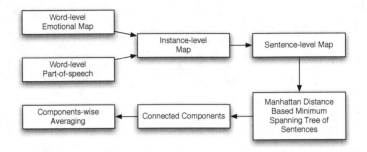

Figure 5.8
The process of producing a sentence group emotion map.

the word was removed from the collection. (b) For each remaining word, the translation that had been suggested by the largest number of translators was chosen as the translation. The second level was created by adding to this collection approximately 2,000 synonyms by looking up a Chinese synonym database. Following the word gathering, they annotated all the words in the database in terms of their Chinese parts-of-speech (POS). This resulted in a two-tier emotion word collection consisting of nearly 3,000 words.

In the second step, the words in the database were mapped into $\mathbf{R}^2$. In the ANEW database every word is rated with respect to its Valence, Arousal, and Dominance levels, each with nine levels. Hu, Chen, and Yang used the first two, Valence and Arousal. That is, if a Chinese word w translated from an English word u in ANEW or if w is a synonym of a Chinese word w' translated from an English word u in ANEW, then its map, $\mu(w)$, is:

$$\mu(w) = (\text{Valence}(u), \text{Arousal}(u)).$$

The range of the map was then extended to the sentence level. Given a sentence s, all the emotion words appearing in s were identified using the POS information attached to them, and then depending on the tense in which the words are used in s and on their modifiers appearing in s, the Valence and Arousal values are scaled (the values may go out their original range of $[-1, 1]$ due to scaling). The map of the sentence s is computed by taking the component-wise average of the maps of the emotion words in s.

Next, Hu, Chen, and Yang collected nearly 1,000 popular music lyrics sung in Chinese along with the timing information of the sentences appearing in the lyrics in their corresponding performance. They had seven people label them into one of the four quadrants of the Thayer map. They kept only those for which at least six labelers agreed. There remained 400 songs. Interestingly, the Calm-Tiredness quadrant had only eight lyrics and the Tension-Energy had as few as 54.

In the next step, using the Manhattan distance on the map they devised a similarity measure between two sentences. Given a lyric, its sentences can be viewed as nodes with the distance between any pair of nodes set to the similarity between the two nodes. By computing the maximum spanning tree of the graph and then cutting the tree into components by eliminating all the edges whose distance value is below a fixed threshold, a set of sentence groups was obtained (a method suggested by Wu and Yang [88]). Each group was then mapped to a point in $\mathbf{R}^2$ by computing the weighted average of the maps where the weight of a sentence was calculated based on which level of the two-level database their emotion words came from and how much time was allocated to the sentences in the singing. Then they selected as the represented group from each lyric the one with the highest total weight.

The above process produced a map of the lyrics to the quadrants of Thayer. The authors compared this against the human based map using F-measure:

$$\frac{\text{Precision} \cdot \text{Recall}}{\text{Precision} + \text{Recall}}. \tag{5.2}$$

Here for given an alignment the Calm-Energy had the F-measure of 0.70, suggesting that there is strong agreement between the human perception of this emotion in Chinese songs and representation of this emotion in Chinese lyrics. As to the remaining three, Tension-Energy had F-measure of 0.44 and the other two had much lower F-measure values.

5.4.3   Mixing Audio and Tag Features for Mood Classification

Bischoff et al. [6] used All Music Guide (allmusic.com) [1] for collecting mood labels. All Music Guide is a Web-based guide book that offers an extensive discography, musicological information, various metadata, and sound samples. It offers theme and mood labels assigned by experts. Bischoff et al. collected from All Music Guide, 73 those expert-assigned theme labels and 178 mood labels for nearly 6,000 songs. The 178 mood labels were organized in two ways: using the two-dimensional diagram of Thayer and using the five-cluster groups according to the aforementioned MIREX mood label clusters [29]. Then from the nearly 6,000 songs, any piece that matched more than one Thayer class was removed, which resulted in a reduction of size to 1,192 songs.

For these songs, audio features and tag features were obtained. As to the audio features, audio files in the MPEG-3 format were used with the 192K bps bit rate. The extracted features included MFCC, Tempo, Loudness, Pitch Classes, chroma, and the Spectral moments. The total number of features was 240. As to the tags, they were collected from Last.fm [2].

Bischoff et al. built a support vector machine classifier based on audio features and a Naive Bayes classifier based on tags. They actually tested various regressions for audio features and concluded that support vector machines performed the best. These classifiers were configured to output a real value between 0 and 1 for each mood class. Then, using a mix ratio parameter β,

$0 < \beta < 1$, the two classifiers were linearly combined into a single classifier. Let c be a class and let f_c and g_c respectively be the classifier for c with respect to audio and the classifier for c with respect to tags. Then the output of mixed classifier, h_c, with β is:

$$\beta f_c(x) + (1 - \beta)g_c(x), \tag{5.3}$$

where x is the input feature vector consisting of the audio and tag features. Both in the case of single-feature set classifiers and in the case of linearly mixed classifiers, the predicted class assignment is the class c for which the output is the largest; that is,

$$\operatorname*{argmax}_c \varphi_c(x), \tag{5.4}$$

where φ is one of f, g, and h.

Both for the Thayer class prediction and for the MIREX mood class prediction, with an appropriately chosen α, the linearly mixed classifier performed better than the single-feature set classifiers. For the MIREX classes, the best accuracy, in terms of F-measure, was 0.572 and that was achieved when $\beta = 0.3$. This was better than the reported accuracies of 0.564 for tag-only features and of 0.432 for audio-only features. As for the Thayer mood classes, the best accuracy, again in terms of F-measure, was 0.569 and that was achieved when $\alpha = 0.8$. This was an improvement from the accuracy of 0.539 for tag-only features and 0.515 for audio-only features.

5.4.4 Mixing Audio and Lyrics for Mood Classification

Hu, Downie, and Ehmann [28] mixed audio signals and lyrics for mood classification of Pop and Rock songs. Ground-truth mood-class (multiple classes for some) identification was done as follows.

To set up experiments, approximately 9,000 audio recordings were obtained such that their corresponding lyrics were available at Lyricwiki.org and such that their tags were available at Last.fm [2]. From those, the songs with lyric length of less than 100 words were removed.

The collected tags were processed to create a vocabulary of emotion words and their grouping as follows. First, from the collection words that were not related to emotion were eliminated. Subject to elimination were: (a) genre/style names (such as "Dance"), (b) terms that were judgmental (such as "Bad" and "Great"), and (c) terms for which the intention of their taggers was ambiguous (such as "Love," which could mean either the tagger loved it or the tagger felt that the song was about love). Second, using WordNet-Affect [74] as a guide words having identical origins (e.g., a name and its adjective form) and their synonyms were grouped together. Finally, these groups were then examined by human experts for further merger. This was the vocabulary of emotion words and their groups.

After this preprocessing, for each emotional word group the songs whose

Calm Comfort	Sad Unhappy	Happy Glad	Romantic	Upbeat Gleeful
Depressed Blue	Anger Fury	Grief Heartbreak	Dreamy	Cheerful Festive
Brooding Contemplative	Aggressive Aggression	Confident Encouraging	Angst Anxiety	Desire Hope
Earnest Heartfelt	Pessimism Cynical	Excitement Exciting		

Table 5.4
The 18 Groups by Hu, Downie, and Ehmann

tags match a word in the group was collected from the song collection. However, if a tag of a song happened to be contained in its artist name (such as "Blue" in Blue Öyster Cult), that particular occurrence of the tag was ignored. Then, emotion word groups of song collection size less than 20 were removed. This resulted in 18 emotion word groups that in large part correspond to the emotions appearing in the Russell diagram over 135 tags and 2,829 songs. A couple of tags from each group are presented in Table 5.4. Note that because these words came from free-form tags they are not necessarily emotion adjectives.

About 43% of the songs were assigned to multiple mood classes, and so the mood class prediction problem was naturally considered to be a multiclass labeling problem. So, it was decomposed as a collection of individual mood classification problems.

As to feature extraction, the lyrics were processed for the following:

- Part-of-Speech (POS): A grammatical classification of words into some types (such as nouns, verbs, and pronouns)

- Function Words (FW): The words that very frequently occur in documents and serve some functions but not so directly related to the contents (such as "about," "the," and "and")

- Content Words (Content): The words that are not function words, may or may not be processed by stemming (removing ending of words to turn them into their stems, for example, changing "removing" to "remov-")

As to the audio features, MFCC and spectral features were extracted using Marsyas [81, 80]. These extracted features were incorporated in support vector machines with a linear kernel. The average prediction accuracy was around 60%. For some categories, the use of both text and audio improved prediction accuracy, but there were categories in which the combined features performed worse than one of the text-only based predictor and the audio-only-based predictor.

5.4.4.1 Further Exploratory Investigations with More Complex Feature Sets

Hu and Downie [26, 27] look deeper into the issue of extracting features from lyrics and combining them with audio features to improve the 18-category binary classification problem. They considered the following five types:

1. They considered bigrams and trigrams of each of the four types of information mentioned earlier, i.e., Part-of-Speech, Function Words, Content Words without Stemming, and Content Words with Stemming, in addition to their unigrams, which had been used earlier.

2. They considered an extensive set of stylistic features of lyrics, such as the number of lines, the number of words per minutes, and the interjection words. There were 19 such features. Some of these features were shown to be effective in artist style classification when used with Bag-of-Words and audio features [48].

3. They used a set of 7,756 affect words constructed from the aforementioned ANEW [8] by adding synonyms using WordNet [16, 54] and by adding words from WordNet-Affect [74]. These additional words were identified with their original words in the ANEW [8] by way of synonym relations. ANEW provides, in addition to the nine-level scores for Valence, Arousal, and Dominance, the standard deviation of their scores. So there are six values assigned to each word in ANEW and its synonyms. By calculating the average and standard deviation of these six values, Hu and Downie generated a 12-feature vector for each lyric.

4. They considered various Bag-of-Words models of the aforementioned General Inquirer. Again, each word appearing in General Inquirer is represented as an 183-dimensional binary vector. The vectors associated with all the words of a lyric that appear in General Inquirer can be summarized in four different ways: frequency, TF-IDF, normalized frequency, and boolean value.

5. For ANEW as well as for General Inquirer, each lyric can be viewed as a collection of words appearing in the vocabulary. Then for each word in the vocabulary, one can compute frequency, TF-IDF normalized frequency, and boolean value.

All these possibilities were individually studied to find their best representation. The best individual performance was achieved by Content Words without Stemming when unigrams, bigrams, and trigrams are combined using their occurrences (accuracy: 0.617) and the second best was achieved by Content Words with Stemming when unigrams, bigrams, and trigrams are combined using TF-IDF (accuracy: 0.613). For Function Words, the best use is to combine their unigrams, bigrams, and trigrams based on occurrences (accuracy:

0.594). When the Context Words without Stemming is combined with Function Words and with either Stylistic Information or the TF-IDF of General Inquirer, the accuracy is further improved to 0.632 or 0.631. Then they mixed these best feature sets with audio features (Timbral and Spectral) to achieve accuracy of more than 0.670.

5.4.5 Exploration of Acoustic Cues Related to Emotions

Eerola, Lartillot, and Toiviainen [13] used the two data sets described in Section 5.3.2 for exploring features that are indicative of the emotions using their feature extraction toolbox [40]. This is in some sense the acoustic-feature-version of the music structure studies of emotion representation in music mentioned in Section 5.2.6.

Eerola, Lartillot, and Toiviainen extracted a total of 29 features from half-overlapping segments in duration of 0.046 seconds. The features extracted pertained to dynamics, timbre, harmony, register, rhythm, and articulation in part using feature extraction methods [17, 24, 39, 60, 77].

These features were then analyzed for their usefulness in prediction of emotional labels through regression. Three regression models were examined: Multiple Linear Regression (MLR) analysis, Principal Component Analysis (PCA), and Partial Least Squares (PLS) regression analysis [87], where five components were used for MLR and PCA and two for PLS. Also, the possibility of using power transformation (i.e., raising to the power of some λ all the components) was considered. The performance of these analysis methods was compared using the R-squared statistics (R^2):

$$1 - \frac{\sum (x - x')^2}{\sum (x - \overline{x})^2},$$

(5.5)

where the sums are over all data points, x' is the predicted value for x, and $\overline{x}$ is the average of all values.

The best results were obtained with PLS with power transform both in the case of three-dimensional emotional space (Valence: 0.72, Activity: 0.85, and Tension: 0.71) and in the case of five emotional categories (Angry: 0.70, Scary: 0.74, Happy: 0.68, Sad: 0.69, and Tender: 0.58).

Multiple linear regression analysis selects components from the original audio feature representation that are the most collectively effective in predicting the values. For "Anger," the five features selected were: Fluctuation peaks, Key clarity, Roughness, Spectral centroid variance, and Tonal novelty; for Tenderness, the five features selected were: Root mean square variance, Key clarity, Majorness, Spectral centroid, and Tonal novelty.

5.4.6 Prediction of Emotion Model Parameters

Targeting the prediction of affect-arousal coordinates from audio, Yang et al. introduced the use of regression for mapping high-dimensional acoustic features to the Barrett-Russell two-dimensional space [91]. Support vector regression (SVR) and a variety of ensemble boosting algorithms, including AdaBoost.RT [70], were applied to the regression problem, using one ground-truth coordinate label for each of 195 music clips. Features were extracted using publicly available extraction tools such as PsySound [9] and Marsyas [81], totaling 114 feature dimensions. To reduce the data to a tractable number of dimensions, PCA was applied prior to regression. This system achieves an R^2 (coefficient of determination) score of 0.58 for arousal and 0.28 for valence.

Schmidt et al. and Han et al. each began their investigation with a quantized representation of the Barrett-Russell affect-arousal space and employed SVMs for classification [69, 22]. Citing unsatisfactory results (with Schmidt obtaining 50.2% on a four-way classification of affect-arousal quadrants, based upon data collected by the *MoodSwings* game, and Han obtaining 33% accuracy in an 11-class problem), both research teams moved to regression-based approaches. Han reformulated the problem using regression, mapping the projected results into the original mood categories, employing SVR and GMM regression methods. Using 11 quantized categories with GMM regression they obtain a peak performance of 95% correct classification.

Schmidt et al. also approached the problem using both SVR and MLR. Their highest performing system obtained 13.7% average error distance in a unit-normalized Barrett-Russell space [69]. Schmidt et al. also introduced the idea of modeling collected human responses in the affect-arousal space as a parameterized stochastic distribution, noting that for most popular music segments the collected points are well-represented by a single 2-D Gaussian. They first perform parameter estimation in order to determine the ground-truth parameters, $\mathcal{N}(\mu, \Sigma)$ and then employ MLR, PLS, and SVR to develop parameter prediction models. They have also performed experiments to evaluate the time-varying (per second) prediction of affect-arousal distributions, achieving average mean errors of 15.4% in a unit-normalized affect-arousal space [68].

5.5   Discussion

Table 5.5 summarizes the work on emotion and mood in music presented in Sections 5.3 and 5.4. Some interesting questions to be addressed are:

- The work of Hu and Downie [26, 27] currently offers the most extensive evaluation of text features extracted from lyrics and acoustic features extracted from audio. The accuracy is around 67%. Can this be improved? Or, has the ceiling been reached [4]?

Ref.	Topic	# of Classes	Features	Methods	Accuracy
Li & Ogihara [46]	Classification	10 binary	Audio (Spectra, Timber, Wavelet)	SVM	≈ 0.44 (F-measure)
Li & Ogihara [47]	Classification	3 binary	Audio (Spectra, Timber, Wavelet)	SVM	70–83% (accuracy)
Liu, Lu, & Zhang [50]	Classification	4	Audio (Spectra, Timber, Rhythm)	GMM	85% (accuracy)
Yang & Lee [90]	Classification	4	Audio (LLD,EDS) + Lyrics (GI)	C4.5	≈80% (accuracy)
Hu, Chen, & Yang [30]	Classification	4	Lyrics (POS, ANEW, Timing)	Fuzzy CMeans	0.70 for 1 class (F-measure)
Bischoff et al. [6]	Classification	4	Tags+ Audio (Timbral, chroma, Pitch, etc.)	SVM, NB	≈0.530 (F-measure)
Hu, Downie, & Ehmann [28]	Classification	18 binary	Lyrics (FW, Content, POS) +Audio (Spectral, Timbral)	SVM	≈0.60 (Average accuracy)
Hu & Downie [28]	Classification	18 binary	Lyrics (FW, Content, POS, GI, ANEW) +Audio (Spectra, Timbral)	SVM	≈0.67 (Average accuracy)
Eerola, Lartillot, & Toiviainen	Correlation Analysis	3 binary + 5	Audio (Spectra, Tempi, Timber, Major/Minor,	PLS, MLS, PCA	N/A

[13]		numeric	Rhythm, etc.)		
Laurier et al. [41]	Cluster-ing	4	Tags (LSA)	KMeans	N/A
Schmidt & Kim [68]	Regres-sion	N/A	Audio (Spectral features)	MLR	N/A

Table 5.5: A Summary of Emotion and Mood Labeling Work Covered in This Chapter (The acronyms are: ANEW = Affective Norms for English Words, Content = Content Words, EDS = Extractor Discovery System, FW = Function Words, GMM = Gaussian Mixture Models, LLD = Low-Level Descriptors, LSA = Latent Semantic Indexing, MLR = Multiple Linear Regression, NB = Naive Bayes, PCA = Principal Component Analysis, PLS = Partial Least Squares, POS = Parts-of-Speech, SVM = Support Vector Machines.)

- Are tags useful in calculating mood labels? This does not apply to data sets like the 18-binary-class data set [26], where the labels were calculated from the tags.

- Of all these possible emotion and mood classes, Hevner, Farnsworth, Watson-Tellegen, Thayer, Schaver, and Russell-Barrett, which one is the easiest to predict?

- How much does emotion and mood vary in music over time, and how difficult is it to predict emotion changes within a single piece of music?

Bibliography

[1] All Music Guide. http://www.allmusic.com.

[2] Last.fm. http://www.last.fm.

[3] *New Oxford American Dictionary,* 2nd edition. Oxford University Press, Oxford, UK, 2005.

[4] J.-J. Aucouturier and F. Pachet. Representing musical genre: A state of art. *Journal of New Music Research,* 32(1):83–93, 2003.

[5] L. Barrington, D. Turnbull, D. O'Malley, and G. Lanckriet. User-centered design of a social game to tag music. *ACM KDD Workshop on Human Computation,* 2009.

[6] K. Bischoff, C.S. Firan, R. Paiu, W. Nejdl, C. Laurier, and M. Sordo. Music mood and theme classification: A hybrid approach. In *Proceedings of the 10th International Society for Music Information Retrieval Conference*, pages 657–662, 2009.

[7] G.H. Bower. How might emotions affect learning? In S. Christianson, editor, *The Handbook of Emotion and Memory: Research and Theory*, pages 3–31. Lawrence Erlbaum Associates, Hillsdale, New Jersey, 1992.

[8] M.M. Bradley and P.J. Lang. Affective norms for English words (ANEW): Stimuli, instruction manual and affective ratings. Technical Report C-1, The Center for Research in Psychophysiology, University of Florida, Gainesville, Florida, 1999.

[9] D. Cabrera, S. Ferguson, and E. Schubert. Psysound3: Software for acoustical and psychoacoustical analysis of sound recordings. In *Proceedings of the International Conference on Auditory Display*, pages 356–363, Montreal, Canada, June 26–29, 2007.

[10] C. Cao and M. Li. Thinkit's submissions for MIREX2009 audio music classification and similarity tasks, MIREX, 2009.

[11] I. Daubechies. The wavelet transform, time-frequency localization and signal analysis. *IEEE Transactions on Information Theory*, 36(5):961–1005, 1990.

[12] J.A. Easterbrook. The effect of emotion on cue utilization and the organization of behavior. *Psychological Review*, 66:183–201, 1959.

[13] T. Eerola, O. Lartillot, and P. Toiviainen. Prediction of multidimensional emotional ratings in music from audio using multivariate regression models. In *Proceedings of the 10th International Society for Music Information Retrieval Conference*, pages 621–626, 2009.

[14] E. Eich, J.T.W. Ng, D. Macaulay, A.D. Percy, and I. Grebneva. Combining music with thought to change mood. In J. Coan and J. Allen, editors, *Handbook of Emotion Elicitation and Assessment*, pages 124–136. Oxford University Press, Oxford, UK, 2007.

[15] P.R. Farnsworth. *The Social Psychology of Music*. Dryden Press, Bel Air, California, 1952.

[16] C. Fellbaum, editor. *WordNet: An Electronic Lexical Database*. MIT Press, Cambridge, Massachusetts, 1998.

[17] J.T. Foote and M.L. Cooper. Media segmentation using self-similarity decomposition. In *Proceedings of the SPIE Conference on Storage and Retrieval for Multimedia*, pages 167–175, 2003.

[18] A. Gabrielsson. The performance of music. In D. Deutsch, editor, *The Psychology of Music*, 2nd edition, Series in Cognition and Perception, pages 501–602, Academic Press, New York, 1999.

[19] A. Gabrielsson. Perceived emotion and felt emotion: Same or different? *Musicae Scientiae*, 6(1):123–148, 2002.

[20] A. Gabrielsson and S. Lindström. The influence of musical structure on emotional expression. In P.N. Juslin and J.A. Sloboda, editors, *Music and Emotion: Theory and Research*, pages 223–248. Oxford University Press, Oxford, UK, 2001.

[21] P.J. Hampton. The emotional element in music. *Journal of General Psychology*, 33:237–250, 1945.

[22] B. Han, S. Rho, R.B. Dannenberg, and E. Hwang. SMERS: Music emotion recognition using support vector machines. In *Proceedings of the 10th International Society for Music Information Retrieval Conference*, pages 651–656, 2009.

[23] D.J. Hargreaves and A.C. North. The functions of music in everyday life: Redefining the social in music psychology. *Psychology of Music*, 27(1):71–83, 1999.

[24] C.A. Harte, M.B. Sandler, and M. Gasser. Detecting harmonic changes in musical audio. In *Proceedings of the 1st ACM Workshop on Audio and Music Computing Multimedia*, pages 21–26, ACM Press, New York, 2006.

[25] K. Hevner. Experimental studies of the elements of expression in music. *American Journal of Psychology*, 48:246–268, 1936.

[26] X. Hu and J.S. Downie. Improving mood classification in music digital libraries by combining lyrics and audio. In *Proceedings of the Joint Conference on Digital Libraries (JCDL)*, 2010.

[27] X. Hu and J.S. Downie. When lyrics outperform audio in music mood classification: A feature analysis. In *Proceedings of the 11th International Society for Music Information Retrieval Conference*, 2010.

[28] X. Hu, J.S. Downie, and A.F. Ehmann. Lyric text mining in music mood classification. In *Proceedings of the 10th International Society for Music Information Retrieval Conference*, pages 411–416, 2009.

[29] X. Hu, J.S. Downie, C. Laurier, M. Bay, and A.F. Ehmann. The 2007 MIREX audio mood classification task: Lessons learned. In *Proceedings of the 9th International Society for Music Information Retrieval Conference*, pages 462–467, 2008.

[30] Y. Hu, X. Chen, and D. Yang. Lyric-based song emotion detection with affective lexicon and fuzzy clustering method. In *Proceedings of the 10th International Society for Music Information Retrieval Conference*, pages 123–128, 2009.

[31] P.N. Juslin. Cue utilization in communication of emotion in music performance: Relating performance to perception. *Journal of Experimental Psychology: Human Perception and Performance*, 26(6):1797–1813, 2000.

[32] P.N. Juslin. Communicating emotion in music performance: A review and theoretical framework. In P.N. Juslin and J. Sloboda, editors, *Music and Emotion: Theory and Research*, pages 309–337. Oxford University Press, Oxford, UK, 2001.

[33] P.N. Juslin and P. Laukka. Expression, perception, and induction of musical emotions: A review and a questionnaire study of everyday listening. *Journal of New Music Research*, 33(3):217–238, 2004.

[34] Y.E. Kim, E. Schmidt, and L. Emelle. MOODSWINGS: A collaborative game for music mood label collection. In *Proceedings of the 9th International Society for Music Information Retrieval Conference*, pages 231–236, 2008.

[35] W.E. Knight and N.S. Rickard. Relaxing music prevents stress-induced increases in subjective anxiety, systolic blood pressure, and heart rate in healthy males and females. *Journal of Music Therapy*, 38(4):254–272, 2001.

[36] M.D. Korhonen. Modeling continuous emotional appraisals of music using system identification. Masters thesis, University of Waterloo, Waterloo, Ontario, Canada, 2004.

[37] M.D. Korhonen, D.A. Clausi, and M.E. Jernigan. Modeling emotional content of music using system identification. *IEEE Transactions on Systems, Man, and Cybernetics, Series B*, 36(3):588–589, 2006.

[38] C.L. Krumhansl. Music: A link between cognition and emotion. *Current Directions in Psychological Science*, 11(2):45–50, 2002.

[39] O. Lartillot, T. Eerola, P. Toiviainen, and J. Fornari. Music-feature modeling of pulse clarity: Design, validation, and optimization. In *Proceedings of the 9th International Society for Music Information Retrieval Conference*, pages 521–526, 2008.

[40] O. Lartillot and P. Toiviainen. MIR in MATLAB: A toolbox for musical feature extraction from audio. In *Proceedings of the 8th International Symposium on Music Information Retrieval*, pages 127–130, 2007.

[41] C. Laurier, M. Sordo, J. Serrà, and P. Herrera. Music mood representation from social tags. In *Proceedings of the 10th International Society for Music Information Retrieval Conference*, pages 381–386, 2009.

[42] E. Law and L. von Ahn. Input-agreement: A new mechanism for collecting data using human computation games. In *Proceedings of the 27th International Conference on Human Factors in Computing Systems*, pages 1197–1206, ACM Press, New York, 2009.

[43] E. Law, L. von Ahn, R. B. Dannenberg, and M. Crawford. TagATune: A game for music and sound annotation. In *Proceedings of the 8th International Symposium on Music Information Retrieval*, pages 361–364, 2007.

[44] M. Leman, V. Vermeulen, L. De Voogdt, D. Moelants, and M. Lesaffre. Prediction of musical affect using a combination of acoustic structural cues. *Journal of New Music Research*, 34(1):39–67, 2005.

[45] T. Lesiuk. The effect of music on work performance. *Psychology of Music*, 33(2):173–191, 2005.

[46] T. Li and M. Ogihara. Detecting emotion in music. In *Proceedings of the 4th International Symposium on Music Information Retrieval*, pages 239–240, 2003.

[47] T. Li and M. Ogihara. Content-based music similarity search and emotion detection. In *Proceedings of the 2004 IEEE International Conference on Acoustic Speech and Signal Processing (ICASSP-04)*, pages V705–V708, 2004.

[48] T. Li and M. Ogihara. Semi-supervised learning from different information sources. *Knowledge and Information Systems*, 7(3):289–309, 2004.

[49] T. Li and M. Ogihara. Toward intelligent music information retrieval. *IEEE Transactions on Multimedia*, 8(3):564–574, 2006.

[50] D. Liu, L. Lu, and H.-J. Zhang. Automatic mood detection from acoustic music data. In *Proceedings of the 4th International Symposium on Music Information Retrieval*, pages 81–87, 2003.

[51] D. Liu, L. Lu, and H.-J. Zhang. Automatic mood detection from acoustic music data. *IEEE Transactions on Acoustics, Speech, and Language Processing*, 14(1):5–18, 2006.

[52] M.I. Mandel and D.P.W. Ellis. A Web-based game for collecting music metadata. In *Proceedings of the 8th International Symposium on Music Information Retrieval*, pages 365–366, 2007.

[53] L.B. Meyer. *Emotion and Meaning in Music*. University of Chicago Press, Chicago, Illinois, 1956.

[54] G.A. Miller. WordNet: A lexical database for English. *Communications of ACM*, 38(11):39–41, 1995.

[55] B.G. Morton, J.A. Speck, E.M. Schmidt, and Y.E. Kim. Improving music emotion labeling using human computation. In *HCOMP 2010: Proceedings of the ACM SIGKDD Workshop on Human Computation*, Washington, DC, 2010.

[56] D.A. Ritossa nd N.S. Rickard. The relative utility of "pleasantness" and "liking" dimensions in predicting the emotions expressed by music. *Psychology of Music*, 32(1):5–22, 2004.

[57] A.C. North, D.J. Hargreaves, and J.J. Hargreaves. Use of music in everyday life. *Music Perception*, 22(1):41–77, 2004.

[58] G.R. Oldham, A. Cummings, L.J. Mischel, J.M. Schmidtke, and J. Zhou. Listen while you work? Quasi-experimental relations between personal-stereo headset use and employee work responses. *Journal of Applied Psychology*, 80(55):547–564, 1995.

[59] F. Pachet, A. La Burthe, A. Zils, and J.-J. Aucouturier. Popular music access: The Sony music browser. *Journal of the American Society for Information Science and Technology*, 55(12):1037–1044, 2003.

[60] E. Pampalk, A. Rauber, and D. Merkl. Content-based organization and visualization of music archives. In *Proceedings of the 10th ACM International Conference on Multimedia*, pages 570–579, ACM Press, New York, 2002.

[61] G. Peeters. A generic training and classification system for MIREX08 classification tasks: Audio music mood, audio genre, audio artist and audio tag, MIREX, 2008.

[62] J.R. Quinlan. *C4.5: Programs for Machine Learning*. Morgan Kaufmann, San Francisco, 1993.

[63] J.A. Russell. A circumplex model of affect. *Journal of Personality and Social Psychology*, 39(6):1161–1178, 1980.

[64] P. Schaver, J. Schwartz, D. Kirson, and C. O'Connor. Emotion knowledge: Further exploration of a prototype approach. *Journal of Personality and Social Psychology*, 52(6):1061–1086, 1987.

[65] K.R. Scherer. Which emotions can be induced by music? What are the underlying mechanisms? And how can we measure them? *Journal of New Music Research*, 33(3):239–251, 2004.

[66] K.R. Scherer, M.R. Zentner, and A. Schacht. Emotional states generated by music: An exploratory study of music experts. *Musicae Scientiae*, 6(1):149–172, 2002.

[67] H. Schlosberg. The description of facial expressions in terms of two dimensions. *Journal of Experimental Psychology*, 44(4):229–237, 1952.

[68] E.M. Schmidt and Y.E. Kim. Prediction of time-varying musical mood distributions from audio. In *Proceedings of the International Society for Music Information Conference*, Utrecht, Netherlands, August 2010.

[69] E.M. Schmidt, D. Turnbull, and Y.E. Kim. Feature selection for content-based, time-varying musical emotion regression. In *MIR '10: Proceedings of the International Conference on Multimedia Information Retrieval*, pages 267–274, Philadelphia, 2010.

[70] D. Shrestha and D. Solomatine. Experiments with AdaBoost.RT, an improved boosting scheme for regression. *Neural Computation*, 18(7):1678–1710, 2006.

[71] J.A. Sloboda. Music structure and emotional response: Some empirical findings. *Psychology of Music*, 19(2):110–120, 1991.

[72] J.A. Sloboda and S.A. O'Neill. Emotions in everyday listening to music. In P.N. Juslin and J.A. Sloboda, editors, *Music and Emotion: Theory and Research*, pages 415–429. Oxford University Press, Oxford, UK, 2001.

[73] P.J. Stone, D.C. Dunphy, M.S. Smith, and D.G. Ogilvie. *The General Inquirer: A Computer Approach to Content Analysis*. MIT Press, Cambridge, Massachusetts, 1966.

[74] C. Strapparava and A. Valitutti. WordNet-Affect: Affective extension of WordNet. In *Proceedings of the 4th International Conference on Language Resources and Evaluation (LREC 2004)*, pages 1083–1086, 2004.

[75] R.E. Thayer. *The Biopsychology of Mood and Activation*. Oxford University Press, Oxford, UK, 1989.

[76] W.F. Thompson and B. Robitaille. Can composers express emotions through music? *Empirical Studies of the Art*, 10:79–89, 1992.

[77] T. Tolonen and M. Karjalainen. A computationally efficient multipitch analysis model. *IEEE Transactions on Speech and Audio Processing*, 8(6):708–716, 2000.

[78] D. Turnbull, R. Liu, L. Barrington, and G. Lanckriet. A game-based approach for collecting semantic annotations of music. In *Proceedings of the 8th International Symposium on Music Information Retrieval*, pages 535–538, 2007.

[79] G. Tzanetakis. Marsyas submissions to MIREX 2007, MIREX, 2007.

[80] G. Tzanetakis and P. Cook. Music genre classification of audio signals. *IEEE Transactions on Speech and Audio Processing*, 10(5):293–302, 2002.

[81] G. Tzanetakis and P.E. Cook. MARSYAS: A framework for audio analysis. *Organized Sound*, 4(3):169–175, 2000.

[82] L. von Ahn. Games with a purpose. *Computer*, 39(6):92–94, 2006.

[83] D. Watson and L.A. Clark. Affects separable and inseparable: On the hierarchical arrangement of the negative affects. *Journal of Personality and Social Psychology*, 62(3):489–505, 1992.

[84] D. Watson, L.A. Clark, and A. Tellegen. Development and validation of brief measures of positive and negative affect: The PANAS scales. *Journal of Personality and Social Psychology*, 54(6):1063–1070, 1988.

[85] D. Watson and A. Tellegen. Toward a consensual structure of mood. *Psychological Bulletin*, 98(2), 1985.

[86] L. Wedin. A multidimensional study of perceptual-emotional qualities in music. *The Scandinavian Journal of Psychology*, 13(4):241–257, 1972.

[87] S. Wold, M. Sjöström, and L. Eriksson. PLS-regression: A basic tool of chemometrics. *Chemometrics and Intelligent Laboratory Systems*, 58:109–130, 2001.

[88] K. Wu and M. Yang. Alternative c-means clustering algorithm. *Pattern Recognition*, 35(10):2267–2278, 2002.

[89] W.M. Wundt. *Introduction to Psychology*. Muller Press, Germany, 2008. (First published in German, 1912; first English translation published in 1924.)

[90] D. Yang and W. Lee. Disambiguating music emotion using software agents. In *Proceedings of the 5th International Society for Music Information Retrieval Conference*, 2004.

[91] Y.-H. Yang, Y.-C. Lin, Y.-F. Su, and H.H. Chen. A regression approach to music emotion recognition. *IEEE Transactions on Acoustics, Speech, and Language Processing*, 16(2):448–457, 2008.

6

Zipf's Law, Power Laws, and Music Aesthetics

Bill Manaris

College of Charleston

Patrick Roos

University of Maryland

Dwight Krehbiel

Bethel College

Thomas Zalonis

College of Charleston

J.R. Armstrong

College of Charleston

CONTENTS

We present results from a decade-long project in the intersection of artificial intelligence, cognitive neuroscience, computer science, and psychology of music. We have extended original research by George Kingsley Zipf to explore connections between power laws (e.g., Zipf's law) and music aesthetics, the latter in part defined by emotional responses of human listeners. Our results suggest a strong connection between music aesthetics (as perceived by humans) and the complexity or entropy of music (as measured by metrics based on Zipf's law). We believe this reflects the fact that both music and the human brain are self-similar, and that our measurements quantify shared aspects of this fractal nature. We introduce Zipf's law and related power laws. We discuss earlier work connecting complexity of artifacts to aesthetics and perceived pleasantness. We provide an algorithmic description of our metrics and identify the various dimensions they measure. We present experimental results, derived with artificial neural networks, which demonstrate the connection between power laws (as captured by our metrics) and music aesthetics (as captured by popularity statistics from a music Web site). We further demonstrate this connection through Armonique, a music similarity engine based on power laws. The aesthetic similarity of Armonique's recommendations is assessed through various psychological experiments involving human listeners. These experiments compare Armonique's music recommendations against human emotional and physiological responses, further demonstrating the connection between power-law metrics and aspects of human emotion and aesthetics.

6.1 Introduction

There is significant available evidence on the relationship between human aesthetics and power laws (e.g., Zipf's law) in the context of sound, images, video, and text. Results connecting power laws with emotional and physiological responses of human listeners in the context of music are presented herein and elsewhere [42, 59]. Similar results exist in the visual domain [52, 54, 55]. Earlier research by George K. Zipf (in the 1940s) suggests a similar connection in text [63]. The same approach may apply to film and video repositories [15]. Finally, power laws (and self-similarity) have been applied to data mining in the context of graph topologies (Web and social networks) as well as astronomical and other large data sets [19]. These results suggest that metrics based on power laws represent a viable approach for data mining in digital archives.

The continued investigation of this relationship (between power laws and human aesthetics) is an emerging research direction with broad and significant implications for the development of automated tools for information retrieval, knowledge discovery, and data mining. Assessment experiments, which document the relationship between human response (both psychological and biological) and power-law patterns in sounds, images, video, and text, raise the question: Do mind and body naturally "resonate" with certain sound (visual, or textual) patterns? Below, we explore some of the ways in which they may do so in music.[1]

6.1.1 Overview

This chapter discusses results from many years of research in the intersection of artificial intelligence, cognitive neuroscience, computer science, and psychology of music. We have developed hundreds of metrics which extract power-law features from Musical Instrument Digital Interface (MIDI) and MP3 audio. Essentially, these metrics capture statistical proportions of music-theoretic and other attributes of music, such as pitch, duration, melodic intervals, harmonic intervals, melodic bigrams, etc. These metrics have been assessed through var-

[1]This material is based upon work supported by the National Science Foundation under grants IIS-0736480, IIS-0849499, and IIS-1049554. Any opinions, findings, and conclusions or recommendations expressed in this material are those of the authors and do not necessarily reflect the views of the National Science Foundation.

ious experiments, including experiments with human subjects. Results from these experiments suggest that power-law metrics model essential aspects of music aesthetics.[2]

Sections 6.2 and 6.3 discuss the history of and some issues related to quantifying music aesthetics. Section 6.4.1 introduces Zipf's law and related power laws. Section 6.5 provides an overview of our power-law metrics for music. Section 6.6 describes automated classification and unsupervised learning tasks used to validate these metrics. Section 6.7 presents Armonique, a music similarity engine utilizing power-law metrics. Section 6.8 presents results from psychological experiments with human subjects assessing Armonique's similarity model with respect to human music aesthetics. Conclusion, acknowledgments, and references follow.

6.2   Music Information Retrieval

There is significant research in quantifying and measuring properties of music. This has been motivated by the interest to automatically classify music in ways meaningful to human aesthetics. The majority of this research has focused on genre and author classification using features extracted from audio signals and/or symbolic (e.g., MIDI) representations [14].

6.2.1   Genre and Author Classification

6.2.1.1   Audio Features

One of the most cited works in music classification at the audio level is by Tzanetakis and Cook [56]. In this study, they use timbre texture, rhythm, and pitch content as classification features. These, in turn, are calculated from statistical features of fast Fourier transform (FFT) and Mel-frequency cepstral coefficients (MFCCs) frequency analyses, statistical features of pitch histograms, and wavelet transforms, respectively. Using these features, Tzanetakis and Cook achieve a classification accuracy of 62% on a corpus of 1,000 songs across 10 music genres: Blues, Classical, Country, Disco, HipHop, Jazz, Metal, Pop, Reggae, and Rock. In a related study on the same corpus, Li et al. [33], report a classification accuracy of 78.5% using amplitude variation statistics based on Daubechies Wavelet Coefficient Histograms (DWCHs).

Other related audio-level classification studies include Dixon et al. [18], who focus only on rhythm classification, and Lidy and Rauber [35]. Both use similar rhythm features plus various psycho-acoustic transformations for the tasks of rhythm and genre classification. Of these, Lidy and Rauber achieve

---

[2]Some results summarized herein have been published more extensively [38, 39, 40, 41, 42, 43, 44, 51].

an accuracy of 74.9% on the Tzanetakis and Cook corpus (10 genres). They also report an accuracy of 70.4% and 84.2% for genre classification on the International Society for Music Information Retrieval (ISMIR) 2004 rhythm and genre classification data sets.

Cano et al. [13] report the development of a content-based (i.e., no metadata) music discovery engine, called MusicSurfer, which is based entirely on audio features related to timbre, tempo, and rhythm patterns. They report an artist identification rate of 24% on a corpus of 273,751 songs from 11,257 artists. They also report an accuracy of 60% (twice as high as the next best system) on the ISMIR 2004 author identification data set.

Aucouturier and Pachet [4] state that timbre features via spectral analysis (MFCCs) are the most common approach to audio-based music similarity studies. They investigate the limits of this approach, suggesting it to be at about 65% R-precision.

6.2.1.2 MIDI Features

While audio-based features mainly capture the timbre and rhythm characteristics of music pieces, working at a higher level (e.g., MIDI) allows the calculation of various music-theoretic and other symbolic features. Given the Aucouturier and Pachet accuracy upper limit (65% R-precision) mentioned above, audio features alone are clearly not enough to achieve human-like classification performance. The development of music theory, in the last few hundred years, indicates that humans recognize and process higher-level musical features. Therefore, higher-level musical features may be useful to model human classification performance along various aesthetic dimensions.

In this domain, Basili et al. [7] calculate features based on melodic intervals, instruments, instrument classes and drum kits, meter/time changes, and pitch range. They report an accuracy of about 70% for genre classification tasks on a corpus of approximately 300 MIDI files from six genres: Blues, Classical, Disco, Jazz, Pop, and Rock.

McKay and Fujinaga [48] use 109 features based on texture, dynamics, pitch statistics, melody and chords, as well as instrumentation and rhythm. They classified 950 pieces from three broad genres (Classical, Jazz, Popular) with an accuracy of 98%. However, according to Karydis et al. [31], "the system requires training for the 'fittest' set of features, a cost that trades off the generality of the approach with the overhead of feature selection."

Karydis et al. [31] work at a MIDI-like level with features based on repeating patterns of pitches and selected properties of pitch and duration histograms. They report an accuracy of approximately 90% on a corpus of 250 music pieces spanning five classical subgenres (i.e., ballads, chorales, fugues, mazurkas, sonatas).

Lidy et al. [36] use features calculated from MIDI transcriptions of audio including attributes of note pitches, durations, and nondiatonic notes combined with typical timbre audio features for the same music pieces. They achieve genre classification accuracies of 76.8% and 90.4% on ISMIR 2004 audio data sets.

Finally, Manaris et al. [40] calculate 156 power-law features to capture the proportions of various melodic features such as pitch, duration, melodic intervals, and chords. With these features they achieve an accuracy of 71.52% on a corpus of close to 2,500 music pieces from nine genres. They also report accuracies of 93.6% to 95% on a five-composer identification task [42].

6.2.2   Other Aesthetic Music Classification Tasks

A smaller but equally interesting body of work exists for various aesthetic music classification tasks, other than genre or author classification. This work is interesting because it requires quantitative measures for music capable of capturing aesthetic dimensions that may vary greatly within genres or authors.

Feng et al. [20] present music classification experiments in terms of mood. They separate musical pieces into four mood categories: happiness, anger, sadness, and fear. They use relative tempo and an articulation feature, based on statistics from the average silence ratio. They train an Artificial Neural Network (ANN) to classify 23 music pieces, given a training set of 330 pieces. They report accuracies of 75% to 86% for the first three mood categories, whereas the accuracy for fear is 25%.

Li and Ogihara [32] present experiments on music similarity search and emotion detection. They work at the audio level and extract features based on FFT, DWCHs, and MFCCs (see Section 6.2.1.1) using 30 seconds of audio. Using Euclidean distance of normalized histograms, they perform similarity retrieval experiments on two corpora: a corpus of 250 vocal Jazz audio files, and a corpus of 288 classical audio files. They report an accuracy of 86%. This is determined based on whether or not pieces retrieved are from the same album as the target. For emotion detection, they rate music using bipolar adjective pairs: (Cheerful, Depressing), (Relaxing, Exciting), and (Comforting, Disturbing). Binary classification experiments, on a corpus of 235 Jazz pieces, achieve accuracies between 70% and 83%.

Manaris et al. [42] report on an experiment using emotional responses from 21 human subjects. These responses were measured as self-reported pleasantness and activation ratings on a standard two-dimensional structure-of-affect instrument (see Barrett and Russell [6]). Using 80 power-law features from a corpus of 210 music excerpts, an ANN achieved an average success rate of 97.22% in predicting (within one standard deviation) human emotional responses to those pieces. Manaris et al. [43] report a related experiment showing high correlation between human emotional responses to music pieces and measured power-law features. Roos and Manaris [51] use a similar set of features to classify between most and least popular classical pieces—as identified by a music service's download statistics—with an accuracy of 90.7%. Finally, Manaris et al. [40] introduce *Armonique*, a music similarity engine, which uses an extended set of power-law features calculated from both symbolic (i.e., MIDI) and timbre (i.e., FFT frequency analysis) data. They also present results from human similarity-judgment experiments demonstrating the effectiveness of the system to distinguish between similar and dissimilar music pieces. In Sections 6.6 and 6.8, we provide previously unpublished results related to these studies.

Other related work in modeling human emotional responses to music includes Yang et al. [60] and Oliveira and Cardoso [49]. Similar to Manaris et al. [42], both projects measure human emotional response using the two-dimensional space of pleasantness (valence) and activation (arousal). While Yang et al. focus on common audio-based features and a regression approach, Oliveira and Cardoso use common MIDI-based features with a Knowledge Base of mappings between emotions and musical features. Both studies find that pleasantness/valence prediction is harder than activation/arousal prediction: 58.3% versus 28.1% R^2 in Yang et al. and 81.6% versus 79.9% R^2 in Oliveira and Cardoso. These findings, compared to the high prediction accuracy for pleasantness reported in Manaris et al. [42], suggest that power laws may be useful in computational modeling of music aesthetics. This possibility is further explored in the next section.

6.3 Quantifying Aesthetics

Webster's defines *aesthetics* as "the study or theory of beauty and of the psychological responses to it; specif., the branch of philosophy dealing with art, its creative sources, its forms, and its effects" [22]. Aesthetics originates from the Greek, αίσθηση – αισθάνομαι, which means to *perceive, feel, sense* (all three notions combined). These notions span the artifact (external), the emotional response (internal), and the sensory organs (interface between external and internal). Over the centuries, use of the term has become less philosophical (i.e., the nature of beauty, art, and taste), and more functional (the analysis, synthesis, and evaluation of artifacts), perhaps reflecting our society's evolution. Schoenberg, among others, promoted this transition in his 1911 "Theory of Harmony" [16, pp. 1–3].

What is the nature of beauty? Where can we find beauty in music? Is it culturally independent (objective) or does it rely on cultural conditioning (subjective)? These are old questions, which are unavoidably raised in the context of this work.

Kahlil Gibran asks: "Where shall you seek beauty, and how shall you find her unless she herself be your way and your guide? And how shall you speak of her except she be the weaver of your speech?" [21, p. 74]. Gibran's perspective raises the intriguing possibility that any potential answers about quantifying aspects of music aesthetics will inevitably also reflect related aspects of human physiology/psychology.

To begin, let's consider two musical pieces, *Song1* and *Song2*, that is,

Figure 6.1
Elias Gottlob Haußmann: Johann Sebastian Bach (1748).
(From Bach-Archiv Leipzig.)

http://tiny.cc/song1 and http://tiny.cc/song2. (It is recommended that you listen to them before reading on. Also, see Figure 6.1 for a hint about their origin.) Assuming you find the pieces at least aesthetically agreeable, then what aspects of these pieces make you feel this way?

This, actually, is a very old exploration. It begins at least 2,500 years ago with the Pythagoreans, who were the first to connect numbers with aesthetics. Aristotle states that "the Pythagoreans were the first to take up mathematics, and ... thought its principles were the principles of all things" [2, pp. 70–71]. They observed that strings exhibit harmonic proportions, that is, they resonate at integer ratios of their length (that is, 1/1, 1/2, 1/3/, 1/4, 1/5, etc.). They also observed that these proportions are aesthetically pleasing to the human ear. Accordingly, they developed musical modes based on these ratios, which formed the basis of our modern-era musical scales.

Aristotle supported the Pythagorean view that "[the interplay] between opposites is the beginning of all beings" [2, pp. 72–73]. Plato, Euclid, and others provided a more precise description of this interplay in the form of proportional analogies (e.g., "A is to B as C is to D"). The apex of this exploration may have been the discovery of the golden mean, or 1.61803399... This special proportion, which humans find aesthetically very pleasing, is found in natural or human-made artifacts [8, 37, 47], [12, pp. 46–57], [26, pp. 91–132], [50, pp. 203–205]. It is also found in the human body (for example, the bones of our hands, the cochlea in our ears, etc.). The golden ratio reflects a place of balance in the structural interplay of opposites.

Considering again our Song1 and Song2, what makes a musical piece aesthetically appealing? Given the Aristotelian/Pythagorean view of opposites, perhaps it is the interplay between silence (rests) and sound (notes). Also, it is the interplay among different sound frequencies occurring concurrently (harmony) and sequentially (melody). Of course, some forms of interplay are more aesthetically pleasing than others. Music theory, which originated with the Pythagorean modes, was developed precisely to codify the aesthetics of this interplay (for example, scales and modes, chords and inversions, cadences, counterpoint, etc.).

Arnheim [3] discusses another kind of interplay—between *disorder* (chaos, randomness) and *order* (monotony). He argues that this interplay affects how aesthetically pleasing artifacts may be. In other words, if the artifact is too chaotic or unpredictable, it will be difficult to comprehend or appreciate (e.g., *12-tone* or *aleatory* music). At the other extreme, if the artifact is too ordered (monotonous) or predictable, it will be uninteresting or boring (e.g., John Cage's *4'33"*).

This theory was experimentally validated by Voss and Clarke [58, 59]. Music was generated through a computer program, which used various random-number generators to control the pitch and duration of successive notes. One piece was created with chaotic (also known as *white noise*) statistical proportions, a piece with somewhat monotonous (also known as *brown noise*) statistical proportions, and a piece with statistical proportions between chaos and monotony (also known as *pink noise* or $1/f$ proportions). As predicted by Arnheim, they observed that the $1/f$ music was much more pleasing to most listeners. The chaotic music was "too random," whereas the brown-noise music was "too correlated." They concluded, "the sophistication of this $1/f$ music (which was 'just right') extends far beyond what one might expect from such a simple algorithm, suggesting that $1/f$ noise (perhaps that in nerve membranes?) may have an essential role in the creative process" [1975, p. 318]. It should be noted that the harmonic proportions observed by the Pythagoreans on strings (that is, $1/1$, $1/2$, $1/3$, $1/4$, $1/5$, etc.) are statistically equivalent to $1/f$ proportions.

In our case, both Song1 and Song2 exhibit near $1/f$ proportions in terms of notes (pitches, durations), melodic intervals, and harmonic intervals, among others. Song2 is J.S. Bach's "Invention No. 13 in A Minor" (BWV784). Song1 was "composed" by a computer program, called NEvMuse, which recombined Song2 notes, while aiming to preserve its $1/f$ proportions. One goal of this experiment was to demonstrate the relationship between music aesthetics and proportions [43]. For comparison, also consider Song3 (i.e., http://tiny.cc/song3), which was created to "counterbalance" the original's $1/f$ proportions, by aiming toward chaotic (white-noise) proportions.

Schroeder [53] explains that the basilar membrane found in the cochlea of the human ear is attuned to sounds with $1/f$ proportions. Since the cochlea is a logarithmic spiral (see Figures 6.2 and 6.3), such sounds stimulate "a constant density of the acoustic nerve endings that report sounds to the brain" [53, p.

122]. Logarithmic spirals exhibit golden ratio proportions (see Figure 6.3). This demonstrates a physiological connection between $1/f$ proportions and the golden ratio, and both to music aesthetics.

6.4 Zipf's Law and Power Laws

George Kingsley Zipf (1902–1950) was a linguistics professor at Harvard University. His seminal book, "Human Behavior and the Principle of Least Effort," contained results from various fields demonstrating the presence of $1/f$ (harmonic) proportions in natural and human-made phenomena [63]. Zipf was the first one (with the possible exception of Johannes Kepler and his 1619 "Harmonices Mundi" work) to hypothesize that there is a universal principle at play, and to propose a mathematical formula to describe it. This formula is known as *Zipf's law*.

Interestingly, aspects of Zipf's universal principle have been observed by others in specific domains. As a result, several related laws exist, such as *Pareto's principle* (80-20 rule in economics), *Lotka's law* (in bibliometrics), *Bradford's law* (in library science), *Benford's law* (in statistics), *Archie's law* (in petrophysics), *Heaps' law* (in linguistics), *Stevens' power law* (in psychophysics), and *inverse-square laws* and other *power laws* (in physics) (e.g., see Li [34]). Finally, Zipf's work influenced the development of Benoit Mandelbrot's concept of Fractal Geometry [45] and Per Bak's concept of Self-Organized-Criticality [5].

6.4.1 Zipf's Law

Informally, Zipf's law describes phenomena where certain types of events are frequent, whereas other types of events are rare. For example, in English, short words (e.g., "a," "the") are very frequent, whereas long words (e.g., "anthropomorphologically") are quite rare. If we compare a word's frequency of occurrence with its statistical rank, we notice an inverse relationship: successive word counts are roughly proportional to $1/1, 1/2, 1/3, 1/4, 1/5$, and so on [10]. In other words, books contain the same type of harmonic proportions as those observed by the Pythagoreans on strings 2,500 years ago.

Zipf generalized this observation to other types of harmonic proportions [10, pp. 130–131]. This is captured by the *Generalized Harmonic Series* equation:

$$F \cdot Sn = \frac{F}{1^p} + \frac{F}{2^p} + \frac{F}{3^p} + ... + \frac{F}{n^p} \qquad (6.1)$$

where F is a constant, n is a positive integer, and p may range from 0 to ∞, with 1 corresponding to Zipf's law. This equation may be best understood by

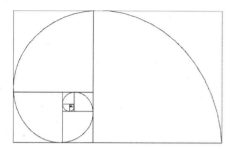

Figure 6.2
A logarithmic spiral (sides of consecutive boxes approximate the golden ratio).

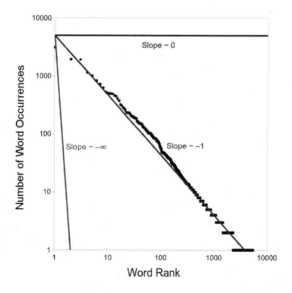

Figure 6.3
Number of unique words (y-axis) ordered by word statistical rank (x-axis) on log scale for Plato's *Phaedo* (slope $= -1.0308$, $R^2 = 0.9551$).

plotting the data (see Figure 6.3). This produces a near straight line whose slope corresponds to the exponent p above. The slope may range from 0 to $-\infty$, with -1.0 denoting Zipf's ideal (also known as *pink-noise, harmonic,* or *1/f* proportions). A slope near 0 indicates a random probability of occurrence (i.e., chaotic or *white-noise* proportions). A slope of -2.0 denotes brown-noise proportions. A slope tending toward $-\infty$ indicates a very monotonous phenomenon, for example, a musical piece consisting mostly of one note (also known as *black-noise* proportions).

In physics, white-noise, pink-noise, brown-noise, and black-noise proportions are known as *power laws*.

Zipf (pink-noise) proportions have been discovered in a wide range of human and naturally occurring phenomena, including music, city sizes, peoples' incomes, subroutine calls, earthquake magnitudes, thickness of sediment

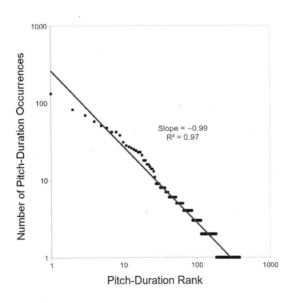

Figure 6.4
Pitch-duration proportions of Chopin's *Nocturne*, Op. 9, No. 1 (slope = -0.9853, $R^2 = 0.9653$).

depositions, clouds, trees, extinctions of species, traffic jams, visits to Web sites, and opening chess moves [9, 45, 53, 58, 59, 63].

In many cases, size may be used instead of statistical rank. According to Salingaros and West, the most pleasing designs in human artifacts exhibit power-law behavior. "The relative multiplicity p of a given design element, that is, the relative number of times it repeats (frequency), is determined by a characteristic scale size x as roughly $px^m = C$, where C is related to the overall size of the structure, and the index m is specific to the structure [52, p. 909]".

A logarithmic plot of p versus x has a slope of m, where $-1 \leq m \leq -2$. Exceptions to this rule correspond to "incoherent, alien structures" [52, p. 909].

As mentioned earlier, Voss and Clarke [58] showed that classical, rock, jazz, and blues music exhibit a power law with slope approximately -1. Later, Voss and Clarke [59] generated music artifacts exhibiting power-law distributions with m ranging from 0 (white noise), to -1 (pink noise), to -2 (brown noise). Pink-noise music was much more pleasing to most listeners, whereas white-noise music sounded "too random," and brown-noise music "too correlated."

In earlier research [44], we have shown that 196 "socially sanctioned" (popular) music pieces exhibit power laws with m near -1 across various music attributes, such as pitch, duration, and melodic intervals. In Section 6.6.1, given a corpus of 14,695 pieces, we observe again that the 1,000 most-popular (aesthetically pleasing?) classical pieces exhibit Zipfian proportions (m near -1); whereas the 1,000 least-popular pieces exhibit more chaotic proportions.

6.4.2 Music and Zipf's Law

Zipf reports results from four musical pieces: Mozart's *Bassoon Concerto in Bb*, Chopin's *Etude in F Minor, Op. 25, No. 2*, Irving Berlin's *Doing What Comes Naturally*, and Jerome Kern's *Who* [63, pp. 336–337]. Since Zipf and his students did not have access to computers, they manually counted notes in music scores. They focused on notes and distances between repeated notes. In both cases, they demonstrated that the above songs exhibit *1/f* proportions similar to the ones observed in natural language.

With the use of a computer and the proper algorithms, this arduous effort may be performed in a few seconds. As mentioned earlier, we have developed hundreds of metrics based on Zipf's law. These metrics capture proportions of music-theoretical and other attributes, such as pitch, duration, melodic intervals, chords, and various proportions of timbre within FFT power spectra.

For example, using a note (pitch) metric, J.S. Bach's *Air on the G String* exhibits a slope of -1.08 and an R^2 of 0.81 (see Figure 6.4). Again, a slope near -1 indicates a Zipf distribution. The R^2 value indicates how well the data points fit the trendline—it may range from 0 (no fit) to 1 (perfect fit). Anything above 0.7 is considered a good fit. We have studied thousands of musical pieces. Our results indicate that most socially sanctioned music, across styles, exhibits near Zipfian distributions across various attributes [42]. Moreover, deviations from ideal Zipfian proportions tend to correlate with composer and style, as we discuss in the next section.

6.5   Power-Law Metrics

As mentioned earlier, we employ metrics based on power laws to extract aesthetically relevant features from musical pieces. We have two categories of metrics, symbolic (MIDI) metrics, and timbre (audio) metrics.

6.5.1   Symbolic (MIDI) Metrics

Each symbolic metric measures the *entropy* of particular music-theoretic or other attribute of musical pieces. For example, in the case of *pitch*, we count each occurrence of each pitch in the piece, for example, 168 C5 notes (i.e., a C note, 5th octave), 86 G5 notes, 53 E5 notes, and so on. Then we calculate the slope and R^2 values of the logarithmic rank-frequency distribution (see Figure 6.4).

In general, the slope may range from 0 to $-\infty$. Again, the R^2 value may range from 0 to 1, with 1 denoting a straight line. This captures the proportion of y-variability of data points with respect to the trendline. It indicates how self-similar the measured attribute is, and, accordingly, how reliable the calculated slope is.

We have three types of symbolic metrics, namely regular, high-order, and local interval variability metrics.

6.5.1.1   Regular Metrics

Regular metrics capture the entropy of a regular attribute or event (an "event" is anything countable, for example, a melodic interval). We currently employ 25 regular metrics related to pitch, duration, harmonic intervals, melodic intervals, harmonic consonance, bigrams, chords, and rests. Table 6.1 provides details for each metric.

6.5.1.2   Higher-Order Metrics

Higher-order metrics capture the entropy of the *difference* between two consecutive regular events. Similar to the notion of derivative in mathematics, for each regular metric one may construct an arbitrary number of higher-order metrics (e.g., the difference of two events, the difference of two differences, and so forth).

It should be noted that these metrics implicitly capture significant aspects of musical hierarchy. Similar to the Schenkerian analysis, music events (e.g.,

Metric	Measures Distribution Of
Pitch	pitches of notes (retains octave)
Chromatic Tone	pitches of notes (ignores octave)
Duration	durations of notes
Quantized Duration	quantized durations of notes
Pitch Duration	combined pitch and duration of notes
Pitch and Quantized Duration	combined pitch and quantized duration of notes
Pitch Distance	distances between same pitches
Duration Distance	distances between same durations
Quantized-Duration Distance	distances between same quantized durations
Duration Bigram	bigrams of note durations
Quantized-Duration Bigram	bigrams of quantized durations
Contour Melody Pitch	pitches of melodic-line notes
Contour Bassline Pitch	pitches of bassline notes
Contour Melody Duration	durations of melodic-line notes
Contour Melody Quantized Duration	same, for quantized durations
Contour Bassline Duration	durations of bass-line notes
Contour Bassline Quantized Duration	same, for quantized durations
Melodic Interval	melodic intervals
Harmonic Interval	harmonic intervals
Melodic Bigram	melodic bigrams
Harmonic Bigram	harmonic bigrams
Melodic Consonance	melodic consonance
Harmonic Consonance	harmonic consonance
Chord	chord progressions
Rest	rests between notes

Table 6.1

List and Description of Regular MIDI Power-Law Metrics (Each metric returns a slope and R^2 value, which capture the entropy and self-similarity of the measured attribute.)

pitch, duration, and so on) are recursively reduced to higher-order ones, capturing long-range structure in pieces [29]. Consequently, pieces without hierarchical structure have significantly different measurements than pieces with structure.

Theoretically, we can calculate as many higher-order levels as the musical data will allow. However, we have discovered that, for classification purposes of normal musical pieces, a few levels (e.g., two to three) suffice, as higher-order metrics tend to be correlated across levels. This has been tested across thousands of musical pieces.

6.5.1.3 Local Variability Metrics

Local variability metrics capture the entropy of the difference of an event from the local average. In other words, local variability, $d[i]$, for the ith event is

$$d[i] = abs(tNN[i] - average(tNN, i))/average(tNN, i) \qquad (6.2)$$

where tNN is the sequence (array) of events, *abs* is the absolute value, and $average(tNN, i)$ returns the average of events within a narrow, say five-event wide window [30]. We provide one local variability metric for each regular and higher-order metric.

6.5.2 Timbre (Audio) Metrics

The extraction of audio features is most commonly based on analyzing the frequency content of audio signals. Common techniques include fast Fourier transform, Mel-frequency cepstral coefficients, and wavelets [62, 28].

Our approach is motivated by the work of Voss and Clarke [58, 59]. By analyzing 12 hours worth of radio recordings from various stations (e.g., Classical, Jazz, Rock, News, and so forth), they discovered that certain properties of the audio signal (e.g., loudness and pitch fluctuation) exhibit Zipfian (*1/f*) distributions.

Based on this observation, we developed power-law metrics that measure the proportions (entropy and self-similarity) of frequencies within audio signals [40].

6.5.2.1 Frequency Metric

Our timbre metrics all rely on a single algorithm. This base algorithm extracts frequency information using FFT on a per window basis. First, we split the signal up into equal size windows. Then, we extract FFT information from each window. We create a histogram of the frequency magnitudes across windows (in the time domain). Finally, we calculate power-law proportions of the summed frequency magnitudes. This generates a measurement of the proportions of frequencies within the signal (i.e., a slope and R^2 value).

6.5.2.2 Signal Higher-Order Metrics

Signal higher-order metrics are based on the frequency metric described above. Similar to the notion of derivatives in mathematics, we calculate the differences between pairs of successive raw signal amplitudes. In other words, this gives us the first higher-order amplitudes of the signal. Then, we split this higher-order signal up into equal size windows. We extract FFT information from each window, and so on.

This is repeated for additional higher levels. Accordingly, *higher-order 0* is based on the original signal, *higher-order 1* is based on the differences between raw signal amplitudes, *higher-order 2* is based on the differences of the differences, and so on. Each higher-order metric calculates the proportions of frequencies within the (corresponding higher-order) signal (i.e., produces a new pair of slope and R^2 features).

6.5.2.3 Intrafrequency Higher-Order Metrics

These timbre metrics capture the entropy of energy change for each frequency over time. These metrics operate on a variation of the above higher-order principle. Instead of calculating differences of signal amplitudes, *intrafrequency higher-order* metrics calculate differences of individual frequencies across pairs of FFT windows. In other words, they calculate differences of magnitudes for each frequency, across two consecutive FFT windows. This gives us the first higher-order frequency magnitudes of the signal (i.e., two consecutive FFT windows are merged into a higher-order one).

Similar to the base frequency metric (see Section 6.5.2.1), we create a histogram of the frequency magnitudes across (higher-order) windows. Finally, we calculate power-law proportions of the summed frequency magnitudes. This generates new measurements of the proportions of frequencies within the signal (i.e., pairs of slope and R^2 values).

6.5.2.4 Interfrequency Higher-Order Metrics

Similar to the intrafrequency metrics above, *interfrequency higher-order* metrics operate on a variation of the higher-order principle. These metrics, however, calculate differences of consecutive frequency bins across a single FFT window. In other words, they calculate differences of magnitudes for each frequency, across pairs of consecutive FFT frequencies. This gives us the first higher-order frequency magnitudes of the signal (i.e., an FFT window's length is reduced by 1).

Again, we create a histogram and calculate proportions of the higher-order frequency magnitudes. This generates new power-law measurements (i.e., pairs of slope and R^2 values).

Discussion

Each of the above metrics can be applied as many times as necessary to improve classification accuracies. Evaluations of this technique using Weka's Principal Components Analysis (PCA) algorithm [23] indicate that, for normal music, there is much correlation (i.e., information overlap) between consecutive higher-order levels. However, the same evaluations also show that additional accuracy is gained per application.

Finally, in addition to base frequency metric and the three approaches to generate higher-order features described above, additional features can be created by varying the FFT window size and the signal's sampling rate. These variations alter the granularity of the signal before applying the base and higher-order metrics to produce even more timbre measurements. Overall, we have experimented with up to a total of 234 audio features.

Since we are interested in power-law distributions within the human hearing range, assuming CD-quality sampling rate (44.1 KHz), we use window sizes up to one second. Interestingly, given our technique, the upper frequencies in this range do not appear to be as important for calculating timbre similarity. The most important frequencies appear to be from 1 kHz to 11 kHz.

6.6 Automated Classification Tasks

Our experiments demonstrate that extracting a large number of power-law metrics serves as a statistical "signature" mechanism, which can help to identify musical pieces and even to automatically classify them in terms of composer or style. As mentioned in Section 6.2, we have trained numerous Artificial Neural Networks (ANNs) on hundreds of values derived from applying our metrics to many music corpora. These ANNs were trained to perform various classification tasks in order to assess our metrics. These tasks included:

- Composer classification: (J.S. Bach, Beethoven, Chopin, Debussy, Purcell, D. Scarlatti) with 93.6% to 95% accuracy [38].

- Style identification: (Medieval, Renaissance, Baroque, Classical, Romantic, Modern, Jazz, Country, Rock) with 71.5% to 96.6% accuracy [40].

- Popularity (pleasantness?) prediction: We used a corpus of 14,695 classical pieces from the Classical Music Archives and a Web access log for one month (1,034,355 downloads). Using this log, we extracted from the corpus the 1,000 most-popular (most-downloaded) pieces and the 1,000 least-popular (least-downloaded) pieces. Trained on a subset of the data, the ANN managed to classify pieces into the proper category (popular versus nonpopular) with 90.7% accuracy. Preliminary results appeared in the work of Roos and Manaris [51].

Of these, the popularity experiment is the most intriguing. Below, we present an expanded version of this experiment, conducted with newer metrics, and providing previously unpublished results.

6.6.1   Popularity Prediction Experiment

The problem with assessing aesthetics is that (similar to assessing intelligence) there seems to be no objective way of doing so. One possibility is to use a variant of the Turing Test, where human subjects may be asked to rate the aesthetics of music pieces, and then check for correlations between those ratings and features extracted using our power-law metrics. In this section, we explore this approach. The experiment reported herein is a larger scale version of the preliminary experiment reported Roos and Manaris [51].

Ideally, our corpus would consist of two types of music, pieces of *high* aesthetic value and pieces of *low* aesthetic value. Also, these pieces should be from the same genre to ensure that we are focusing on aesthetics as opposed to genre preferences. However, it is difficult to find music of low aesthetic quality, since such music would be unlikely to survive the test of time. For instance, all surviving classical music is considered to be of reasonable-to-high aesthetic value—each and every surviving classical piece has to have been enjoyed by enough people to be played, written down, archived, and so forth.

Given the difficulty of finding enough pieces of low aesthetic quality, another option is to have a large collection of known pieces ranked by human listeners. We were fortunate enough to be given access to the Classical Music Archive corpus, which at the time consisted of 14,695 classical pieces encoded in MIDI (http://www.classicalarchives.com). We were also given access to a download log of a total of 1,034,355 downloads for the month of November 2003. Through this log, we identified the 1,000 most popular (most downloaded) and 1,000 most unpopular (least downloaded) pieces. (Tables 6.2 and 6.3 show names and composers of pieces from each group.) We then performed several binary classification tasks using these equally sized sets.

Given this configuration of our corpus, with over 10,000 music pieces separating the two classes by popularity, we believe the hypothesis of a general correlation between popularity and aesthetics has merit. It should also be noted that in the unpopular class all pieces were accessible and playable, ruling out the possibility that pieces received few download requests due to inaccessibility and MIDI errors.

6.6.1.1   ANN Classification

We conducted several ANN classification tasks, using the power-law metrics described earlier, between the popular and unpopular classes.

All classification tasks involved feed-forward ANNs trained via backpropagation. Training ran for 500 epochs, with a value of 0.2 for momentum and 0.3 for learning rate. The ANNs contained a number of nodes in the input

Composer	Piece	Count
BEETHOVEN, Ludwig van	Bagatelle No. 25 in A Minor "Für Elise"	9965
VIVALDI, Antonio	No. 1. La Primavera (Spring) in E, Op. 8, No. 1	6382
MOZART, Wolfgang Amadeus	Divertimento for strings in D	6190
BACH, Johann Sebastian	Toccata and Fugue in D, Toccata	5576
CHOPIN, Frédéric François	Etude in C–, Op. 10, No. 12	4723
TCHAIKOVSKY, Pyotr Ilich	The Nutcracker, Op. 7	3948
CHOPIN, Frédéric François	Polonaise in Ab, Op. 53	3564
DEBUSSY, Achille-Claude	Arabesque No. 1 in E	3545
CHOPIN, Frédéric François	Nocturne in Ab, Op. 32, No. 2	3166
BACH, Johann Sebastian	Prelude and Fugue No. 1 in C	2914
CHOPIN, Frédéric François	Fantaisie-impromptu in C#, Op. 66	2913
MOZART, Wolfgang Amadeus	Serenade in G for strings, 1. Allegro	2827
BACH, Johann Sebastian	Toccata and Fugue in D–, Fugue	2691
BEETHOVEN, Ludwig van	Symphony No. 5 in C–, Op. 67	2504
BEETHOVEN, Ludwig van	Piano Sonata No. 14 in C#, Op. 27, No. 2	2192
MOZART, Wolfgang Amadeus	Divertimento for strings in D, Andante	2132
BACH, Johann Sebastian	Brandenburg Concerto No. 1 in F, 3. Allegro	2129
BRAHMS, Johannes	Waltzes, Op. 39—No. 15 in Ab ("Lullaby")	2075
HANDEL, George Frideric	Harpsichord Suite in E, 4. Air and Variations	2039
PACHELBEL, Johann	Canon in D	1965

Table 6.2
Top 20 (Most Popular) of the 14,695 Pieces of the Classical Music Archives Corpus

Composer	Piece	Count
HOLBORNE, Antony	Alman: The Honeysuckle	10
MARINI, Biagio	La Ponte	10
MORLEY, Thomas	Sweet Nymph, Come to Thy Lover	10
CATO, Diomedes	Galliarde I	9
MILAN, Luys	Six Pavanes for Guitar	9
CERTON, Pierre	Chanson Parisienne "Frère Thibault"	9
HOLBORNE, Antony	The Widow's Mite	9
LASSUS, Orlande de	12 Fantasies for 2 Parts—No. 2	8
BYRD, William	Pavans and Galliards—Pavan: 2 Parts in One	8
GORZANIS, Giacomo	Che Giova far Morir—Napolitana a 3 v.p.	8
GUERRERO, Francisco	Todo Quanto Pudo Dar	8
GASTOLDI, Giovanni Giácomo	Balletti a Tre Voci, 1594—Il Tedesco	7
MAIER, Michael	Atalanta Fugiens—Atala-48	7
GOUDIMEL, Claude	150 Pseaumes de David— Pseaume XXXIII	7
GROTTE, Nicolas de la	Douce Maistresse Touche (Poème de Ronsard)	7
BYRD, William	Dances—Alman	6
MAIER, Michael	Atalanta Fugiens Atala-44	6
MAIER, Michael	Atalanta Fugiens Atala-12	6
MIKOLAJ, of Cracow	Ach, Hilf Mich Leid	5
MIELCZEWSKI, Marcin	Canzona Seconda a 2	4

Table 6.3
Bottom 20 (Most Unpopular) of the 14,695 Pieces of the Classical Music Archives Corpus

Classification Experiment	Success (%)
228 features—symbolic and timbre combined	91.65%
157 features—symbolic only	90.90%
73 features—timbre only	85.25%
30 selected features (CfsSubsetEval-Bestfit)	88.40%
Randomly assigned classes (control)	47.70%

Table 6.4
Success Rates of Different ANN Popularity Classification Experiments (Ten-fold cross validation.)

Value	Average	Std
Slope (30 selected features)	0.9569	0.3436
R^2 (30 selected features)	0.7801	0.1991
Slope (228 features)	0.8644	0.4855
R^2 (228 features)	0.7772	0.1956

Table 6.5
Popular Pieces: Average and Standard Deviation (Std) of Slope and R^2 Values

layer equal to the features used for training, 2 nodes in the output layer and $(inputnodes + outputnodes)/2$ nodes in the hidden layer. For evaluation, we used 10-fold cross validation. The results for these experiments are listed in Table 6.4. Details on the listed experiments are described below.

First, we conducted a classification task using 228 features per piece to train an ANN. These features consisted of regular symbolic metrics, two higher-orders for each, and a local variability metric for each regular and higher-order metric. The remaining features are made up of timbre metrics, including five higher-orders and five sampling-rate reductions.

For control purposes, we conducted a classification task identical to the first, but with classes assigned randomly for each piece. Finally, we conducted a classification task identical to the first, but using only the 30 most relevant features to train the ANN. These attributes were selected via Weka's CfsSubsetEval-Bestfit attribute selection algorithm on the combined metric set [23].

Tables 6.5 and 6.6 list the average and standard deviation of slope and R^2 values for the popular and unpopular pieces, respectively. It is interesting to note how the most meaningful slopes for popular pieces approximate Zipf's ideal (i.e., -0.9569). On the other hand, the most meaningful slopes for unpopular pieces are more chaotic (i.e., -0.8399).

Value	Average	Std
Slope (30 selected features)	0.8399	0.3800
R^2 (30 selected features)	0.7665	0.2452
Slope (228 features)	0.7854	0.4716
R^2 (228 features)	0.7620	0.2237

Table 6.6
Unpopular Pieces: Average and Standard Deviation (Std) of Slope and R^2
Values

6.6.2 Style Classification Experiments

This section presents results from various style classification experiments using symbolic (MIDI) and timbre (audio) power-law metrics on a corpus of 8,370 MP3 pieces from Magnatune (http://magnatune.com). The corpus includes the following genres: Rock, Folk, Punk, Blues, Electronica, Classical, World, New Age, Jazz, Metal, Pop, Ambient, and Children. We created two different data sets:

- **10-genre set**: This data set retains as many different genres as possible, but removes and combines some genres for which the original data set only had very few pieces. Specifically, the "Folk" and "Children" pieces are removed and the few "Punk" pieces are combined with the "Rock" pieces to form a "Rock/Punk" genre class. The resulting data set has the following genre classes and class sizes: Classical (2849), Rock/Punk (1239), Blues (102), Electronica (1047), World (1028), New Age (577), Jazz (194), Metal (313), Pop (364), and Ambient (657).

- **6-genre set**: This data set combines more of the genres in an attempt to reduce stylistic overlap between the different genre classes. The resulting data set has the following genre classes and class sizes: Classical (2849), Rock/Punk/Metal (1552), Blues/Jazz/Pop (660), Electronica (1047), Ambient/NewAge (1244), and World (1028).

For all our classification experiments, since many of our metrics are highly correlated, we first applied Weka's PCA tool to transform and reduce the metrics used. This aims to reduce the metrics into a set of uncorrelated metrics.

6.6.2.1 Multiclass Classification

For both data sets, PCA generated a total of 116 reduced features. A majority class classifier would achieve a classification accuracy of $2849/8370 = 34.04\%$. We conducted 10-fold stratified cross-validation experiments using Weka's Multilayer Perceptron ANN.

With the 10-genre set, the classification accuracy was 65.07%. Below is the ANN confusion matrix:

a	b	c	d	e	f	g	h	i	j	<-- classified as
2649	100	38	4	0	5	27	20	3	3	a = Classical
158	514	61	16	15	93	79	60	26	6	b = World
63	66	225	9	0	51	69	69	18	7	c = New Age
16	11	18	79	3	11	32	10	8	6	d = Jazz
4	12	1	1	183	27	71	4	9	1	e = Metal
14	97	48	12	27	573	147	71	54	4	f = Electronica
41	55	63	37	65	124	717	37	95	5	g = Rock/Punk
42	56	72	10	3	81	33	350	8	2	h = Ambient
11	23	8	11	14	69	137	6	82	3	i = Pop
4	8	3	1	0	1	7	2	2	74	j = Blues

With the 6-genre set, the classification accuracy was 68.39%. Below is the ANN confusion matrix:

a	b	c	d	e	f	<-- classified as
2628	99	70	14	26	12	a = Classical
142	514	128	47	100	97	b = World
113	130	718	57	96	120	c = Ambient/NewAge
37	45	46	259	185	88	d = Blues/Jazz/Pop
37	80	106	161	1018	150	e = Rock/Punk/Metal
16	107	125	68	144	587	f = Electronica

6.6.2.2 Multiclass Classification (Equal Class Sizes)

In order to explore genre classification on a data set with equal class sizes, we reduced the data sets to N randomly chosen pieces of each of the classes, where N is the smallest class size of the original data set.

Reducing the 10-genre set to consist of equal classes resulted in 102 instances in each class. PCA reduced the metrics down to 101 attributes. A majority class classifier would achieve a classification accuracy of $102/102 * 10 = 10.00\%$. The Weka ANN (Multilayer Perceptron), achieves a classification accuracy of 52.06% in a 10-fold stratified cross validation experiment. Below is the ANN confusion matrix:

a	b	c	d	e	f	g	h	i	j	<-- classified as
36	9	0	3	25	9	3	8	2	7	a = Rock/Punk
8	72	1	6	5	4	3	1	0	2	b = Metal
1	4	88	1	3	1	1	1	1	1	c = Blues
4	3	4	60	5	2	6	10	2	6	d = Jazz
24	7	1	5	30	15	1	7	4	8	e = Pop
8	6	1	7	12	49	10	4	0	5	f = Electronica
3	1	4	4	3	3	57	13	6	8	g = Ambient
6	2	4	11	9	8	15	35	5	7	h = New Age
1	3	2	4	1	0	10	4	70	7	i = Classical
2	4	7	7	9	8	10	10	11	34	j = World

Reducing the 6-genre set to consist of equal classes resulted in 660 instances in each class. PCA reduced the metrics down to 114 attributes. A majority class classifier would achieve a classification accuracy of $660/660 * 6 = 16.67\%$. The Weka ANN (Multilayer Perceptron), achieves a classification accuracy of 58.28% in a 10-fold stratified cross validation experiment. Below is the ANN confusion matrix:

```
  a   b   c   d   e   f   <-- classified as
340 127  75  55  16  47 |   a = Rock/Punk/Metal
133 339  79  44  23  42 |   b = Blues/Jazz/Pop
 80  80 356  68  11  65 |   c = Electronica
 40  49  70 371  40  90 |   d = Ambient/NewAge
  8  15   3  36 543  55 |   e = Classical
 43  54  56  89  59 359 |   f = World
```

6.6.2.3 Binary-Class Classification (Equal Class Sizes)

Using the reduced 6-genre set consisting of equal classes (660 pieces in each), we performed a number of binary classification experiments. For each of these experiments, we chose one genre class and combined all other genres into a genre called "Other." Here are the results:

- Rock/Punk/Metal versus Other: ANN accuracy 80.09% (1552 instances per class)

- Classical versus Other: ANN accuracy 93.56% (2849 instances per class)

- Electronica versus Other: ANN accuracy 80.42% (1047 instances per class)

- Blues/Jazz/Pop versus Other: ANN accuracy 72.96% (660 instances per class)

- Ambient/NewAge versus Other: ANN accuracy 76.82% (1244 instances per class)

- World versus Other: ANN accuracy 73.64% (1028 instances per class)

It is interesting to note the high accuracy of the Classical versus Other classification task (93.56%). One interpretation is that our metrics are better "tuned" to classical music. Another possibility is that this high accuracy reflects a stronger stylistic distinction between classical music and the other genres of the experiment. This possibility is supported by the fact that many of the nonclassical pieces in the Magnatune corpus (e.g., Jazz, Rock, Electronica, and so forth) have multiple genre labels (e.g., a Jazz-Rock piece, or an

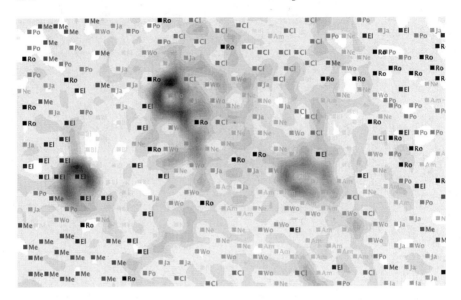

Figure 6.5
MIDI metrics ESOM U-Matrix visualization. (**Ro** = Rock/Punk, **Bl** = Blues,
El = Electronica, **Cl** = Classical, **Wo** = World, **Ne** =New Age, **Ja** = Jazz,
Me = Metal, **Po** = Pop, **Am** = Ambient)

Electronica-Ambient piece, and so forth). In other words, with the exception
of Classical, the other Magnatune genres have "fuzzier" boundaries, even for
human listeners.

6.6.3 Visualization Experiment

In order to better understand the high-dimensional space defined by our metrics, we generated various visualizations.

6.6.3.1 Self-Organizing Maps

Emergent self-organizing maps (ESOMs) are an unsupervised learning technique for visualizing the structure and organization of high-dimensional data
through a low-dimensional representation.

We created separate visualizations for MIDI versus audio metrics. This was
done to observe the differences in genre classification ability between the two
types of metrics (for example, see Figures 6.5 and 6.6). We also created visualizations for the combined metrics (for example, see Figure 6.7). These graphs
were generated using the *Databionics ESOM Tools* by training a toroid 50 x
82 ESOM with our power-law timbre and MIDI features [57]. We utilized a U-
Matrix display, which superimposes a coloring on the usually two-dimensional

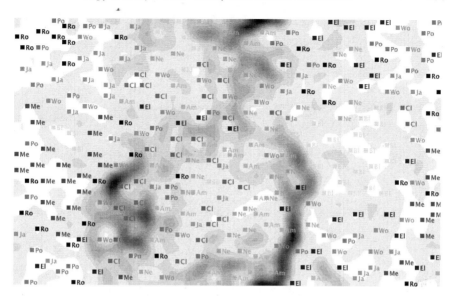

Figure 6.6
Audio metrics ESOM U-Matrix visualization. (**Ro** = Rock/Punk, **Bl** = Blues,
El = Electronica, **Cl** = Classical, **Wo** = World, **Ne** =New Age, **Ja** = Jazz,
Me = Metal, **Po** = Pop, **Am** = Ambient)

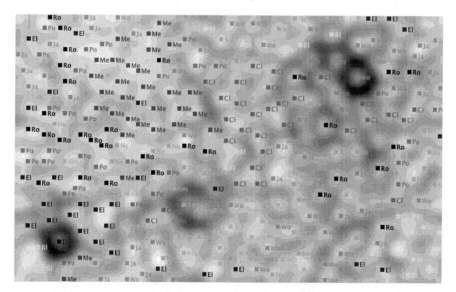

Figure 6.7
Combined (MIDI + audio) metrics ESOM U-Matrix visualization. (**Ro** =
Rock/Punk, **Bl** = Blues, **El** = Electronica, **Cl** = Classical, **Wo** = World, **Ne**
=New Age, **Ja** = Jazz, **Me** = Metal, **Po** = Pop, **Am** = Ambient)

grid of the ESOM. The grid visualizes the projection of the higher-dimensional space to a lower-dimensional map space (in this case a toroid). The U-Matrix display uses color to visualize the original high-dimensional distances of the input space. The colors represent U-matrix values of nodes and can be interpreted as height. The U-matrix value of a particular node is the average distance between the node and its closest neighbors. Light shades (mountains) represent large distances in the original data space, dark shades (valleys) imply similarity with respect to the extracted features.

In order to create practical visualizations, we reduced our data set to 40 pieces from each genre class, for a total of 400 pieces. While using these smaller data sets decreased the amount of training samples available, we believe the visualizations are representative of the complete data set. We trained ESOMs as described above using the metrics derived through PCA reduction.

Discussion

The visualizations from our ESOM experiments provide useful information on the genre classification ability of power-law metrics.

For instance, Figure 6.5 suggests that MIDI metrics alone can reasonably identify several genres, including Ambient, Classical, and Metal. Figure 6.6 suggests that timbre metrics can identify Blues better than MIDI metrics alone. Also, timbre metrics can identify the Metal genre as well as MIDI metrics. Finally, Figure 6.7 suggests that the combined timbre and MIDI metrics (as was to be expected) perform better; various genres are clustered overall more clearly than in Figures 6.5 and 6.6.

Again, it should be noted that the Magnatune corpus does not necessarily follow a consistent, or highly accurate genre-labeling scheme. For instance, Magnatune genres are assigned at the level of artist or album. Also many artists/albums have multiple genre labels (e.g., Jeff Wahl is a New Age/Jazz guitarist, whereas Kirsty Hawkshaw's *The Ice Castle* album is Ambient/Electronica). We labeled pieces in our data sets using the first genre identified. We believe this may have negatively affected cluster cohesion in the above visualizations. This belief is supported by our results with Armonique, discussed in the next two sections.

6.7 Armonique—A Music Similarity Engine

To further explore and assess the connection between power-law metrics and music aesthetics, we developed a music similarity engine, called *Armonique* [40, 39, 51]. This system started as a demo for power-law metrics. However, with National Science Foundation (NSF) funding, it has evolved into an effective, scalable system for music information retrieval (see http://armonique.org).

The majority of existing music search engines (50+) focus on context/metadata (e.g., text input, social networking, and/or users' listening habits). This includes systems such as iTunes Genius, Last.fm, and Pandora, which involve either musicologists listening and carefully tagging every new song across numerous dimensions (e.g., Pandora), or collaborative filtering techniques based on user preferences and ratings (e.g., Genius).

Armonique utilizes approximately 250 power-law metrics. These metrics extract features shown to correlate with aspects of human aesthetics (e.g., see Section 6.6). Since this extraction does not require interaction by humans (musicologists or listeners), Armonique can handle large and/or rapidly increasing music collections. Also, this allows users to discover songs of interest that are rarely listened to and are hard to find otherwise.

Currently, Armonique incorporates 10,000+ MP3 songs from Magnatune. These songs span Ambient, Classical (Baroque, Renaissance, Medieval, Contemporary, Minimalism), Electronica, Jazz and Blues, Metal and Punk Rock, New Age, Rock and Pop, and World (Indian, Celtic, Arabic, Tango, Eastern-European, Native-American) music.

As a search query, the user inputs a musical piece. This may be a piece already in the system, or a piece uploaded in real time by the user (in MP3). Armonique extracts timbre metrics from this piece (see Section 6.5.2). Then it converts the piece to MIDI. To do so, it utilizes an efficient audio-to-MIDI transcription algorithm, which handles polyphony and captures harmonic, vocal and percussive instrumentation. This algorithm involves calculating Fourier components of a signal at specific frequencies, and using variable window durations based on specific frequency bands (under publication). Then, Armonique extracts symbolic (MIDI) metrics (see Section 6.5.1). Thus, a piece is mapped to hundreds of slopes and R^2 features. Armonique utilizes a mean-square error (MSE) calculation to find the "closest" pieces. We have experimented with various techniques to make the search real time. Currently, we use a binary-search approach that we present elsewhere (under publication). Finally, Armonique outputs pieces that it "considers" most similar, according to its power-law aesthetics model (see Figure 6.8).

The next section presents results assessing the aesthetic relevance of Armonique's recommendations.

6.8 Psychological Experiments

We have conducted various assessment and validation experiments which compare the computational aesthetics model described above to emotional and physiological responses of human subjects. These experiments further

the music you like, every time

Armonique is a music similarity engine. Unlike other systems, it does not require humans to tag songs.
It analyzes music using a computational model of aesthetics. Listeners find that this approach works well.

If you have an iPhone, also try out Armonique Lite, our portable music similarity engine.

(Get new songs)

▶ Philharmonia Baroque - Recitativo Oime che pene
Handel - Atalanta CD2
Not only is this the only in-print recording of Handel's magnificent and rarely performed opera, but as the Wall
Street Journal observed in April 2006, this two CD release by ... (more)

Results 1-10 of 132

Philharmonia Baroque
Handel - Atalanta CD2
Not only is this the only in-print recording of Handel's magnificent and rarely performed opera, but as the Wall
Street Journal observed in April 2006, this two CD release by... (more)
▶ Recitativo Sono Irene oppur sogno   - 96.11% similar - Similar Songs
▶ Recitativo Oh del crudo mio bene   - 93.38% similar - Similar Songs
▶ Recitativo Il mio caro pastore   - 90.84% similar - Similar Songs
▶ Recitativo Oh forz del Destin   - 88.83% similar - Similar Songs

Philharmonia Baroque
Handel - Atalanta CD1
Not only is this the only in-print recording of Handel's magnificent and rarely performed opera, but as the Wall
Street Journal observed in April 2006, this two CD release by... (more)
▶ Recitativo Sempre ti lagni oh Tirsi   - 94.11% similar - Similar Songs
▶ Recitativo Amarilli Oh Dei qui Tirsi   - 91.25% similar - Similar Songs
▶ Recitativo Cerchi invano la morte   - 90.90% similar - Similar Songs
▶ Recitativo Perche sospesa o figlia   - 90.72% similar - Similar Songs

▶ The Sarasa Ensemble - Cantata Il Delirio Amoroso - Recitativo
A Baroque Mosaic
Hailed for its "great clarity" and "irresistible energy," The Sarasa Ensemble performs Vivaldi, Bach, Handel and
other compositions from the early Baroque through the Romantic... (more)
     - 89.27% similar - Similar Songs

 ▶ The Sarasa Ensemble - Ich Habe Genug Recitative
Bach Cantatas
This deeply beautiful recording presents two of Bach's cantatas for solo voice, one—Weichet nur—a celebratory
wedding ode and the other his celebrated masterpiece for... (more)
     - 88.69% similar - Similar Songs

Figure 6.8
The Armonique user interface.

demonstrate the connection between power-law metrics and aspects of human
emotion and aesthetics.

6.8.1 Earlier Assessment and Validation

We have approached the validation of power-law metrics from several perspectives. In this section we discuss experiments where we search for correlations
of power-law features with various aspects of human response to music.

6.8.1.1 Artificial Neural Network Experiment

When it became clear that an Artificial Neural Network (ANN) could perform well employing these metrics to classify music by composer and style [42], we examined whether an ANN could also predict average ratings by human listeners of their own emotional responses to music.

This experiment utilized 210 excerpts of 12 classical compositions. Half of these received ratings, averaged across 21 human participants, of "pleasant," while the remaining were rated "unpleasant." Using power-law metrics, the ANN had a success rate of 98.41% in classifying the 210 excerpts into these two human response categories. In addition, even the time course of the changes in average pleasantness ratings during the music could be somewhat predicted by the ANN [42, p. 66].

6.8.1.2 Evolutionary Computation Experiment

A second strategy was to use power-law metrics to create a software system for analyzing and composing music and then obtain emotional-response ratings of human listeners to the resulting computer-composed music.[3]

An experiment employing this strategy examined the continuously recorded self-ratings of pleasantness and activation by 23 human participants obtained during listening to J.S. Bach's *Invention No. 13 in A Minor (BWV 784)*, as well as to 17 computer-composed variations of this piece [43]. Fifteen of these computer-composed variations were judged by the system's music critic (fitness function) to be pleasant, while the other two were judged as unpleasant. The results showed that the time course of the both pleasantness and activation ratings for these 18 music pieces could be predicted by a combination of the MSE criterion of the music critic and elapsed time. The time courses of the human ratings to the two unpleasant variations contrasted sharply with those of the original Bach composition (lower pleasantness and higher activation than for the original).

6.8.1.3 Music Information Retrieval Experiment

A third strategy is the one employed in our most recent experiments. These experiments entailed measurement of human psychological and physiological responses to music selected by Armonique (http://armonique.org). This music discovery engine utilizes numerous power-law metrics as its search criteria.

Our first experiment employing this approach was designed to validate an early version of the search engine [40]. That version worked only with MIDI files and did not incorporate any measures of timbre. The search engine was used to find three similar and three dissimilar MIDI pieces in relation to a piece of classical music that each participant said they enjoyed. The music selected

---

[3]This evolutionary computing system, called *NEvMuse*, employs power-law metrics as fitness functions [44, 41, 43]. Section 6.3 presented music generated by NEvMuse to illustrate the connection between power laws and music aesthetics.

came from the Classical Music Archives 14,695-piece corpus. The results indicated that our 21 human participants rated the similar and dissimilar music differently, with ratings of similar music being closer to those of the original piece. However, these differences were generally quite small, and physiological measures did not clearly differentiate between similar and dissimilar music, perhaps because of inadequate sample size. Additional factors contributing to the relatively small differences may include the lack of timbre measures in the search engine and the fact that participants were listening to MIDI files rather than more naturalistic music.

An interesting feature of the ratings data in the experiment just described was that ratings of liking were significantly higher for the song chosen by the participant than for either the similar or dissimilar songs. This finding contrasts with the absence of a difference in liking between a particular song and similar songs chosen by Armonique when the participant plays no role in the choice of the original song (see Figure 6.9).

6.8.2　Armonique Evaluation Experiments

This section describes experiments on the latest version of Armonique, which works with MP3 pieces.

6.8.2.1　Methodology

A more comprehensive study was undertaken to evaluate a second-generation version of Armonique, which takes timbre into account. This version presents the music as ordinary audio (MP3 or WAV) files, thereby creating a more normal music listening experience for participants.

Since the corpus of music employed came from Magnatune (http://magnatune.com), it did not include much widely known music. Consequently, this study relied upon favorite genres of participants rather than particular pieces that they identified as favorites.

Two sets of musical excerpts (one minute in length, all instrumental music) were presented:

(a) One set of seven pieces (henceforth, *Set A*), which was presented to all 40 participants. This common set consisted of an original piece, the three pieces most similar to the original from the corpus, and the three pieces most dissimilar (according to Armonique). This set represented a search based upon a classical composition chosen by (liked by) the experimenter.

(b) A second set of five excerpts (henceforth, *Set B*), which was unique for each participant. Each of these 40 different sets consisted of an original piece, two most similar, and two most dissimilar pieces (again, according to Armonique). Each set came from a different search. This search was initiated using a random piece from one of the participant's three favorite genres (based on Magnatune's genre taxonomy).

Therefore, each of the 40 participants was presented with 12 pieces. These pieces were presented in a different random order every time. While listening, participants had no knowledge of which piece was the starting point for any of the searches. The original was revealed only at the end of the experiment, when we asked participants to rate the similarity (to the original piece) of the pieces recommended by Armonique.

Each participant listened to the music through speakers, in a separate experimental session of about one hour conducted in a small laboratory room. The session began with attachment of electrodes and sensors for measuring a variety of responses (ActiveTwo system, BioSemi B.V., Amsterdam, The Netherlands)—heart rate, skin conductance (left fore and middle fingers), skin temperature (left thumb), left forehead (corrugator supercilii) electromyographic (EMG) activity, and 32 channels of electroencephalographic (EEG) activity.

Digitized, amplified physiological signals were passed through a fiber-optic cable into an adjacent room where signals were recorded and saved into data files. Music presentation was controlled by a second computer (LabVIEW software, National Instruments, Austin, Texas), which also signaled the beginning and end of each excerpt to the recording computer and prompted the participant to make various ratings immediately after each excerpt.

An additional excerpt was presented at the beginning of the recording session to allow participants to practice these ratings. Ratings were not made during the music to avoid possible contamination of physiological responses by the conscious process of rating the music.

Four types of ratings were made by participants using a mouse to move sliders on the LabVIEW software—*pleasantness response* to the music, *activation response* to the music, *liking* of the music, and *perceived familiarity* of the music. Pleasantness and activation sliders employed the Self-Assessment Manikin of Bradley and Lang [11].

After the experimental session, participants rated the similarity of each piece within Set A and Set B, in relation to the original piece for each respective search.

6.8.2.2   Results—Psychological Ratings

In contrast to the rather small differences demonstrated in our previous study (see Section 6.8.1.3), robust differences were obtained on a variety of psychological measures in the present study. Here we summarize results on the comparison of music identified in the searches as similar to an original piece with that identified as dissimilar. Repeated-measure analyses of variance were followed by a contrast of the similar and dissimilar excerpts. The similarity ratings conducted at the conclusion of the experimental sessions showed large and statistically significant ($p < 0.001$) differences between similar and dissimilar excerpts, for both Set A and Set B, though somewhat more consistently in Set A (as would be expected since all participants were listening to the same

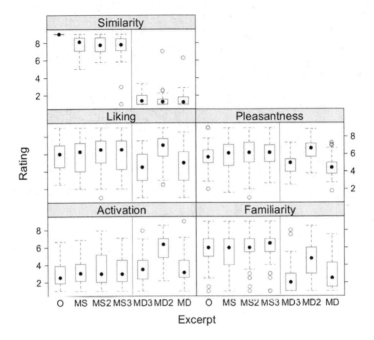

Figure 6.9
Set A: Responses (self-ratings) from 40 participants to music recommended by
Armonique. (O = original piece; MS, MS2, MS3 = 1st, 2nd, 3rd most similar
piece; MD3, MD2, MD = 3rd, 2nd, 1st most dissimilar piece)

music in this set). Figure 6.9 shows the results for Set A, and Figure 6.10 for
Set B.

The affective response ratings may be more relevant to the question of
aesthetic response. These data also showed statistically significant though
somewhat less dramatic differences (Figures 6.9 and 6.10). Self-ratings of participant pleasantness responses were significantly higher in similar excerpts
than in dissimilar ones in both Set A ($p = 0.005$) and Set B ($p < 0.001$).
Liking ratings displayed much the same pattern—significantly higher liking of
similar excerpts in both Set A ($p = 0.034$) and Set B ($p < 0.001$). Self-ratings
of participant activation were significantly lower in similar excerpts than in
dissimilar ones in Set A ($p = 0.001$), but no such differences were found in Set
B ($p > 0.5$).

Familiarity ratings may provide a partial explanation for the affective response ratings. As shown in Figures 6.9 and 6.10, the pattern of differences
in these ratings is quite similar to those for pleasantness and liking; also

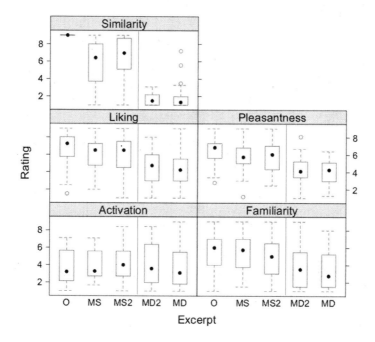

Figure 6.10
Set B: Responses (self-ratings) from 40 participants to music recommended
by Armonique. (O = original piece; MS, MS2 = 1st, 2nd most similar piece;
MD3, MD2, MD = 2nd, 1st most dissimilar piece)

differences between similar and dissimilar excerpts are significant for both
sets of excerpts ($p < 0.001$).

An obvious exception to the above appears in Figure 6.9 (Set A) for the
second most dissimilar excerpt (MD2). This piece was predominantly percussion music. Although participants judged it to be very dissimilar from the
original, they reported high pleasantness and activation responses to it. They
also reported liking it and being familiar with it. No such departure from the
main pattern of differences is observed in Set B. Perhaps this is because Set
B actually represents an average across 40 searches, whereas Set A represents
a single search.

6.8.2.3 Results—Physiological Measures

A variety of measures of physiological response were extracted. Data files were
read with EEGLAB software (Swartz Center for Computational Neuroscience,
University of California, San Diego). Separate data sets were created for each
song and the EEG channels; and for each song and the remaining psychophys-

iological measures. EEG data were corrected for eye-movement artifacts. Then they were subjected to Fourier analysis to extract power in the alpha frequency band (8–13 Hz) as a measure of activation at each electrode site (high-alpha power represents low-brain activation). Data from two participants had to be dropped from the analysis because of excessive recording artifacts.

Data from symmetrically located pairs of electrodes were employed to create measures of hemispheric activation asymmetry. This process entailed subtracting alpha powers between homologous locations on the left and right hemispheres. Differences on these asymmetry measures between musical pieces were analyzed, using repeated-measures analysis, as described in Section 6.8.2.2. These asymmetry measures are of special interest because of evidence that they correlate with affective responses [17, 1, 46]; but see also Harmon-Jones et al. [25].

The following pairs of electrodes (10-20 system) produced significantly higher asymmetry measures in similar than in dissimilar songs of Set A: $F8 - F7$ ($p = 0.044$), $FC2 - FC1$ ($p = 0.008$), $FC6 - FC5$ ($p = 0.042$), $C4 - C3$ ($p = 0.028$). Differences for one of these electrode pairs are shown in Figure 6.11.

Thus, as illustrated in Figure 6.11, similar songs activated the posterior and lateral portion of the left frontal lobe more than the right. This difference was not as pronounced for dissimilar songs. This observation might be associated with the more positive affective response to the similar songs. Such a relationship is consistent with previous findings of an association between left hemisphere activation and positive affect [46]. However, these differences were not significant for Set B.

Data for heart rate were analyzed by extracting the *interbeat intervals* using Open AnsLab software (Department of Psychology, University of Basel, Switzerland). Average interbeat intervals were calculated for each participant and each excerpt in spreadsheet software. Excerpts were compared using repeated-measure analyses as described in Section 6.8.2.2. Mean interbeat intervals were significantly greater for dissimilar than for similar excerpts in both Set A ($p = 0.016$, Figure 6.12) and Set B ($p = 0.001$, Figure 6.13).

The remaining psychophysiological measures did not display consistent differences between musical pieces. However, it is interesting to return to the psychological ratings of Set A for guidance in the physiological analyses. Recall the high ratings for piece MD2. This finding is somewhat paralleled by the skin conductance data for Set A (Figure 6.14). A repeated-measure analysis of variance, followed by a contrast of MD2 with the two other dissimilar pieces, indicated that skin conductance was significantly higher ($p = 0.016$) during MD2 than during the other two pieces.

6.8.2.4 Discussion

These results indicate that similar music (i.e., pieces recommended by Armonique as similar to a given piece) produces different responses than

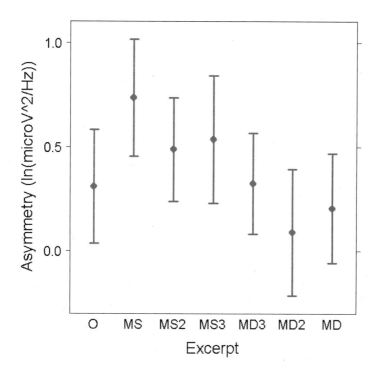

Figure 6.11
Set A: Hemispheric asymmetry (mean ± standard error) of alpha EEG activity at the $FC2$ (right hemisphere) and $FC1$ (left hemisphere) electrodes over the frontal lobes of 38 participants for the seven excerpts of Set A (see code legend in Figure 6.10).

dissimilar music. The most obvious differences are in the similarity judgments that participants make, which are generally consistent with the search engine ratings. Participants also differentiate the two categories of music in their reported affective responses, though there is evidence of inconsistencies in the ratings within the dissimilar category. It appears that there may be more than one underlying dimension of similarity that determines affective response.

Reported familiarity with the music displays a pattern of differences much like that of the pleasantness and liking aspects of affect. This finding agrees with the frequent claim that people like a piece of music because they are familiar with it.

To investigate this possibility, we employed hierarchical linear modeling to examine the relationship among these variables. Both familiarity and pleasantness were strong predictors of liking ($p < 0.001$), explaining different portions

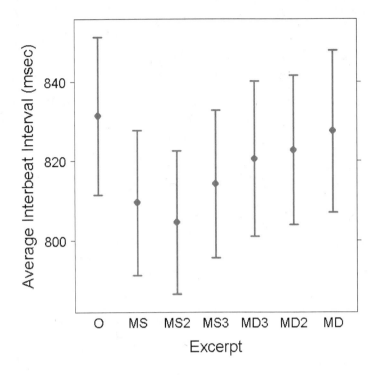

Figure 6.12
Set A: Intervals between heartbeats (mean ± standard error) of 40 participants
for the seven excerpts of Set A (see code legend in Figure 6.10).

of the overall variance in liking ratings. This relationship was present in Set
A; it was also confirmed in separate analyses on Set B. However, since the
participants were aware of their affect ratings when they made their familiarity ratings, these data cannot exclude the possibility that participants were
simply trying to make their ratings agree with each other.

Physiological responses also differentiated the two categories of music, similar and dissimilar. Hemispheric asymmetry measures derived from spectral
analyses were greater for similar than for dissimilar pieces, as the affective
rating differences would lead us to expect. However, this occurred only in
Set A, in spite of the fact that strong affective response differences were also
found in Set B. That the hemispheric asymmetry differences only emerged in
response to the music heard by all participants suggests that these differences
may be more directly linked to the processing of the physical stimuli than
to higher-level conscious affective responses. On the other hand, differences
in heart interbeat interval were present for both sets of music, suggesting a

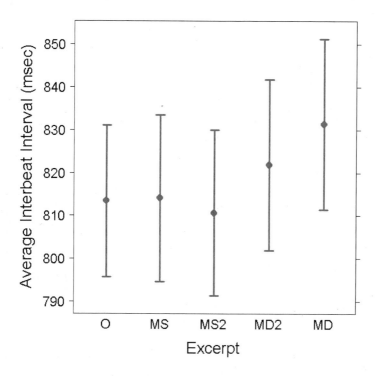

Figure 6.13
Set B: Intervals between heartbeats (mean ± standard error) of 40 participants
for the five excerpts of Set B (see code legend in Figure 6.10).

response that is more closely associated to the conscious affective response.
Other physiological responses are apparently not strongly determined by these
musical differences. A difference not predicted by the search engine, the high
activation ratings produced by MD2 in Set A, also had a physiological component, relatively high skin conductance.

A peculiarity of the results is evident especially in Figures 6.11 and 6.12:
On some measures, for Set A, the dissimilar songs produced responses more
like those of the original than those of the similar songs. The explanation for
this outcome is not apparent. However, note that results for Set A are based
only on a single search by Armonique, whereas results for Set B are based on
a different search per participant. It may be that some unknown idiosyncrasy
of the original song in Set A produced this pattern. Such an outcome for Set
B was presumably prevented by the fact that the results were averaged over
40 searches.

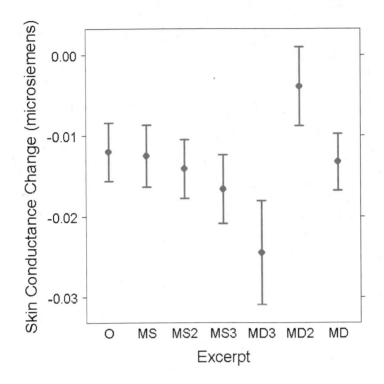

Figure 6.14
Set A: Skin conductance change from baseline (mean ± standard error) of 40
participants for the seven excerpts of Set A (see code legend in Figure 6.10).

6.8.2.5   Final Thoughts

The primary practical purpose of a music similarity engine is to find music
that listeners like. From that perspective, the ratings of liking may provide
especially important information about Armonique's performance (and, respectively, the promise of power-law metrics for music information retrieval).

The data from the evaluation experiments described above (see Figures 6.9
and 6.10) provide evidence for high ratings of liking in similar songs. This is
also evident in unpublished results of several other experiments. However, as
noted earlier, when the starting point for a search was a piece of music chosen
by the listener, liking was generally higher for this song than for similar songs
found by Armonique. This finding is consistent with those of many studies
in the literature demonstrating individual differences in preferences for music
[24].

In closing, evidence from a variety of different psychological assessment approaches indicates that power-law metrics capture fundamental aspects of music that determine human responses, including affective ones. However, the evidence also indicates wide individual variation in self-ratings to music and in physiological response. These variations may imply:

1. Deficiencies in measurement methods, such that error variance is large and obscures fundamental patterns;

2. Individual differences in response produced by factors outside the music, such as participants' musical experience or social group [24]; or

3. Properties of the music that are not captured by the employed power-law metrics.

The ratings methods employed in these experiments were based upon dimensional and categorical models of emotions that are not specific to music. Methods that are more specifically tailored to music-elicited emotions [61] might help reduce error variance and make differences even clearer.

The findings with physiological methods were helpful in demonstrating correlations with power-law metrics, but the differences between similar and dissimilar songs seemed relatively weak and poorly correlated with self-ratings. While physiological methods have now been widely used in measuring responses to music, the evidence is mixed regarding how useful they are as indicators of fundamental emotional response [27]. It may be that the emotional effects elicited in these experiments are too subtle to produce robust physiological differences. Thus, further research with more refined methods may lead to stronger evidence of the predictive value of power-law metrics.

6.9 Conclusion

This chapter presented results from applying power laws to music information retrieval and data mining tasks. These results suggest that power-law metrics represent a promising new approach for automatic extraction of metadata from musical archives. Essentially, these metrics capture statistical proportions (e.g., entropy and self-similarity) of music-theoretic and other attributes. These metrics were assessed in various ways, including through Armonique, a music similarity engine (http://armonique.org). Evaluation experiments suggest a connection between power laws and music aesthetics (e.g., emotional and physiological responses of listeners).

Our adaptation of Zipf's law in regular and higher-order metrics across many dimensions, essentially, captures the structure of a musical piece economically across different levels of resolution. Musical pieces are self-similar to a certain extent; this means that the entropy (e.g., information content, grammatical structure, and so forth) is similar across different levels of hierarchy. Therefore, it is not necessary to calculate all possible higher-order levels (or local interval variability metrics), since, after a certain point, not much new information is being extracted. Where that point of *saturation* occurs is subject to experimentation, and probably varies across musical genres and styles. But, once discovered, it can probably be applied safely across musical artifacts of the same genre/style. In our work with thousands of pieces from diverse musical genres and styles (e.g., Ambient, Baroque, Classical, Impressionist, Renaissance, Medieval, Contemporary, Minimalist, Electronica, Jazz and Blues, Metal and Punk Rock, New Age, Rock and Pop, Indian, Celtic, Arabic, Tango, and Eastern European), we find that a few higher-order levels (two to three) most often suffice.

Alternatively, this suggests that any type of metric, which captures the entropy of an artifact, as long as it is applied at different levels of granularity, may have similar information-extraction effectiveness and aesthetic relevance as the power-law metrics described here.

In closing, we live in a self-similar (fractal) world. This self-similarity is reflected across phenomena, events, objects, artifacts, and living entities, at different levels of granularity, dimensions, and time scales [5]. This reflection, of course, is not *mirror-perfect*—it comes with fluctuations. Our work suggests that it is precisely these fluctuations that provide the identity and uniqueness of different types of phenomena; and within one type of phenomenon, the identity and uniqueness of individual events/objects/artifacts. (For instance, one rainstorm is different from another, but statistically they are all rainstorms, hence the word "rainstorm." Accordingly, the recorded sound of one rainstorm is different from that of another, but all are recognizable for what they are.) The same applies to music, in terms of genres (e.g., Baroque versus Jazz) or individual pieces (e.g., J.S. Bach's *Toccata and Fugue in D Minor* versus his *Invention No. 13 in A Minor*). Hopefully, we have provided enough information and evidence to encourage further exploration of Zipf's Law [63], and related power laws for music data mining in specific, and computational modeling of aesthetics in general.

Acknowledgments

Our work has been supported by the U.S. National Science Foundation (Grants IIS-0736480, IIS-0849499, and IIS-1049554). We also received generous donations of music corpora from Pierre R. Schwob, Classical Music Archives (www.classicalarchives.com), John Buckman, Magnatune (www.magnatune.com), and Stephanie Merakos, Music Library of Greece (www.mmb.org.gr). The following have contributed to research in power-law metrics and Armonique development: Dana Hughes, Brys Sepulveda, Luca

Pellicoro, Timothy Hirzel, Perry Spyropoulos, Clayton McCauley, Penousal Machado, Juan Romero, Brian Muller, William Daugherty, Dallas Vaughan, Christopher Wagner, Charles McCormick, Tarsem Purewal, Valerie Sessions, and James Wilkinson. The following have contributed to assessment and validation experiments involving emotional and physiological responses of human subjects: Aimee Siebert, José Rojas, Sonia Barrera, and Brittany Baker. We finally thank Yiorgos Vassilandonakis, Blake Stevens, and Renée McCauley for their invaluable comments on earlier drafts of this chapter. Dana Hughes helped develop some of the figures.

Bibliography

[1] J.J.B. Allen, J.A. Coan, and M. Nazarian. Issues and assumptions on the road from raw signals to metrics of frontal EEG asymmetry in emotion. *Biological Psychology*, pages 183–218, 2004.

[2] Aristotle. *Complete Works, Vol. 10, "Metaphysics I."* Hatzopoulos, O., (ed.), Kaktos, Athens, Greece (in Greek), 1992.

[3] R. Arnheim. *Entropy and Art: An Essay on Disorder and Order.* University of California Press, Berkeley, 1971.

[4] J.-J. Aucouturier and F. Pachet. Improving timbre similarity: How high is the sky? *Journal of Negative Results in Speech and Audio Sciences*, 1(1), 2004.

[5] P. Bak. *How Nature Works: The Science of Self-Organized Criticality.* Copernicus, New York, 1996.

[6] L.F. Barrett and J.A. Russell. The structure of current affect: Controversies and emerging consensus. *Current Directions in Psychological Science*, 8(1):10–14, 1999.

[7] R. Basili, A. Serafini, and A. Stellato. Classification of musical genre: A machine learning approach. In *Proceedings of the 5th International Conference on Music Information Retrieval (ISMIR '04)*, pages 505–508, Barcelona, Spain, 2004.

[8] M. Beer. Mathematics and music: Relating science to arts? *Mathematical Spectrum*, 41(1):36–42, 2008.

[9] B. Blasius and R Tönjes. Zipf's law in the popularity distribution of chess openings. *Physics Review Letters*, 103(21):218701-1–218701-4, 2009.

[10] Bogomolny, A. Benford's Law and Zipf's Law, http://www.out-the-knot.org/do_you_know/zbflaw.html.

[11] M.M. Bradley and P.J. Lang. Measuring emotion: The self-assessment manikin and the semantic differential. *Journal of Behavioral Therapy and Experimental Psychiatry*, 25(1):49–59, 1994.

[12] P.A. Calter. *Content-Based Music Information Retrieval: Current Directions and Future Challenges.* John Wiley & Sons, New York, 2008.

[13] P. Cano, M. Koppenberger, and N. Wack. Content-based music audio recommendation. In *Proceedings of the 13th Annual ACM International Conference on Multimedia (MULTIMEDIA'05)*, pages 211–212, ACM Press, New York, 2005.

[14] M.A. Casey, R. Veltkamp, M. Goto, M. Leman, C. Rhodes, and M. Slaney. Content-based music information retrieval: Current directions and future challenges. *Proceedings of the IEEE*, 96(4):668–696, March 2008.

[15] J.E. Cutting, J.E. DeLong, and C.E. Nothelfer. Attention and the evolution of Hollywood film. *Psychological Science*, 21(7):e–print, 2010.

[16] C. Dahlhaus. *Esthetics of Music.* Austin, W.W. (trans.), Cambridge University Press, Cambridge, UK, 1982.

[17] R.J. Davidson. EEG measures of cerebral asymmetry: Conceptual and methodological issues. *International Journal of Neuroscience*, 39:71–89, 1988.

[18] S. Dixon, F. Gouyon, and G. Widmer. Toward characterisation of music via rhythmic patterns. In *Proceedings of the 5th International Conference on Music Information Retrieval (ISMIR'04)*, pages 509–516, Barcelona, Spain, 2004.

[19] C. Faloutsos. Next generation data mining tools: Power laws and self-similarity for graphs, streams and traditional data. *Lecture Notes in Computer Science, Knowledge Discovery in Databases, PKDD 2003*, LNCS 2838:10–15, 2003.

[20] Y. Feng, Y. Zhuang, and Y. Pan. Music information retrieval by detecting mood via computational media aesthetics. In *Proceedings of the 2003 IEEE/WIC International Conference on Web Intelligence (WI'03)*, pages 235–241, IEEE Computer Society, Washington, DC, 2003.

[21] K. Gibran. *The Prophet*, 9th edition. Alfred A. Knopf, New York, 1973.

[22] D.B. Guralnik, editor. *Webster's New World Dictionary of the American Language*, 2nd college edition. William Collins, Ohio, 1979.

[23] M. Hall, E. Frank, G. Holmes, B. Pfahringer, P. Reutemann, and I.H. Witten. The WEKA data mining software: An update. *SIGKDD Explorations*, 11(1):10–18, 2009.

[24] D.J. Hargreaves and A.C. North. Experimental aesthetics and liking for music. In P.N. Juslin and J.A. Sloboda, editors, *Handbook of Music and Emotion: Theory, Research, Applications.* Oxford University Press, Oxford, UK, 2010.

[25] E. Harmon-Jones, L. Lueck, M. Fearn, and C. Harmon-Jones. The effect of personal relevance and approach-related action expectation on relative left frontal cortical activity. *Psychological Science*, 17(5):434–440, 2006.

[26] P. Hemenway. *Divine Proportion: Φ (Phi) in Art, Nature, and Science.* Sterling Publishing, New York, 2005.

[27] D.A. Hodges. Psychophysiological measures. In P.N. Juslin and J.A. Sloboda, editors, *Handbook of Music and Emotion: Theory, Research, Applications.* Oxford University Press, Oxford, UK, 2010.

[28] J.H. Jensen, M.G. Christensen, D.P.W. Ellis, and S.H. Jensen. Quantitative analysis of a common audio similarity measure. *IEEE Transactions on Audio, Speech and Language Processing*, 17(4):693–703, 2009.

[29] O. Jonas. *Introduction to the Theory of Heinrich Schenker.* Longman, New York, 1982.

[30] J. Kalda, M. Sakki, M. Vainu, and M. Laan. Zipf's Law in human heartbeat dynamics, http://arxiv.org/abs/physics/0110075, 2001.

[31] I. Karydis, A. Nanopoulos, and Y. Manolopoulos. Symbolic musical genre classification based on repeating patterns. In *Proceedings of the 1st ACM Workshop on Audio and Music Computing Multimedia (AMCMM'06)*, pages 53–58, ACM Press, New York, 2006.

[32] T. Li and M. Ogihara. Content-based music similarity search and emotion detection. In *Proceedings of the IEEE International Conference on Acoustics, Speech, and Signal Processing (ICASSP'04)*, pages V705–V708, IEEE Computer Society, Los Alamitos, California, 2004.

[33] T. Li, M. Ogihara, and Q. Li. A comparative study on content-based music genre classification. In *Proceedings of the 26th Annual International ACM SIGIR Conference on Research and Development in Information Retrieval (SIGIR'03)*, pages 282–289, ACM Press, New York, 2003.

[34] Li, W. Information on Zipf's law, http://www.nslij-genetics.org/wli/zipf.

[35] T. Lidy and A. Rauber. Evaluation of feature extractors and psycho-acoustic transformations for music genre classification. In *Proceedings of the 6th International Conference on Music Information Retrieval (ISMIR 2005)*, pages 34–41, London, September 11–15, 2005.

[36] T. Lidy, A. Rauber, A. Pertusa, and J.M. Iñesta. Improving genre classification by combination of audio and symbolic descriptors using a transcription system. In *Proceedings of the 8th International Conference on Music Information Retrieval (ISMIR 2007)*, pages 61–66, Vienna, Austria, 2007.

[37] M. Livio. *The Golden Ratio.* Broadway Books, New York, 2002.

[38] P. Machado, J. Romero, M. L. Santos, A. Cardoso, and B. Manaris. Adaptive critics for evolutionary artists. *Lecture Notes in Computer Science, Applications of Evolutionary Computing*, LNCS 3005:437–446, 2004.

[39] B. Manaris, J.R. Armstrong, T. Zalonis, and D. Krehbiel. Armonique: A framework for Web audio archiving, searching, and metadata extraction. *International Association of Sound and Audiovisual Archives (IASA) Journal*, 35:57–68, 2010.

[40] B. Manaris, D. Krehbiel, P. Roos, and T. Zalonis. Armonique: Experiments in content-based similarity retrieval using power-law melodic and timbre metrics. In *Proceedings of the 9th International Conference on Music Information Retrieval (ISMIR 2008)*, pages 343–348, Philadelphia, 2008.

[41] B. Manaris, P. Machado, C. McCauley, J. Romero, and D. Krehbiel. Developing fitness functions for pleasant music: Zipf's law and interactive evolution systems. *Lecture Notes in Computer Science, Applications of Evolutionary Computing*, LNCS 3449:498–507, 2005.

[42] B. Manaris, J. Romero, P. Machado, D. Krehbiel, T. Hirzel, W. Pharr, and R.B. Davis. Zipf's law, music classification and aesthetics. *Computer Music Journal*, 29(1):55–69, 2005.

[43] B. Manaris, P. Roos, P. Machado, D. Krehbiel, L. Pellicoro, and J. Romero. A corpus-based hybrid approach to music analysis and composition. In *Proceedings of the 22nd Conference on Artificial Intelligence (AAAI-07)*, pages 839–845, AAAI Press, Vancouver, BC, 2007.

[44] B. Manaris, D. Vaughan, C. Wagner, J. Romero, and R.B. Davis. Evolutionary music and the Zipf-Mandelbrot law: Progress toward developing fitness functions for pleasant music. *Lecture Notes in Computer Science, Applications of Evolutionary Computing*, LNCS 2611:522–534, 2003.

[45] B. Mandelbrot. *Fractal Geometry of Nature.* Freeman and Company, New York, 1977.

[46] D. Mathersul, L.M. Williams, P.J. Hopkinson, and A.H. Kemp. Investigating models of affect: Relationships among EEG alpha asymmetry, depression, and anxiety. *Emotion*, 8(4):560–572, 2008.

[47] M. May. Did Mozart use the golden section? *American Scientist*, 84(1):118–119, 1996.

[48] C. McKay and I. Fujinaga. Automatic genre classification using large high-level musical feature sets. In *Proceedings of the 5th International Conference on Music Information Retrieval (ISMIR'04)*, pages 525–530, Barcelona, Spain, 2004.

[49] A. Oliveira and A. Cardoso. Modeling affective content of music: A knowledge base approach. In *Proceedings of the 5th Sound and Music Computing Conference (SMCC'08)*, pages e1–e6, Berlin, Germany, 2008.

[50] C.A. Pickover. *Computers and the Imagination*. St. Martin's Press, New York, 1991.

[51] P. Roos and B. Manaris. A music information retrieval approach based on power laws. In *Proceedings of 19th IEEE International Conference on Tools with Artificial Intelligence (ICTAI-07)*, vol. 2, pages 27–31, Patras, Greece, 2007.

[52] N.A. Salingaros and B.J. West. A universal rule for the distribution of sizes. *Environment and Planning B: Planning and Design*, 26(6):909–923, 1999.

[53] M. Schroeder. *Fractals, Chaos, Power Laws: Minutes from an Infinite Paradise*. W.H. Freeman and Co., New York, 1991.

[54] B. Spehar, C.W.G. Clifford, B.R. Newell, and R.P. Taylor. Universal aesthetic of fractals. *Computers & Graphics*, 27:813–820, 2003.

[55] R.P. Taylor, A.P. Micolich, and D. Jonas. Fractal analysis of Pollock's drip paintings. *Nature*, 399:422, 1999.

[56] G. Tzanetakis and P. Cook. Musical genre classification of audio signals. *Speech and Audio Processing, IEEE Transactions*, 10(5):293–302, 2002.

[57] A. Ultsch and F. Mörchen. *ESOM-Maps: Tools for Clustering, Visualization, and Classification with Emergent SOM*. CiteSeer (Online), 2005.

[58] R.F. Voss and J. Clarke. 1/F noise in music and speech. *Nature*, 258:317–318, 1975.

[59] R.F. Voss and J. Clarke. 1/F noise in music: Music from 1/F noise. *Journal of the Acoustical Society of America*, 63(1):258–263, 1978.

[60] Y.-H. Yang, Y.-C. Lin, Y.-F. Su, and H.-H. Chen. A regression approach to music emotion recognition. *IEEE Transactions on Audio, Speech and Language Processing (TASLP)*, 16(2):448–457, 2008.

[61] M. Zentner and T. Eerola. Self-report measures and models. In P.N. Juslin and J.A. Sloboda, editors, *Handbook of Music and Emotion: Theory, Research, Applications*. Oxford University Press, Oxford, UK, 2010.

[62] X. Zhang and W.R. Zbigniew. Analysis of sound features for music timbre recognition. In *Proceedings of the 2007 International Conference on Multimedia and Ubiquitous Engineering (MUE'07)*, pages 3–8, IEEE Computer Society, Washington, DC, 2007.

[63] G.K. Zipf. *Human Behavior and the Principle of Least Effort*. Hafner Publishing, New York, 1949.

Part III

Social Aspects of Music Data Mining

7

Web-Based and Community-Based Music Information Extraction

Markus Schedl

Johannes Kepler University

CONTENTS

There exists a tremendous and ever growing amount of digital music files in today's music collections, both in private and commercial repositories. This trend has been caused by various developments in the past two decades. Some of them were new audio compression techniques, high-speed Internet access, cheap storage devices, and novel channels for digital music distribution.

This remarkable increase in the size of music collections requests for intelligent methods to analyze, structure, and visualize them, as finding desired music is obviously becoming more difficult with growing repository size. To address this issue, a key task in the fields of music information extraction and retrieval is elaborating methods to uncover music-related information from various sources, as diverse as digital manifestations of audio signals (e.g., MP3 files), Web pages about music artists, or file sharing data in peer-to-peer net-

works. The traditional approach to this problem relies on analyzing the audio
signal of a piece of music to derive various features that describe some of its
musical properties, such as loudness, rhythmic structure or timbral aspects.
Such techniques are commonly referred to as "signal-based," "audio-based," or
"music content-based" approaches. In cases where audio files are not available,
however, these content-based approaches have obviously restricted applicability.

On the other hand, an incredible amount of information is available in
today's largest data source, the World Wide Web. Focusing only on the music
domain, there exists a wealth of information provided in the form of classical
"Web 1.0" pages, which can be either static or dynamically created. Typical
examples are fan pages, artists' or bands' personal Web pages, but also music
information systems such as allmusic.com [4], formerly known as the All Music
Guide (AMG), that provide music-related information in abundance. Moreover, the emergence of platforms and services that are commonly referred to as
the "Web 2.0" contributed considerably to the amount of user-generated content available today. The term "Web 2.0" was coined in 1999 by DiNucci [24],
but became popular only later in 2004, when O'Reilly launched the first Web
2.0 conference [71]. Web 2.0 applications include blogging services, social networks, platforms to share user-generated content and attach tags to items such
as images, videos, or songs. In the music domain, we find services specialized
on providing structured information about music, music recommendations,
or personalized Web radio, such as Last.fm [51] or Pandora [74]. What makes
such services particularly appealing for information retrieval tasks, in addition
to the abundance of information provided, is the ease of information gathering since they typically offer a developer's application programming interface
(API) to retrieve desired pieces of information.

With all those new platforms and services it has become possible to obtain
information on almost all musical artists. However, the information's quality
has not necessarily improved. In fact, quite the opposite observation can be
made as it has never been easier to make spurious information public using
Web 2.0 systems. One of the largest challenges of Web-based music information extraction is therefore how to overcome the noise in the data [50].

In this chapter, we look into different methods to automatically determine music-related information using the Web as data source. Such techniques
are commonly referred to as "Web-based," "community-based," or "music
context-based" approaches. The chapter is split into two parts: the first one
describes approaches to gather specific types of music-related information,
the second one presents techniques that address one of the key challenges in
music information research, the development of *similarity measures*—in this
case, based on information mined from Web and community data sources.

7.1 Approaches to Extract Information about Music

Metadata describing properties of a music entity, such as an artist, an album, or a song, plays an important role in a variety of music-related applications. Such applications include music information systems, recommender systems, and user interfaces to explore music collections.

Music information systems are services that offer various kinds of music-related information, typically for different music entities. For example, information commonly offered for an artist or a band includes biographies, discographies, band members and instruments, tour and concert dates, and photographs. For the entity album, images of album cover artwork are usually provided, as well as release date, album reviews, and links to online stores. Song- or track-specific information commonly encompasses lyrics, similar tracks (determined by content-based analysis [17, 8, 90], or by collaborative filtering, [18]), and sometimes preview snippets. Popular examples of music information systems are Last.fm [51], allmusic.com [4], and Discogs [25]; but also the lesser known systems, such as *Ishkur's Guide to Electronic Music* [38], *Map of Metal* [67], or the *Automatically Generated Music Information System* (AGMIS) [79], are also interesting examples.

Given a set of seed artists or tracks, music recommender systems provide recommendations to their users, which they hopefully like. These recommendations may either be based on content-based (CB) feature extraction from audio files and subsequent application of a similarity measure to find songs or artists similar to the seeds, or they may rely on collaborative filtering (CF) techniques that predict the taste of a user by comparing his or her preferences with like-minded users. To this end, a CF-based system keeps track of all items assigned to or liked by a user (e.g., products purchased, Web pages visited, or songs played). Based on a new user's seeds or her actual interaction with the system, similar users can be determined, and recommendations of the form "similar users also liked item X" can be made. A more detailed elaboration of recommender systems and CF can be found [47, 18, 57, 78].

User interfaces for exploring music collections typically also make use of metadata to support browsing. For example, the nepTune interface [44] to explore music repositories in a virtual, three-dimensional landscape extracts terms and images from artist-related Web pages and displays them in the regions to which songs of the corresponding artists are mapped. This technique facilitates recognizing areas that contain specific music styles, for example, "blues" or "vocal" music. The songs themselves are grouped based on content-based features which are clustered using a Self-Organizing Map [46]. Lübbers and Jarke [60] present a similar user interface. In this case, images of album covers serve as representations for individual songs. The user interface (UI) further allows the user to adapt the landscape by creating or destroying hills that separate groups of songs and accordingly adjust the underlying similarity measure. Another application that uses album cover images as identifiers for music pieces is the MusicGalaxy interface [87] that organizes a music collection by applying the similarity-preserving data projection technique multidimensional scaling (MDS) [49, 22] to a similarity measure defined on content-based features. The MusicSun interface [73] provides a means of discovering new artists by letting the user first define a set of seed artists and then select from a list of descriptive terms that are retrieved from the seed artists' Web pages, using different dictionaries of terms related to music, such as genres and styles, instruments, moods, and countries. Subsequently, artists that are similar to the seeds and/or important to the selected term are recommended using different similarity definitions (content based and Web based).

Metadata about music can be coarsely divided into "editorial metadata" and "cultural metadata," depending on its source. The former typically originates from a music editor or another music expert, most often from a record label, whereas the latter makes use of the wisdom of the crowd in that it reflects the knowledge and opinions of a large number of people. In the Web 2.0 era, mining such cultural metadata therefore involves analyzing Web pages, blogs, music-related Web platforms, social networks, and similar services.

In the following, a selection of approaches to extract some types of music-related metadata from Web sources are presented. Given the abundance of available literature on Web-based extraction of various kinds of music-related metadata, the following sections can only illustrate a small fraction. More precisely, approaches to extract song lyrics, an artist's country of origin, band members and instrumentation, and images of album cover artwork are presented. These information categories have been selected since they focus on different music entities: songs, artists, bands, and albums.

7.1.1   Song Lyrics

The lyrics of a song represent an important piece of music-related information as they usually give indication of the semantics of a piece of music. They are furthermore capable of revealing aspects such as the artist's/songwriter's cultural background (e.g., use of a certain language or presence of slang words), political orientation, or musical style. There exists a bunch of Web sites that offer song lyrics, for example, Lyrics.com [62]. However, song coverage of these sites is usually limited. To overcome this issue, applications such as Evil-Lyrics [28] offer semiautomated retrieval of song lyrics by consulting different Web sources. To this end, EvilLyrics queries search engines with artist and track name and filters the results for known lyrics pages. The content of these pages is then fetched, and the lyrics are extracted using predefined filters. The user can eventually select between the different results.

Applications such as EvilLyrics suffer from several drawbacks. First, only Web pages of known lyrics sites are considered, which limits the scope of the retrieval process. For example, lesser known artists whose lyrics are only present on their own Web page are not considered. Second, since predefined filters are used to parse lyrics pages, structural changes of a lyrics provider's page necessitate an update of the corresponding filter. Moreover, the choice of the correct or best version of a song's lyrics from the candidates must be effected by the user as the application does not try to estimate the quality of the results.

In general, lyrics extracted from dedicated lyrics pages are usually prone to certain inconsistencies or ambiguities. Some examples are simple typos, different spellings for the same word, different versions caused by misheard lyrics, or structural differences (e.g., repetitions are sometimes indicated as "repeat chorus," sometimes as "chorus 2x," sometimes the complete chorus just occurs twice). Addressing these issues, some approaches to automatically determine the correct version of a song's lyrics have been proposed.

Knees et al. [45] first query Google using the scheme "artist name" "track name" lyrics to obtain a set of Web pages likely containing the sought lyrics. Subsequently, the retrieved pages are cleaned by removing all HTML tags and converting the resulting plain text to lowercase (a process also known as "casefolding"). A page selection step is then performed, which uses either keywords in the title of the HTML documents or correlation between term weight vectors to figure out similar pages that are hence likely to contain the same lyrics. Hereafter, abbreviations for repetitions such as "repeat chorus" or "chorus (x2)" are sought and expanded. In order to align the lyrics present on the different Web pages and therefore determine the most correct version of a song's lyrics, a technique from bioinformatics, namely multiple sequence alignment (MSA), is employed. It aims at finding optimally matching word sequences over the different pages. To this end, the Needleman–Wunsch algorithm [69] is applied to pairs of word sequences. This algorithm makes use of dynamic programming to find a globally optimal alignment of two sequences with respect to a given scoring scheme. Knees et al. propose to assign a score of 10 to matching word pairs, a score of -1 to the insertion of a gap, and a score of 0 to mismatching word pairs. Iterative alignment of the highest scoring pairs of sequences and selecting for each position in the different alignments the most frequently occurring word yields an output string comprising the lyrics aligned from the various pages. Since the output of this procedure often contains additional words before and after the actual lyrics, such as the artist or track name or copyright information, a smoothing and confidence estimation step eventually aims at removing such parts [45].

Knees et al. evaluate their approach using 258 song lyrics taken from booklets of compact discs as reference. They report on having achieved a median recall value of 98% for this collection, that is, for half of the songs in the test collection, at least 98% of the words constituting the lyrics were correctly determined. The authors, however, also report on some drawbacks of their approach. First, it relies on the existence of at least two pages containing lyrics for each song under consideration. Second, the output does not contain any punctuation or line breaks. User acceptance may therefore be restricted. Finally, in some cases the alignment process discards important words. This should be resolved since completeness of the results seems to be more important than precision [45].

Korst and Geleijnse [48] present an alternative, more efficient approach to automatically determine a song's lyrics using Web retrieval techniques. First, similar to Knees et al.'s approach [45], Korst and Geleijnse issue search requests to Google. However, they use the query scheme allinanchor: "artist name" "track name" lyrics, which returns only pages whose links from other pages have artist name, track name, and the word "lyrics" in their anchor text. If this search does not result in a predefined number of pages, the constraints are successively weakened (e.g., by omitting the *allinanchor*-constraint, the word "lyrics," and the artist name). Unlike Knees et al. [45] who discard all tags and retain only the plain text content of the retrieved pages, Korst and Geleijnse [48] do make explicit use of the HTML documents' structure. Exploiting the fact that lyrics are usually organized in stanzas, the authors analyze `<pre>`...`</pre>` and `<br>` tags to distill likely occurrences of lyrics within the retrieved pages, by defining corresponding parsing rules. Since the resulting text fragments contain a large amount of noise, that is, text that does not represent lyrics, in a next step a "fingerprint" of the text segments of each page is created, for subsequent clustering of the pages. To this end, the five longest words in each parsed page are used as the page's representatives, since long words in general tend to be more discriminative than short ones. The pages are then clustered based on the criterion that the fingerprints of all pages within a certain cluster must share at least three words. Retaining only the largest cluster effectively eliminates outliers. Alignment is then performed using a dynamic programming approach that minimizes the edit distance on the word level between the text fragments of each pair of pages. To maximize the number of matching words, insertions, deletions, and substitutions of words are assigned a score of -1, while matches receive a positive score. For computational reasons, Korst and Geleijnse do not align all pairs of text fragments. Instead they select the fragment of maximum length as reference, assuming that this sequence contains (at least) the complete lyrics of the song under consideration, and align this reference with all other fragments. The resulting alignments are then combined to a single multiple alignment.

Evaluation is performed on the same data set as used by Knees et al. [45]. Korst and Geleijnse report average recall values of 93% (percentage of words in the reference, that is, lyrics from CD booklets, that were correctly found by the approach) and average precision values of 86% (percentage of words correctly output by the approach among all words output). Korst and Geleijnse [48] note that these results are comparable to those achieved by Knees et al., pointing out that their approach is computationally more efficient.

7.1.2   Country of Origin

As a second type of music-related metadata, we focus on the place of birth of an artist or the place of foundation of a band. Determining this information is not as straightforward as it might seem. Consider for example Farrokh Bulsara, also known as Freddie Mercury. He was born in Zanzibar, United Republic of Tanzania. However, he relocated to the United Kingdom at the age of 17, where he later became world famous as cofounder of the band Queen. Mercury's country of origin is nevertheless Tanzania, whereas Queen's is the United Kingdom, where the band was founded by Mercury, Brian May, and Roger Taylor in April 1970, [cf., 56]. This illustrates the problem of determining the country of origin in cases where the main country of musical activity differs from the place of birth.

Basically, Web-based approaches to find the country of origin can be categorized into two groups: methods that mine data from specific Web sources, such as Wikipedia or Last.fm, and methods that try to distill the country of origin from arbitrary Web pages. As representative for the first category, Govaerts and Duval's work [33] will be summarized in the following. Approaches belonging to the latter category were proposed by Schedl et al. [84, 83] and will be presented thereafter.

Govaerts and Duval use different data sources and heuristics to determine an artist's or band's country of origin. More precisely, they look into artist pages and biographies available on Last.fm, Freebase [30], and Wikipedia. Freebase is a collaborative database, and therefore provides information in a more structured manner than, for example, Wikipedia. On Last.fm the authors search for occurrences of geographical locations to predict the country of origin. On Freebase they either look into the "origin" property of the artist's database entry or gather all nationalities, birthplaces, and residences and subsequently predict the most frequently occurring country. Biographies extracted from Wikipedia are also sought for geographical locations, and three heuristics are proposed to predict the country of origin using this data source. All heuristics rely on the number of country occurrences. The first one simply predicts the country that occurs most frequently in the artist's biography. The other two heuristics favor early occurrences of country names in the biography.

For evaluation Govaerts and Duval use a set of more than 11,000 artists from Aristo Music [5], which has been manually annotated by music experts. The set is rather unevenly distributed with respect to continents since more than 95% of the artists originate from Europe or North America. The authors report that they were able to determine the origin for 59% of the test set by at least one of the analyzed methods. A comparison among the three data sources showed that Wikipedia performed best in terms of coverage, with a recall value of 56%. Coverage was 7% for Last.fm and 26% for Freebase. Accuracy values varied between 70% (Wikipedia) and 90% (Last.fm and Freebase). Combining different methods by chaining them in decreasing order of accuracy, 77% accuracy at 59% coverage could be achieved.

Schedl et al. [84, 83] propose three alternative approaches to predict an artist's country of origin, based on Web pages indexed by a search engine. The first one relies solely on the search engine's estimate of an artist's number of Web pages that contain specific country terms. To this end, the search engine is queried for all pairs of artist names and country names. For each artist, the country with the highest page count estimate is then predicted. The second and third approach analyze artist-related Web pages that have been determined using the search engine. After having fetched the corresponding Web content, the second approach applies term weighting measures commonly used in text-based information extraction and retrieval research [11, 76], to the retrieved Web pages. The third approach employs heuristics based on the text distance between country names and key terms in the retrieved Web pages.

As for the term weighting measures employed in the second approach, Schedl et al. analyzed the following, where term t represents a term indicating a country name.

Document frequency (DF): $df_{t,a}$ is the total number of Web pages retrieved for artist a on which term t occurs at least once.

Term frequency (TF): $tf_{t,a}$ is the total number of occurrences of term t in all pages retrieved for a.

Term frequency–inverse document frequency (TF-IDF): The basic idea of the $tf \cdot idf_{t,a}$ measure is to increase the weight of t if t occurs frequently in the set of Web pages retrieved for a, while at the same time decrease t's weight if t occurs in a large number of documents among the whole set of pages (retrieved for all artists), and is thus not very discriminative for a. The authors employ the logarithmic formulation of the TF-IDF weighting, as given in Equation 7.1, where n equals the total number of Web pages retrieved, and df_t is the total number of Web pages containing term t.

$$tf \cdot idf_{t,a} = \ln\left(1 + tf_{t,a}\right) \cdot \ln\left(1 + \frac{n}{df_t}\right) \tag{7.1}$$

Using the set of country names C as input, Schedl et al. [83, 84] calculate the weight for all terms $t \in C$ applying each term weighting function. Predicting the country for an artist is then simply performed by selecting the most important country term as determined by the term weighting measure.

The third approach applies text distance measures between country names and origin-related key terms, such as "born," "founded," "origin," or "country," to the set of retrieved Web pages. Based on the distance between the positions of the country terms and the origin-related key terms on artist a's pages, a model of a's most likely country of origin is constructed. It comprises two different functions: first, a distance measure on the intra-page-level to determine the distances within a Web page of a; second, an aggregation function to combine the intra-page-level distances for all pages retrieved for a. The authors experimented with the minimum and the arithmetic mean functions.

Evaluation is performed on a manually compiled collection of 578 artists, including 69 distinct countries of origin. The authors report precision values of up to 71% at a recall level of 100%, employing the term weighting approach with the TF-IDF measure. Another finding of Schedl et al. [83, 84] is that the term weighting heuristic outperforms the other two approaches. In fact, their first approach, using page count estimates, achieves only 23% precision (at 100% recall). The approach relying on text distances yields 37% precision (at 100% recall).

7.1.3   Band Members and Instrumentation

On the level of musical bands, automated approaches to determine the members of a band and the instruments they play is an interesting task as it enables creating relations between individual artists and bands. Combined with temporal information, that is, when a certain member joined or left a band, such information may also serve to derive similarities between bands or influences a particular artist had on a particular band.

Schedl and Widmer [85] propose a method to find members and their roles, that is, instruments they play, for a given band based on natural language processing (NLP) techniques applied to Web pages. Their NLP approach basically comprises three steps: named entity detection, rule-based pattern analysis, and data aggregation to predict band members. After having determined (and fetched) band-related Web pages via a search engine, the authors use a named entity detection technique [15], to find potential members. For this purpose, 2, 3, and 4 grams, that is, consecutive sequences comprising of 2, 3, and 4 words, are extracted. Then various filtering steps are performed. For example, only n-grams whose tokens are all capitalized are retained, and n-grams consisting of common speech words are discarded. In a next step, a set of rules tailored to the specific task is applied to the extracted n-grams and the surrounding text. Examples for such rules are "M plays the I," "R M," or "M is the R," where M represents the potential member, I represents the instrument (the authors use a set of predefined instruments and synonyms), and R is the corresponding role, e.g., "drummer" or "guitarist." Schedl and Widmer count the number of Web pages analyzed for the band under consideration each rule can be applied to. Subsequently, these counts are summed up for all pairs of members and instruments. Noisy and unlikely correct assignments are filtered using a dynamic threshold for the minimum number of aggregated rule applications. Finally, all remaining members and roles are predicted.

Schedl and Widmer [85] evaluate their approach on a collection of 51 rock and metal bands, for which the current 240 members as well as the current and past 499 members were determined manually by consulting various sources. By varying the threshold used for noise reduction it is possible to adjust precision and recall. For the set of current members, the authors report a precision value of 44% (at 36% recall level), when optimizing both precision and recall. Using all members (current and past ones), a precision of 61% is achieved at 26% recall level. Although these numbers certainly leave room for improvement, the authors note that there exists an upper limit for the recall since artist names that do not occur on the Web pages retrieved for the band under consideration obviously cannot be detected by the approach. This upper limit is reported to lie between 50% and 60%, depending on the scheme used to query the search engine and on the collection (all members versus current members).

7.1.4   Album Cover Artwork

Images of an album's cover artwork play an essential role for identifying and recognizing a musical work or an artist. Due to the steadily increasing importance of online music distribution and consumption, which is effected without the physical presence of an album, techniques to automatically extract album cover artwork are highly requested. Like for the task of lyrics retrieval, which has already been discussed, there exists some applications for gathering album cover artwork from the Web, for example, Album Cover Art Downloader [1]. However, again they offer only semiautomated retrieval, where the task of selecting the correct cover from a set of candidates is left to the user.

In contrast, fully automated approaches are presented by Schedl et al. [81] and Schedl [79]. Based on a set of artist-related Web pages determined via a search engine, first, a full inverted index or word-level index [94] is constructed, that is, a list of occurrence and position information for each term within all pages is stored in a database. This indexation also includes HTML tags and their properties. Given the name of an artist and an album under consideration, the index is then used to determine the text distance between <img> tags referring to potential album covers and artist/album names. These distances are calculated on the level of characters and on the level of tags. The approach assumes that <img> tags of images containing actual album covers have the corresponding artist and album names either in their src or alt attribute, or at least at a nearby position. Therefore, by fetching a fixed number of images with minimal summed distances *<img> tag – artist name* and *<img> tag – album name*, the authors construct a set of potential album cover images. Thereafter, content-based filtering is performed to remove erroneous images. Exploiting the fact that images of album covers almost always have a quadratic shape, all images that do not fulfill this constraint are discarded. Preliminary studies revealed that potential album cover images frequently show scanned discs instead of the actual cover. Hence, the authors propose to apply a simple circle detection technique to the set of possible album covers. To this end, rectangular areas along a circular path that may represent the border of a scanned compact disc are examined. Since images of scanned discs usually show a strong contrast between the inner region (illustrating the disc) and the outer region (the image's background), comparing the color histograms [29] of either type of region it is possible to filter out images that likely represent scanned discs. From the set of candidates that remain after the filtering step, the image with minimal distance between its <img> tag and the textual identifiers is selected and predicted as the correct album cover.

Evaluation is performed using a commercial collection of 3,311 albums by artists from all around the world. In general, calculating the text distance at the tag level outperforms using the distance at character level. The best results are achieved by computing tag-level distances and employing filtering of nonquadratic images and scanned discs. In this case, 59% of the album covers are correctly identified. Among the negative results found are images of other albums by the correct artist (6%), images of other artist-related material, for example, portraits (2%), and completely unrelated images (14%). Moreover, for 20% of the albums, not a single cover image could be determined. This large amount of missing data, however, could be caused by the challenging test collection, as previous experiments on a private collection of 255 albums resulted in 83% correctly found images and only 7% of cases for which no prediction could be made [81].

7.2 Approaches to Similarity Measurement

Elaborating accurate musical similarity measures that are capable of capturing aspects of music similarity that relate to real, perceived similarity is one of the main challenges in music information research. Such similarity measures can help to understand why two music pieces or artists are perceived (dis)similar by the listener. Similarity measures are furthermore a key ingredient of various music-related applications. Examples are systems to automatically generate playlists [7, 75], music recommender systems [20, 93], semantic music search engines [41], and intelligent user interfaces [73, 43].

This section reviews work that exploits context-based data, more precisely, Web-based and community-based data sources, to define similarity between artists and between tracks. The presented techniques to context-based similarity estimation can be categorized into two main groups: text-based and co-occurrence-based methods. The former group includes approaches that make use of texts extracted from Web pages, of tags, and of lyrics. The co-occurrence approaches exploit, for example, playlists, page counts, and peer-to-peer networks as data source.

7.2.1 Text-Based Approaches

This section presents work that exploits textual, music-related information originating from Web pages, user tags, and song lyrics. The methods employed are hence strongly related to traditional text-based information retrieval (IR) and information extraction (IE) techniques. Most approaches to text-based IR and IE rely on some principal assumptions and models, which are detailed in the following. The *bag of words* model, which can be traced back at least to Luhn [61], represents a document as an unordered set of its words, ignoring structure and grammar rules. Words can be generalized to terms, where a *term* may be a single word or a sequence of n words (*n-grams*), or correspond to some grammatical structure, like a noun phrase. Using such a bag of words representation, each term t describing a particular document d is commonly assigned a weight $w_{t,d}$ that estimates the frequency or importance of t in/for d. Each document can then be described by a *feature vector* comprising the single weights. When considering a whole corpus of documents, each document can be thought of as a representation of its feature vector in a *feature space* or *vector space* whose dimensions correspond to the particular term weights. This so-called vector space model is a fundamental model in information retrieval and was originally described by Salton et al. [77].

When it comes to deriving artist-related information from the Web, usually all Web pages returned for a particular artist are regarded as one large, virtual document describing the artist under consideration. This aggregation seems reasonable since, in Web-based music information retrieval, the usual entity of interest is the music artist, not a single Web page. Furthermore, it is easier to cope with very small, or even empty, pages if they are part of a larger virtual document.

7.2.1.1   Term Profiles from Web Pages

One of the most comprehensive sources of cultural data is the vast amount of available Web pages. The majority of the presented approaches uses a Web search engine to retrieve relevant documents and create artist term profiles from unstructured text extracted from Web pages. To focus the search on Web pages relevant to music, different query schemes are employed. Such schemes may comprise the artist's name augmented by the keywords "music review" [92, 12] or "music genre style" [40]. Additional keywords are particularly important for artists whose names have another meaning outside the music context, such as "Bush," "Kiss," and "Air." A comparison of different query schemes can be found in Knees et al. [42].

Whitman and Lawrence [92] extract different term sets (unigrams, bigrams, noun phrases, artist names, and adjectives) from up to 50 artist-related Web pages obtained via a search engine. After downloading the pages, the authors apply parsers and a part-of-speech (POS) tagger [14] to assign each word to its suited test set(s). Individual term profiles are then created for each artist by employing a version of the TF-IDF measure, which assigns a weight to each term t in the context of each artist A_i. In general, the TF-IDF weighting assigns higher weights to terms that occur often within a certain document (here, the Web pages of an artist), but rarely in other documents (other artists' Web pages). Equation 7.2 shows the weighting used by Whitman and Lawrence, where the term frequency $tf(t, A_i)$ is defined as the percentage of retrieved pages for artist A_i containing term t. The document frequency

$df(t)$ is the percentage of artists (in the whole collection) whose set of retrieved Web pages contains term t at least once.

$$w_{simple}(t, A_i) = \frac{tf(t, A_i)}{df(t)} \qquad (7.2)$$

The authors further propose another variant in which also rarely occurring terms, that is, terms with a low DF, are weighted down to emphasize terms in the middle IDF range. Equation 7.3 shows this variant, where μ and σ represent parameters manually set to 6 and 0.9, respectively.

$$w_{gauss}(t, A_i) = \frac{tf(t, A_i)e^{-(log(df(t))-\mu)^2}}{2\sigma^2} \qquad (7.3)$$

Computing the TF-IDF weights for all terms in each term set yields individual feature vectors or term profiles for each artist. The overlap between the term profiles of two artists, that is, the sum of weights of all terms that occur in both term profiles, is then used as an estimate for the artists' similarity (Equation 7.4).

$$sim_{overlap}(A_i, A_j) = \sum_{\{\forall k | a_{i,k} > 0, a_{j,k} > 0\}} a_{i,k} + a_{j,k} \qquad (7.4)$$

For evaluation, the authors compare these similarities to two other sources of artist similarity information, which serve as ground truth (similar-artist relations from allmusic.com and user collections from the music sharing service OpenNap, see Section 7.2.2.3). The test collection comprises about 400 artists. Remarkable differences between the individual term sets could be found. The unigram, bigram, and noun phrase sets perform considerably better than the other two sets, regardless of the ground-truth definition. The authors further note that the expert-based similarity judgments as provided by *allmusic.com* tend to be strongly influenced by a subjective bias of the respective music editors.

Extending the work presented by Whitman and Lawrence [92], Baumann and Hummel [12] introduce filters to prune the set of retrieved Web pages. First, they remove all Web pages with a size of more than 40 kilobytes (after parsing). They also try to filter out advertisements by ignoring text in table cells comprising more than 60 characters, but not forming a correct sentence. Finally, Baumann and Hummel perform keyword spotting in the URL, the title, and the first text part of each page. Each occurrence of the initial query parts (artist name, "music," and "review") contributes to a page score. Pages that score too low are filtered out. In contrast Whitman and Lawrence [92], Baumann and Hummel use a logarithmic IDF weighting in their TF-IDF formulation. With these modifications the authors are able to outperform the approach presented by Whitman and Lawrence [92].

Knees et al. [40] present an approach similar to Whitman and Lawrence [92]. Unlike Whitman and Lawrence who experiment with different term sets,

Knees et al. use only one list of unigrams. For each artist, a weighted term profile is created by applying a TF-IDF variant. Equation 7.5 illustrates the TF-IDF formulation, where n is the total number of Web pages retrieved for all artists in the collection, $tf(t, A_i)$ is the number of occurrences of term t in all Web pages retrieved for artist A_i, and $df(t)$ is the number of pages in which t occurs at least once. In case $tf(t, A_i)$ equals zero, $w_{ltc}(t, A_i)$ is defined to be zero.

$$w_{ltc}(t, A_i) = (1 + \log_2 tf(t, A_i)) \cdot \log_2 \frac{n}{df(t)} \qquad (7.5)$$

Calculating the similarity between the term profiles of two artists A_i and A_j is performed using the cosine similarity according to Equation 7.6. Here, T denotes the set of all terms, and θ gives the angle between A_i's and A_j's feature vectors in the Euclidean space.

$$sim_{cos}(A_i, A_j) = \cos\theta = \frac{\sum_{t \in T} w(t, A_i) \cdot w(t, A_j)}{\sqrt{\sum_{t \in T} w(t, A_i)^2} \cdot \sqrt{\sum_{t \in T} w(t, A_j)^2}} \qquad (7.6)$$

Knees et al. evaluate their approach in a genre classification setting using k-Nearest Neighbor (k-NN) classifiers on a test collection of 224 artists (14 genres, 16 artists per genre). They achieve accuracy values of up to 77%. Employing a term selection technique and a Support Vector Machine (SVM) [91] as classifier, accuracy increases to 87%.

Other approaches derive term profiles from more specific Web resources. For example, Celma et al. [19] propose a music search engine that crawls audio blogs via RSS feeds and calculates TF-IDF features. Hu et al. [37] extract TF-based features from music reviews gathered from Epinions.com [27]. Schedl [80] presents an approach that extracts user posts associated with music artists from the microblogging service Twitter [89]. Subsequently, term profiles are created, using term lists specific to the music domain. Various term weighting measures are then applied (TF, DF, and TF-IDF), and the pair-wise artist similarity between the resulting feature vectors is estimated using cosine similarity. In a genre classification task on the 224-artist set from Knees et al. [40], accuracy values of up to 72% are reported.

7.2.1.2 Collaborative Tags

In the era of the "Web 2.0," platforms and services offer their users various means to contribute user-generated data. One popular example is the assignment of short descriptions to a specific item (often some kind of multimedia object, such as an image, a video, or a music piece). This labeling process is commonly referred to as tagging. The more people are labeling an item with a tag, the more the tag is assumed to be relevant to the item. Although the content of a tag is usually not restricted, in the music domain most tags represent specific properties of an artist, an album, or a music piece, for example, genre and style descriptions, nationalities, epochs, and instruments. One of the most popular music-related platforms that offers tagging functionality is Last.fm. Since Last.fm provides a comprehensive developer's API [52], it represents a valuable source for context-based information about music.

Geleijnse et al. [32] gather tags from Last.fm to generate a "tag ground truth" on the artist level. The authors first filter redundant and noisy tags using the set of tags associated with tracks by the artist under consideration. Similarity between two artists is then estimated as the number of overlapping tags. Evaluation against Last.fm's similar artist function shows that the number of overlapping tags between similar artists is much larger than the average overlap between arbitrary artists (about 10 versus 4 after filtering).

Levy and Sandler [55] retrieve tags from Last.fm and MusicStrands [68] to construct a semantic space for music pieces. To this end, all tags found for a specific piece are tokenized, and a document-term matrix based on TF-IDF weighting is created. As a result, each track is represented by a term vector. Three different extensions to TF weighting are explored: weighting by the number of users that applied the tag, restriction to adjectives, and no weighting at all. Optionally, the dimensionality of the vectors is reduced by applying latent semantic analysis (LSA) [23]. The similarity between feature vectors is calculated via the cosine measure, Equation 7.6.

As for evaluation, for each genre or artist term t, each track labeled with t serves as query, and the mean average precision (MAP) over all queries is computed. It was found that filtering for adjectives considerably worsens the performance. Levy and Sandler [55] further indicate that weighting of term frequency by the number of users may improve genre precision. Without LSA, that is, using the full term vectors, genre precision reaches 80% and artist precision 61%. Using LSA, genre precision is up to 82% and artist precision 63%.

Compared with the Web page-based approaches presented in the last section, the tag-based approaches offer some advantages. First, the used vocabulary, that is, tag set, is much smaller and more focused to the music domain, which may reduce noise. Second, Last.fm not only provides tags on the artist level, but also on the level of individual tracks, making it possible to calculate similarities between tracks. On the other hand, tag-based approaches also suffer from certain restrictions. For example, the tagging of comprehensive collections requires a large and active user community. Another problem is that coverage of artists or tracks from the lesser known "long tail" is usually very low.

Alleviating some of these problems, the idea of gathering tags via "games with a purpose" has recently become popular [88, 65, 54]. Such games aim at solving problems that are hard to solve for computers, for example, capturing emotions evoked when listening to a song. By encouraging users to play such games, a large number of songs can be efficiently labeled with semantic descriptors.

7.2.1.3   Song Lyrics

In Section 7.1.1, it has been shown how song lyrics can be determined using Web mining techniques. Here, approaches that define similarities between artists or songs based on lyrics are presented.

Logan et al. [59] use song lyrics for tracks by 399 artists to determine artist similarity. First, the authors apply Probabilistic Latent Semantic Analysis (PLSA) [35] to a collection of more than 40,000 song lyrics in order to build N models of topics typically present in lyrics. Subsequently, all lyrics by an artist are processed using each topic model to create N-dimensional vectors of which each dimension gives the likelihood of the artist's tracks to belong to the corresponding topic. Artist vectors are then compared by calculating the L_1 distance (also known as Manhattan distance), as shown in Equation 7.7.

$$dist_{L_1}(A_i, A_j) = \sum_{k=1}^{N} |a_{i,k} - a_{j,k}| \qquad (7.7)$$

This approach is evaluated against human similarity judgments, more precisely, the "survey" data for the *uspop2002* set, [13]. It yields worse results than similarities obtained via acoustic features, irrespective of the chosen N, the usage of stemming, or the filtering of lyrics-specific stop words. Since lyrics-based and audio-based approaches make different errors, however, a combination of both is suggested.

Mahedero et al. [64] demonstrate the usefulness of lyrics for four tasks: language identification, song structure detection, thematic categorization, and similarity measurement. For similarity estimation, TF-IDF term weighting is performed, and cosine similarities are calculated. A song's representation is then obtained by aggregating the similarities to all songs in the collection into a new vector. These representations are compared using an unspecified algorithm. Exploratory experiments indicate some potential for cover version identification and plagiarism detection.

Other approaches do not explicitly aim at determining similar songs with respect to lyrical content, but rather at revealing conceptual clusters [39] or at classifying songs into genres [66] or mood categories [53, 36]. Since the extracted features can also be used for similarity estimation, these approaches are nevertheless related to the subject of this section.

Laurier et al. [53] classify songs into four mood categories by means of lyrics and audio content. From a song's lyrics, TF-IDF features are derived, and the cosine similarity measure is applied. A 10-fold cross validation with a k-NN classifier yields accuracies slightly above 60%, using lyrics as data source. Signal-based features perform better than the lyrics-based features. However, a combination of both gave the best results.

Hu et al. [36] experiment with TF, TF-IDF, and Boolean vectors and investigate the impact of stemming, part-of-speech tagging, and function words for soft categorization into 18 mood clusters. Best results are achieved with TF-

IDF weights on stemmed terms. The authors further report that lyrics-based features alone can outperform audio-based features.

7.2.2   Co-Occurrence–Based Approaches

The principal assumption underlying all approaches presented in this section is that the occurrence of two music pieces or artists within the same context indicates some kind of similarity. The context is given by the used data source. In the following, approaches relying on Web pages (or page counts returned by a search engine), playlists, and peer-to-peer (P2P) networks are presented.

7.2.2.1   Web-Based Co-Occurrences and Page Counts

The context of a music entity may be defined by related Web pages. Determining and using such music-related pages as data source for music information retrieval tasks was probably first performed by Cohen and Fan [21]. Cohen and Fan automatically extract lists of artist names from Web pages. To determine pages relevant to the music domain, they query Altavista [2] and Northern Light[1] [70]. The resulting HTML pages are then parsed according to their DOM tree, and all plain text content with minimum length of 250 characters is further analyzed for occurrences of entity names. Pages with multiple occurrences of artists are then treated as "pseudo-users" that "rate" tracks by the contained artists positively. Based on this data, a CF system is constructed, which is used for artist recommendation.

Some co-occurrence-based approaches rely on page count estimates returned to search engine requests. Querying a search engine for pages that contain both of two artist names and retrieving the corresponding page count estimate, it is possible to derive similarity information. For example, Zadel and Fujinaga [93] investigate the usability of two Web services to extract co-occurrence information and consecutively derive artist similarity. More precisely, the authors propose an approach that, given a seed artist as input, retrieves a list of potentially related artists from the Amazon [3] Web service Listmania!. Based on this list, artist co-occurrences are derived by querying the Google Web API[2] and storing the returned page counts of artist-specific queries. Google was queried for `"artist name i"` and for `"artist name i"+"artist name j."` Thereafter, the so-called "relatedness" of each Listmania! artist to the seed artist is calculated as the ratio between the combined page count, that is, the number of Web pages on which both artists co-occur, and the minimum of the single page counts of both artists, Equation 7.8. The minimum is used to account for different popularities of the two

---

[1]Northern Light, formerly providing a meta search engine, in the meantime has specialized on search solutions tailored to enterprises.

[2]Google no longer offers this Web API. It has been replaced by several other APIs, mostly devoted to Web 2.0 development.

artists.

$$sim_{pc\_min}(A_i, A_j) = \frac{pc(A_i, A_j)}{\min\left(pc(A_i), pc(A_j)\right)} \qquad (7.8)$$

Recursively extracting artists from Listmania! and estimating their relatedness to the seed artist via Google page counts allows to construct lists of similar artists. Although the article shows that Web services can be efficiently used to find artists similar to a seed artist, it lacks a thorough evaluation of the results.

Analyzing Google page counts as a result of artist-related queries is also performed by Schedl et al. [82]. Unlike the method presented by Zadel and Fujinaga [93], Schedl et al. derive complete similarity matrices from artist co-occurrences. This offers additional information since it can also be predicted which artists are *not* similar.

Schedl et al. [82] define the similarity of two artists as the conditional probability that one artist is to be found on a Web page that is known to mention the other artist. Since the retrieved page counts for queries like "artist name i" or "artist name i"+"artist name j" indicate the relative frequencies of this event, they are used to estimate the conditional probability. Equation 7.9 gives a formal representation of the symmetrized similarity function.

$$sim_{pc\_cp}(A_i, A_j) = \frac{1}{2} \cdot \left(\frac{pc(A_i, A_j)}{pc(A_i)} + \frac{pc(A_i, A_j)}{pc(A_j)} \right) \qquad (7.9)$$

In order to restrict the search to Web pages relevant to music, different query schemes are proposed by Schedl et al. [82] (see also Section 7.2.1.1). Otherwise, queries for artist names that equal common speech words would unjustifiably lead to higher page counts, hence, distort the similarity relations.

Schedl et al. perform two evaluation experiments on the same 224-artist-data-set as used by Knees et al. [40]. They estimate the homogeneity of the genres defined by the ground truth by applying the similarity function to artists within the same genre and to artists from different genres. To this end, the authors relate the average similarity between two arbitrary artists from the same genre to the average similarity of two arbitrary artists from different genres. The results show that the co-occurrence approach can be used to clearly distinguish between most of the genres. The second evaluation experiment is an artist-to-genre classification task using a k-NN classifier. In this setting, the approach yields accuracy values of about 85%, averaged over all genres.

A shortcoming of the approaches proposed by Zadel and Fujinaga [93] and Schedl et al. [82] is that the number of involved search engine requests is quadratic in the number of artists. If a complete similarity matrix is to be created, these approaches therefore scale poorly to real-world music collections.

Quadratic computational complexity can be avoided with an alternative strategy to co-occurrence analysis, as described by Schedl [79, Chapter 3]. This method resembles Cohen and Fan's [21], presented in the beginning of this section. First, for each artist A_i, a certain amount of top-ranked Web pages returned by the search engine is retrieved. Subsequently, all pages fetched for artist A_i are searched for occurrences of all other artist names A_j in the collection. The number of page hits represents a co-occurrence count, which equals the document frequency of the artist term "A_j" in the corpus given by the Web pages for artist A_i. Relating this count to the total number of pages successfully fetched for artist A_i, a similarity function is constructed. Employing this method, the number of issued queries grows linearly with the number of artists in the collection. The formula for the symmetric artist similarity is given in Equation 7.10.

7.2.2.2 Playlists

An early approach to derive similarity information from the context of a music entity can be found in the research by Pachet et al. [72]. Pachet et al. consider radio station playlists from a French radio channel and compilation CDs from CDDB[3] to extract co-occurrences between tracks and between artists. The authors count the number of co-occurrences of two artists (or pieces of music) A_i and A_j in the radio station playlists and compilation CDs. They define the co-occurrence of an entity A_i to itself as the number of A_i's occurrences in the considered data source. To account for different frequencies, that is, popularities, of songs or artists, the co-occurrence counts are normalized. Assuming that co-occurrence is a symmetric function, the complete similarity measure used by the authors is given in Equation 7.10.

$$sim_{pl\_cooc}(A_i, A_j) = \frac{1}{2} \cdot \left[\frac{cooc(A_i, A_j)}{cooc(A_i, A_i)} + \frac{cooc(A_j, A_i)}{cooc(A_j, A_j)} \right] \quad (7.10)$$

This similarity formulation is incapable of capturing indirect links that an entity may have with others. For example, given three artists A, B, and C and assuming that A and B often co-occur and B and C often co-occur, Equation 7.10 cannot express that A and C are probably also similar. In order to capture such indirect links, the complete co-occurrence vectors of two entities A_i and A_j (i.e., a vector comprising, for a specific entity, the co-occurrence count with all other entities) are constructed and their statistical correlation is computed, cf. Equation 7.11.

$$sim_{pl\_corr}(A_i, A_j) = \frac{Cov(A_i, A_j)}{\sqrt{Cov(A_i, A_i) \cdot Cov(A_j, A_j)}} \quad (7.11)$$

These co-occurrence and correlation functions are used as similarity measures on the track level and on the artist level. Pachet et al. evaluate them on two data sets, one comprising 12 tracks, the other one consisting of 100 artists. As ground truth the authors use similarity judgments by music experts from Sony

---

[3]CDDB is a Web-based album identification service that returns, for a given unique disc identifier, metadata like artist and album name, tracklist, or release year. This service is offered in a commercial version operated by Gracenote [34] as well as in an open source implementation named freeDB [31].

Music. The main finding is that artists or tracks that appear consecutively in radio station playlists or on CD samplers indeed show a high similarity. The co-occurrence function generally outperforms the correlation function (70% to 76% versus 53% to 59% agreement with the ground truth).

Another approach that uses playlists for music similarity estimation is presented by Cano and Koppenberger [16]. Cano and Koppenberger create a similarity network by extracting playlist co-occurrences of more than 48,000 artists retrieved from Art of the Mix [6] in early 2003. Art of the Mix is a Web service that allows users to upload and share their mixed tapes or playlists. The authors analyze more than 29,000 playlists. They subsequently create a similarity network, where a connection between two artists is made if they co-occur in a playlist. The article reveals some interesting properties of the constructed artist similarity network. First, each artist is only connected to a small number of other artists. Thus, a similarity measure constructed from such data is likely to capture only (strong) positive similarities between two artists. In spite of this data sparsity, the network shows one large cluster of nodes connecting more than 99% of the artists. Moreover, the average length of shortest path between two artists is remarkably small (3.8). So is the clustering coefficient that estimates the probability of indirect links, that is, the probability that two neighboring artists of a given one are connected themselves. Thus, given that artist A is similar to B and to C, the probability for B and C being similar is quite small (0.1). Analyzing the average degree of a node shows that each artist is on average connected to 12.5 other artists.

In a more recent article that exploits playlists to derive artist similarity information, Baccigalupo et al. [10] analyze co-occurrences of artists in playlists shared by members of a Web community. The authors look at more than 1 million playlists made publicly available by MusicStrands [68], a Web service (no longer in operation) that allowed users to share playlists. The authors extract the 4,000 most popular artists from the playlist set, measuring popularity as the number of playlists in which each artist occurs. They further take into account that two artists that consecutively occur in a playlist are probably more similar than two artists that occur farther away in a playlist. To this end, the authors define a distance function $d_h(A_i, A_j)$ that counts how often a song by artist A_i co-occurs with a song by A_j at a distance of h. Thus, h is a parameter that defines the number of songs in between the occurrence of a song by A_i and the occurrence of a song by A_j in the same playlist. Baccigalupo et al. define the distance between two artists A_i and A_j as in Equation 7.12, where the playlist counts at distances 0 (two consecutive songs by artists A_i and A_j), 1, and 2 are weighted with β_0, β_1, and β_2, respectively. The authors empirically set the values to $\beta_0 = 1$, $\beta_1 = 0.8$, $\beta_2 = 0.64$.

$$d_{pl\_d}(A_i, A_j) = \sum_{h=0}^{2} \beta_h \cdot [d_h(A_i, A_j) + d_h(A_j, A_i)] \qquad (7.12)$$

To account for the popularity bias, that is, very popular artists co-occur with a lot of other artists in many playlists simply because of their popularity, the authors perform normalization according to Equation 7.14, where $\widehat{dist}_{pl\_d}(A_i)$ denotes the average distance between A_i and all other artists (Equation 7.13), and X is the set of the $n-1$ artists other than A_i. The authors do not report any evaluation dedicated to artist similarity.

$$\widehat{d_{pl\_d}}(A_i) = \frac{1}{n-1} \cdot \sum_{j \in X} d_{pl\_d}(A_i, A_j) \tag{7.13}$$

$$dist_{|pl\_d|}(A_i, A_j) = \frac{d_{pl\_d}(A_i, A_j) - \widehat{d_{pl\_d}}(A_i)}{\left| max\left(d_{pl\_d}(A_i, A_j) - \widehat{d_{pl\_d}}(A_i) \right) \right|} \tag{7.14}$$

7.2.2.3   Peer-to-Peer Networks

Peer-to-peer (P2P) networks represent a rich source of music-related data since their users are commonly willing to reveal metadata about the shared content. In the case of music files, file names and ID3 tags are usually disclosed.

Early work makes use of data extracted from P2P networks [92, 26, 58, 13]. All these studies use, among other sources, data extracted from the P2P network OpenNap to derive music similarity information.[4] Logan et al. [58] and Berenzweig et al. [13] report on having determined the 400 most popular artists on OpenNap in mid-2002. The authors gathered metadata on shared content, which yielded about 175,000 user-to-artist relations from about 3,200 shared music collections. Logan et al. [58] especially highlight the sparsity in the OpenNap data, in comparison with data extracted from the audio signal. Logan et al. compare similarities defined by artist co-occurrences in OpenNap collections, by expert opinions from allmusic.com, by playlist co-occurrences from Art of the Mix, by data gathered from a Web survey, and by audio feature Mel-frequency cepstral coefficients (MFCCs) [9]. To this end, they calculate a "ranking agreement score" by comparing the top N most similar artists according to each data source and calculating the pair-wise overlap between the sources. The main findings are that the co-occurrence data from OpenNap and from Art of the Mix show a high degree of overlap, the experts from allmusic.com and the participants of the Web survey agree moderately, and the signal-based measure has a rather low agreement with all other sources (except compared to the allmusic.com data).

In research conducted by Whitman and Lawrence [92], a software agent is used to retrieve from OpenNap a total of 1.6 million user-song relations over a period of three weeks in August 2001. To alleviate the popularity bias, Whitman and Lawrence use a similarity measure as shown in Equation 7.15, where $C(A_i)$ denotes the number of users that share songs by artist A_i, $C(A_i, A_j)$

---

[4]It is not clear whether the four mentioned publications make use of exactly the same data set. In any case, the authors emphasize that they only extract metadata from OpenNap, but do not download any files.

is the number of users that have both artists A_i and A_j in their shared collection, and A_k is the most popular artist of the whole data set. The second factor (in the right-hand part of the equation) down weights the similarity between two artists if one of them is very popular and the other is not.

$$sim_{p2p\_wl}(A_i, A_j) = \frac{C(A_i, A_j)}{C(A_j)} \cdot \left(1 - \frac{|C(A_i) - C(A_j)|}{C(A_k)}\right) \qquad (7.15)$$

Ellis et al. [26] use the same artist set that Whitman and Lawrence [92] used. The aim is to build a ground truth for artist similarity estimation. The authors report on having extracted from OpenNap about 400,000 user-to-song relations, covering about 3,000 unique artists. Again, the co-occurrence data is compared with artist similarity data gathered by a Web survey and with allmusic.com data. In contrast to Whitman and Lawrence [92], Ellis et al. [26] take indirect links in allmusic.com's similarity judgments into account. To this end, Ellis et al. propose a transitive similarity function on similar artists from the allmusic.com data, called "Erdös distance." More precisely, the distance $d(A_1, A_2)$ between two artists A_1 and A_2 is measured as the minimum number of intermediate artists needed to form a path from A_1 to A_2. As this procedure also allows to derive information on dissimilar artists (those with a high minimum path length), it can be employed to obtain a complete distance matrix. Furthermore, the authors propose an adapted distance measure, the so-called "Resistive Erdös measure," which takes into account that there may exist more than one shortest path of length l between A_1 and A_2. Assuming that two artists are more similar if they are connected via many different paths of length l, the Resistive Erdös similarity measure equals the electrical resistance in a network, see Equation 7.16, where each path from A_i to A_j is modeled as a resistor whose resistance equals the path length $|p|$. However, this adjustment does not improve the agreement of the similarity measure with the data from the Web-based survey, as it fails to overcome the popularity bias, that is, many different paths between popular artists unjustifiably lower the total resistance.

$$dist_{p2p\_res}(A_i, A_j) = \left(\sum_{p \in Paths(A_i, A_j)} \frac{1}{|p|}\right)^{-1} \qquad (7.16)$$

A recent approach that derives similarity information on the artist and on the song level from the Gnutella P2P file sharing network is presented by Shavitt and Weinsberg [86]. Shavitt and Weinsberg collected metadata of shared files from more than 1.2 million Gnutella users in November 2007. They restricted their search to music files (MP3 and WAV). The crawl yielded a data set of 530,000 songs. Information on both users and songs are represented via a two-mode graph showing users and songs. A link between a song and a user is created when the user shares the song. One finding of analyzing the resulting network is that most users in the P2P network share similar files.

The authors use the gathered data for artist recommendation. To this end, they construct a user-to-artist matrix V, where $V(i,j)$ gives the number of songs by artist A_j that user U_i shares. Shavitt and Weinsberg then perform direct clustering on V using the k-means algorithm [63] with the Euclidean distance metric. Artist recommendation is then performed using either data from the centroid of the cluster to which the seed user U_i belongs or by using the nearest neighbors of U_i within the cluster to which U_i belongs.

In addition, Shavitt and Weinsberg [86] address the problem of song clustering. Accounting for the popularity bias, the authors define a distance function that is normalized according to song popularity, as shown in Equation 7.17, where $uc(S_i, S_j)$ denotes the total number of users that share songs S_i and S_j. C_i and C_j denote, respectively, the popularity of songs S_i and S_j, measured as their total occurrence in the data set.

$$dist_{p2p\_pop}(S_i, S_j) = -\log_2\left(\frac{uc(S_i, S_j)}{\sqrt{C_i \cdot C_j}}\right) \tag{7.17}$$

Evaluation experiments are carried out for song clustering. The authors report an average precision of 12.1% and an average recall of 12.7%, which they judge as quite good considering the vast amount of songs shared by the users and the inconsistency in the metadata (ID3 tags).

7.3   Conclusion

This chapter gave an overview of Web- and community-based music information extraction techniques and of approaches to similarity measurement based on music context data. In comparison with signal-based or content-based approaches, techniques that make use of the vast amount of information available on the Web offer certain advantages. First, they do not require having access to the actual music files. They are potentially capable of determining high-level semantic descriptors, for example, via tags. Furthermore, community-based data sources can reflect people's opinions, feelings, and other subjective aspects of how music is perceived—information which cannot be derived from the pure audio signal alone.

Although context-based approaches show the great potential of Web- and community-based data sources, they suffer from particular shortcomings. First, data sparsity is obviously a problem, in particular for artists in the "long tail." Even if data is available, there is usually an imbalance between the amount of existent information for popular artists and for lesser known ones, a fact usually referred to as popularity bias. Moreover, approaches that rely on user data often include only participants of specific communities. For example, the average Last.fm user does not necessarily correspond to the whole population's average music listener. It is indeed known that users of certain communities tend to have similar music tastes. This phenomenon is referred to as community bias or population bias. Finally, filtering spurious and erroneous information poses a challenge for all approaches.

Acknowledgments

The author would like to thank Peter Knees for his contributions to this chapter. Furthermore, he wishes to acknowledge all researchers whose contributions to the field could not be included here.

Bibliography

[1] http://www.unrealvoodoo.org/hiteck/projects/albumart (accessed: August 2010).

[2] http://www.altavista.com (accessed: February 2008).

[3] http://www.amazon.com (accessed: January 2008).

[4] http://www.allmusic.com (accessed: November 2007).

[5] http://www.aristomusic.com (accessed: January 2010).

[6] http://www.artofthemix.org (accessed: February 2008).

[7] J.-J. Aucouturier and F. Pachet. Scaling up music playlist generation. In *Proceedings of the IEEE International Conference on Multimedia and Expo (ICME 2002)*, pages 105–108, Lausanne, Switzerland, August 2002.

[8] J.-J. Aucouturier and F. Pachet. Improving timbre similarity: How high is the sky? *Journal of Negative Results in Speech and Audio Sciences*, 1(1), 2004.

[9] J.-J. Aucouturier, F. Pachet, and M. Sandler. "The way it sounds": Timbre models for analysis and retrieval of music signals. *IEEE Transactions on Multimedia*, 7(6):1028–1035, December 2005.

[10] C. Baccigalupo, E. Plaza, and J. Donaldson. Uncovering affinity of artists to multiple genres from social behaviour data. In *Proceedings of the 9th International Conference on Music Information Retrieval (ISMIR'08)*, Philadelphia, September 14–18, 2008.

[11] R. Baeza-Yates and B. Ribeiro-Neto. *Modern Information Retrieval.* Addison-Wesley, Upper Saddle River, New Jersey, 1999.

[12] S. Baumann and O. Hummel. Using cultural metadata for artist recommendation. In *Proceedings of the 3rd International Conference on Web Delivering of Music (WEDELMUSIC 2003)*, Leeds, UK, September 15–17, 2003.

[13] A. Berenzweig, B. Logan, D.P.W. Ellis, and B. Whitman. A large-scale evaluation of acoustic and subjective music similarity measures. In *Proceedings of the 4th International Conference on Music Information Retrieval (ISMIR 2003)*, Baltimore, October 26–30, 2003.

[14] E. Brill. A simple rule-based part of speech tagger. In *Proceedings of the 3rd Conference on Applied Natural Language Processing*, pages 152–155, 1992.

[15] J. Callan and T. Mitamura. Knowledge-based extraction of named entities. In *Proceedings of the 11th International Conference on Information and Knowledge Management (CIKM 2002)*, pages 532–537, (McLean, VA), ACM Press, New York, 2002.

[16] P. Cano and M. Koppenberger. The emergence of complex network patterns in music artist networks. In *Proceedings of the 5th International Symposium on Music Information Retrieval (ISMIR 2004)*, pages 466–469, Barcelona, Spain, October 10–14, 2004.

[17] M.A. Casey, R. Veltkamp, M. Goto, M. Leman, C. Rhodes, and M. Slaney. Content-based music information retrieval: Current directions and future challenges. *Proceedings of the IEEE*, 96:668–696, April 2008.

[18] Ò. Celma. *Music recommendation and discovery in the long tail.* PhD thesis, Universitat Pompeu Fabra, Barcelona, Spain, 2008.

[19] Ò. Celma, P. Cano, and P. Herrera. SearchSounds: An audio crawler focused on weblogs. In *Proceedings of the 7th International Conference on Music Information Retrieval (ISMIR 2006)*, Victoria, Canada, October 8–12, 2006.

[20] Ò. Celma and P. Lamere. ISMIR 2007 Tutorial: Music Recommendation. http://mtg.upf.cdu/~ocelma/MusicRecommendationTutorial-ISMIR2007 (accessed: December 2007), September 23 27, 2007.

[21] W.W. Cohen and W. Fan. Web-collaborative filtering: Recommending music by crawling the Web. *WWW9/Computer Networks*, 33(1–6):685–698, 2000.

[22] T.F. Cox and M.A.A. Cox. *Multidimensional Scaling.* Chapman & Hall, New York, 1994.

[23] S. Deerwester, S.T. Dumais, G.W. Furnas, T.K. Landauer, and R. Harsh-
man. Indexing by latent semantic analysis. *Journal of the American
Society for Information Science*, 41:391–407, 1990.

[24] D. DiNucci. Fragmented future. *Design & New Media*, 53(4), 1999.

[25] http://www.discogs.com (accessed: July 2010).

[26] D.P.W. Ellis, B. Whitman, A. Berenzweig, and S. Lawrence. The quest
for ground truth in musical artist similarity. In *Proceedings of 3rd International Conference on Music Information Retrieval (ISMIR 2002)*,
Paris, France, October 13–17, 2002.

[27] http://www.epinions.com/music (accessed: August 2007).

[28] http://www.evillabs.sk/evillyrics (accessed: August 2007).

[29] D.D. Feng, W.-C. Siu, and H.-J. Zhang. *Multimedia Information Retrieval and Management: Technological Fundamentals and Applications*.
Springer, New York, 2003.

[30] http://www.freebase.com (accessed: January 2010).

[31] http://www.freedb.org (accessed: February 2008).

[32] G. Geleijnse, M. Schedl, and P. Knees. The quest for ground truth in
musical artist tagging in the social Web era. In *Proceedings of the 8th
International Conference on Music Information Retrieval (ISMIR 2007)*,
Vienna, Austria, September 2007.

[33] S. Govaerts and E. Duval. A Web-based approach to determine the origin
of an artist. In *Proceedings of the 10th International Society for Music
Information Retrieval Conference (ISMIR 2009)*, Kobe, Japan, October
2009.

[34] http://www.gracenote.com (accessed: February 2008).

[35] T. Hofmann. Probabilistic latent semantic analysis. In *Proceedings of
Uncertainty in Artificial Intelligence (UAI)*, Stockholm, Sweden, 1999.

[36] X. Hu, J.S. Downie, and A.F. Ehmann. Lyric text mining in music mood
classification. In *Proceedings of the 10th International Society for Music
Information Retrieval Conference (ISMIR'09)*, Kobe, Japan, 2009.

[37] X. Hu, J.S. Downie, K. West, and A. Ehmann. Mining music reviews:
promising preliminary results. In *Proceedings of the 6th International
Conference on Music Information Retrieval (ISMIR 2005)*, London, UK,
September 11–15, 2005.

[38] http://techno.org/electronic-music-guide (accessed: August 2010).

[39] F. Kleedorfer, P. Knees, and T. Pohle. Oh oh oh whoah! Toward automatic topic detection in song lyrics. In *Proceedings of the 9th International Conference on Music Information Retrieval (ISMIR 2008)*, pages 287–292, Philadelphia, September 2008.

[40] P. Knees, E. Pampalk, and G. Widmer. Artist classification with Web-based data. In *Proceedings of the 5th International Symposium on Music Information Retrieval (ISMIR 2004)*, pages 517–524, Barcelona, Spain, October 10–14, 2004.

[41] P. Knees, T. Pohle, M. Schedl, and G. Widmer. A music search engine built upon audio-based and Web-based similarity measures. In *Proceedings of the 30th Annual International ACM SIGIR Conference on Research and Development in Information Retrieval (SIGIR 2007)*, Amsterdam, Netherlands, July 23–27, 2007.

[42] P. Knees, M. Schedl, and T. Pohle. A deeper look into Web-based classification of music artists. In *Proceedings of 2nd Workshop on Learning the Semantics of Audio Signals (LSAS 2008)*, Paris, France, June 2008.

[43] P. Knees, M. Schedl, T. Pohle, and G. Widmer. An innovative three-dimensional user interface for exploring music collections enriched with meta-information from the Web. In *Proceedings of the 14th ACM International Conference on Multimedia (MM 2006)*, Santa Barbara, California, October 23–27, 2006.

[44] P. Knees, M. Schedl, T. Pohle, and G. Widmer. Exploring music collections in virtual landscapes. *IEEE MultiMedia*, 14(3):46–54, July–September 2007.

[45] P. Knees, M. Schedl, and G. Widmer. Multiple lyrics alignment: Automatic retrieval of song lyrics. In *Proceedings of 6th International Conference on Music Information Retrieval (ISMIR 2005)*, pages 564–569, London, UK, September 11–15, 2005.

[46] T. Kohonen. *Self-Organizing Maps, Volume 30*, 3rd edition. Springer Series in Information Sciences. Springer, Berlin, Germany, 2001.

[47] Y. Koren, R. Bell, and C. Volinsky. Matrix factorization techniques for recommender systems. *Computer*, 42:30–37, August 2009.

[48] J. Korst and G. Geleijnse. Efficient lyrics retrieval and alignment. In W. Verhaegh, Emile A., W. ten Kate, J. Korst, and S. Pauws, editors, *Proceedings of the 3rd Philips Symposium on Intelligent Algorithms (SOIA 2006)*, pages 205–218, Eindhoven, Netherlands, December 6–7, 2006.

[49] J.B. Kruskal and M. Wish. *Multidimensional Scaling*. Paper Series on Quantitative Applications in the Social Sciences. Sage Publications, Newbury Park, California, 1978.

[50] P. Lamere. Social tagging and music information retrieval. *Journal of New Music Research: Special Issue: From Genres to Tags: Music Information Retrieval in the Age of Social Tagging*, 37(2):101–114, 2008.

[51] http://last.fm (accessed: December 2010).

[52] http://last.fm/api (accessed: August 2010).

[53] C. Laurier, J. Grivolla, and P. Herrera. Multimodal music mood classification using audio and lyrics. In *Proceedings of the International Conference on Machine Learning and Applications*, San Diego, California, 2008.

[54] E. Law, L. von Ahn, R. Dannenberg, and M. Crawford. Tagatune: a game for music and sound annotation. In *Proceedings of the 8th International Conference on Music Information Retrieval (ISMIR 2007)*, Vienna, Austria, September 2007.

[55] M. Levy and M. Sandler. A semantic space for music derived from social tags. In *Proceedings of the 8th International Conference on Music Information Retrieval (ISMIR 2007)*, Vienna, Austria, September 2007.

[56] http://www.last.fm/music/Queen (accessed: January 2010).

[57] G. Linden, B. Smith, and J. York. Amazon.com recommendations: Item-to-item collaborative filtering. *IEEE Internet Computing*, 4(1), 2003.

[58] B. Logan, D.P.W. Ellis, and A. Berenzweig. Toward evaluation techniques for music similarity. In *Proceedings of the 26th Annual International ACM SIGIR Conference on Research and Development in Information Retrieval (SIGIR 2003): Workshop on the Evaluation of Music Information Retrieval Systems*, (Toronto, Canada), ACM Press, New York, July–August 2003.

[59] B. Logan, A. Kositsky, and P. Moreno. Semantic analysis of song lyrics. In *Proceedings of the IEEE International Conference on Multimedia and Expo (ICME 2004)*, Taipei, Taiwan, June 27–30, 2004.

[60] D. Lübbers and M. Jarke. Adaptive multimodal exploration of music collections. In *Proceedings of the 10th International Society for Music Information Retrieval Conference (ISMIR 2009)*, Kobe, Japan, October 2009.

[61] H.P. Luhn. A statistical approach to mechanized encoding and searching of literary information. *IBM Journal*, pages 309–317, October 1957.

[62] http://www.lyrics.com (accessed: August 2010).

[63] J. MacQueen. Some methods for classification and analysis of multivariate observations. In L.M. Le Cam and J. Neyman, editors, *Proceedings of the 5th Berkeley Symposium on Mathematical Statistics and Probability, Volume I of Statistics*, pages 281–297, University of California Press, Berkeley, 1967.

[64] J. P.G. Mahedero, Á. Martínez, P. Cano, M. Koppenberger, and F. Gouyon. Natural language processing of lyrics. In *Proceedings of the 13th ACM International Conference on Multimedia (MM 2005)*, pages 475–478, Singapore, November 6–11, 2005.

[65] M.I. Mandel and D.P.W. Ellis. A Web-based game for collecting music metadata. In *Proceedings of the 8th International Conference on Music Information Retrieval (ISMIR 2007)*, Vienna, Austria, September 2007.

[66] R. Mayer, R. Neumayer, and A. Rauber. Rhyme and style features for musical genre classification by song lyrics. In *Proceedings of the 9th International Conference on Music Information Retrieval (ISMIR'08)*, Philadelphia, 2008.

[67] http://www.mapofmetal.com (accessed: December 2010).

[68] http://music.strands.com (accessed: November 2009).

[69] S.B. Needleman and C.D. Wunsch. A general method applicable to the search for similarities in the amino acid sequence of two proteins. *Journal of Molecular Biology*, 48(3):443–453, 1970.

[70] http://www.northernlight.com (accessed: February 2008).

[71] T. O'Reilly. What Is Web 2.0—Design Patterns and Business Models for the Next Generation of Software. http://oreilly.com/web2/archive/what-is-web-20.html (accessed: March 2010).

[72] F. Pachet, G. Westerman, and D. Laigre. Musical data mining for electronic music distribution. In *Proceedings of the 1st International Conference on Web Delivering of Music (WEDELMUSIC 2001)*, Florence, Italy, November 23–24, 2001.

[73] E. Pampalk and M. Goto. MusicSun: A new approach to artist recommendation. In *Proceedings of the 8th International Conference on Music Information Retrieval (ISMIR 2007)*, Vienna, Austria, September 23–27, 2007.

[74] http://www.pandora.com (accessed: August 2010).

[75] T. Pohle, P. Knees, M. Schedl, E. Pampalk, and G. Widmer. "Reinventing the wheel": A novel approach to music player interfaces. *IEEE Transactions on Multimedia*, 9:567–575, 2007.

[76] G. Salton and M.J. McGill. *Introduction to Modern Information Retrieval*. McGraw-Hill, New York, 1983.

[77] G. Salton, A. Wong, and C.S. Yang. A vector space model for automatic indexing. *Communications of the ACM*, 18(11):613–620, 1975.

[78] B. Sarwar, G. Karypis, J. Konstan, and J. Reidl. Item-based collaborative filtering recommendation algorithms. In *WWW'01: Proceedings of 10th International Conference on World Wide Web*, pages 285–295, 2001.

[79] M. Schedl. *Automatically extracting, analyzing, and visualizing information on music artists from the World Wide Web*. PhD thesis, Johannes Kepler University Linz, Linz, Austria, 2008.

[80] M. Schedl. On the use of microblogging posts for similarity estimation and artist labeling. In *Proceedings of the 11th International Society for Music Information Retrieval Conference (ISMIR 2010)*, Utrecht, Netherlands, August 2010.

[81] M. Schedl, P. Knees, T. Pohle, and G. Widmer. Toward automatic retrieval of album covers. In *Proceedings of the 28th European Conference on Information Retrieval (ECIR 2006)*, London, April 2–5, 2006.

[82] M. Schedl, P. Knees, and G. Widmer. A Web-based approach to assessing artist similarity using co-occurrences. In *Proceedings of the 4th International Workshop on Content-Based Multimedia Indexing (CBMI 2005)*, Riga, Latvia, June 21–23, 2005.

[83] M. Schedl, C. Schiketanz, and K. Seyerlehner. Country of origin determination via Web mining techniques. In *Proceedings of the IEEE International Conference on Multimedia and Expo (ICME 2010): 2nd International Workshop on Advances in Music Information Research (AdMIRe 2010)*, Singapore, July 19–23, 2010.

[84] M. Schedl, K. Seyerlehner, D. Schnitzer, G. Widmer, and C. Schiketanz. Three Web-based heuristics to determine a person's or institution's country of origin. In *Proceedings of the 33rd Annual International ACM SIGIR Conference on Research and Development in Information Retrieval (SIGIR 2010)*, Geneva, Switzerland, July 19–23, 2010.

[85] M. Schedl and G. Widmer. Automatically detecting members and instrumentation of music bands via Web content mining. In *Proceedings of the 5th Workshop on Adaptive Multimedia Retrieval (AMR 2007)*, Paris, France, July 5–6, 2007.

[86] Y. Shavitt and U. Weinsberg. Songs clustering using peer-to-peer co-occurrences. In *Proceedings of the IEEE International Symposium on Multimedia (ISM2009): International Workshop on Advances in Music Information Research (AdMIRe 2009)*, San Diego, California, December 16, 2009.

[87] S. Stober and A. Nürnberger. MusicGalaxy: An adaptive user-interface for exploratory music retrieval. In *Proceedings of the 11th International Society for Music Information Retrieval Conference (ISMIR 2010)*, Utrecht, Netherlands, August 2010.

[88] D. Turnbull, R. Liu, L. Barrington, and G. Lanckriet. A game-based approach for collecting semantic annotations of music. In *Proceedings of the 8th International Conference on Music Information Retrieval (ISMIR 2007)*, Vienna, Austria, September 2007.

[89] http://www.twitter.com (accessed: August 2010).

[90] G. Tzanetakis and P. Cook. Musical genre classification of audio signals. *IEEE Transactions on Speech and Audio Processing*, 10(5):293–302, 2002.

[91] V.N. Vapnik. *The Nature of Statistical Learning Theory.* Springer, New York, 1995.

[92] B. Whitman and S. Lawrence. Inferring descriptions and similarity for music from community metadata. In *Proceedings of the 2002 International Computer Music Conference (ICMC 2002)*, pages 591–598, Göteborg, Sweden, September 16–21, 2002.

[93] M. Zadel and I. Fujinaga. Web services for music information retrieval. In *Proceedings of the 5th International Symposium on Music Information Retrieval (ISMIR 2004)*, Barcelona, Spain, October 10–14, 2004.

[94] J. Zobel and A. Moffat. Inverted files for text search engines. *ACM Computing Surveys*, 38:1–56, 2006.

8

Indexing Music with Tags

Douglas Turnbull

Ithaca College

CONTENTS

8.1 Introduction

In this chapter, we introduce the concept of a *semantic music discovery engine*. We refer to such a system as *discovery* engine (as opposed to a *search* engine) because it is designed to help people discover novel music, as well as uncover new connections between familiar songs and artists. The term *semantic* refers to the fact that our system is built around a *query-by-text description* paradigm where a person can make use of a large, diverse set of musically relevant *tags* to specify the type of music that he or she wishes to hear. For example, a semantic music discovery engine lets one find music by asking for "mellow classic rock featuring slide guitar." From this query phrase, the system first identifies tags like "mellow," "classic rock," and "slide guitar" and then retrieves songs that are semantically associated with these tags. Finally, the engine presents the most relevant music as a streamable playlist of songs.

The core data structure of a semantic music discovery engine is a *music index*. The music index is represented as a large *tag-song matrix* where each element of the matrix reflects the strength of semantic association between a tag and a song. We will begin this chapter by further developing the notion of a music index using concepts from the field of (text-based) Information Retrieval (Section 8.2). We will then describe a number of sources of music information that are useful for constructing a music index (Sections 8.3 and 8.4). These data sources are often complementary and can be combined to improve the music discovery engine both in terms of accuracy and scalability (Section 8.5). We will end the chapter with a few examples of both commercial and academic semantic music discovery engines (Section 8.6).

8.2   Music Indexing

Like traditional Internet search engines (e.g., Google, Yahoo!), the core data structure of a semantic music discovery engine is an *index*. As such, we will briefly describe some of the fundamental concepts from text-based information retrieval (IR) and then further develop these ideas for music discovery. The reader may wish to refer to a recommended textbook on IR for additional background information [30, 3].

8.2.1   Indexing Text

An index is often represented as a large *term-document* matrix where each row represents a short text-token and each column represents a document. In the context of Internet search, a *term* may be any word. A *document* is a Web page found on the Internet. To index a Web page, we count the number of times each word appears in that Web page. For example, in Figure 8.2, the term "apple" appears three times in the first document, zero times in the second document, two times in the third document, and so on.

Once we have indexed all of our documents, a user enters a query and the search engine quickly finds relevant Web pages using the index. For the example query "apple crisp," the most relevant document is the third document

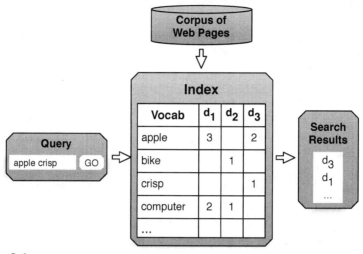

Figure 8.1
Architecture for an Internet search engine.

since it contains both of the terms "apple" and "crisp." The first document is the second-most relevant Web page since one of the terms ("apple") appears frequently in the document (Figure 8.1).

In practice, we modify the raw counts in the term-document matrix to improve our IR system. For example, we will up-weight terms that appear frequently in a document (term frequency) and down-weight terms that appear in many different documents (inverse document frequency). This weighting scheme is referred to as term frequency–inverse document frequency (TF-IDF). We may also normalize each document vector by length so that longer documents are fairly compared with shorter documents when considering a query.

More formally, we construct a term-document matrix $\mathbf{X}$ where each column vector $\mathbf{v}_d = [\mathbf{X}]_{.,d}$ corresponds to a document d. When a user enters a string of words as a query, we think of the query string as a (very short) document and represent it as a document vector $\mathbf{v}_q$. We then calculate the similarity between $\mathbf{v}_q$ and each of the document vectors in our index. The most commonly used similarity function is *cosine similarity*:

$$sim(\mathbf{v}_q, \mathbf{v}_d) = \frac{\mathbf{v}_q \cdot \mathbf{v}_d}{|\mathbf{v}_q||\mathbf{v}_d|} \tag{8.1}$$

where the numerator is the dot product between the two vectors and the denominator is the product of the Euclidean lengths for the two vectors. We note that many other functions of similarity are also used (e.g., Kullback-Leibler [KL]-divergence, L_1-norm) as alternatives to cosine similarity. The search engine will rank order documents based on the similarity function and return documents that are most similar to the query vector.

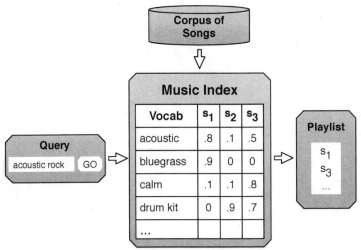

Figure 8.2
Architecture for a semantic music discovery engine.

Once we have an index and a similarity function, we can also compute the pair-wise similarity between each pair of documents in our corpus. This allows us to recommend similar Web pages for a given seed Web page (*query-by-similarity*) and cluster similar Web pages (*topic discovery*[1]).

8.2.2   Indexing Music

In the context of music, we will refer to terms as *tags* to be consistent with the Music-IR literature. A tag is a short phrase like "happy," "hand drums," or "New Orleans jazz." Our vocabulary will consist of thousands of tags that are related to instruments, genres, emotions, moods, usages, geographic origins, musicological terms, and other music-related concepts. Likewise, for a music index, a document can be a song, an album, an artist, a record label, etc. To simplify our discussion, we will consider songs to be our musical documents throughout this chapter.

To create a music index, we need to construct a *tag-song* matrix $\mathbf{X}$ where each element $x_{t,s} = [\mathbf{X}]_{t,s}$ represents the strength of association between each tag t and each song s. Unfortunately, this is somewhat less straightforward than building an index for Web pages because our documents are now audio files. That is, previously we simply needed to count (text-based) terms in (text-based) documents. For music, we need to find a way to relate (text-based) tags to (audio-based) songs.

In the next section, we will describe how we can create a tag-song music

---

[1] A good example of topic discovery is News Aggregator (e.g., Google News) which automatically clusters news article from many different news outlets into a small set of "stories."

index from various sources of data. However, once constructed, we can use the music index much like we use an index for text documents. That is, we create a query vector $\mathbf{v}_q$ from the user's query string and calculate the (cosine) similarity between it and each of the column vectors (e.g., $\mathbf{v}_s = [\mathbf{X}]_{.,s}$ for song s) in our tag-song matrix. We can then rank-order the songs based on this similarity in order to generate a semantically meaningful playlist of songs (Figure 8.2). In addition, we can use our music index to calculate music similarity between pairs of songs or to cluster our music into automatically created genres.

8.3 Sources of Tag-Based Music Information

In this section, we describe five data sources that can provide us with semantic information about music. Three data sources (surveys, social tags, games) rely on human participation, and as such, are expensive in terms of financial cost and human labor. Two sources (text mining, autotagging) rely on automatic methods that are computationally intense but require less direct human involvement.

There are a number of key concepts to consider when comparing these approaches. The *cold-start problem* refers to the fact songs that are not annotated cannot be retrieved. This problem is related to *popularity bias* in that popular songs (in the *short head*) tend to be annotated more thoroughly than unpopular songs (in the *long tail*) [1, 22]. This often leads to a situation in which a short-head song is ranked above a long-tail song despite the fact that the long-tail song may be more semantically relevant. We prefer an approach that avoids the cold-start problem (e.g., autotagging). If this is not possible, we prefer approaches in which we can explicitly control which songs are annotated (e.g., survey, games), rather than an approach in which only the more popular songs are annotated (e.g., social tags, Web documents).

A *strong labeling* [8] is when a song has been explicitly labeled or not labeled with a tag depending on whether or not the tag is relevant. This is opposed to a *weak labeling* in which the absence of a tag from a song does not necessarily indicate that the tag is not relevant. For example, a song may feature drums but is not explicitly labeled with the tag "drum." Weak labeling is a problem if we want to design a music IR system with high recall (e.g., find every song which features drums), or if our goal is to collect a training data set for a supervised autotagging system that uses discriminative classifiers [29, 14, 47].

It is also important to consider the size, structure, and extensibility of the tag vocabulary. In the context of text-based music retrieval, the ideal vocabulary is a large and diverse set of semantic tags, where each tag describes some meaningful attribute or characterization of music. In this chapter, we limit our focus to tags that can be used consistently by a large number of individuals when annotating novel songs based primarily on the listening experience. In general, this does not include tags that are personal (e.g., "seen live") or judgmental (e.g., "horrible") in nature [47].

A tag vocabulary can be *fixed* or *extensible*, as well as *structured* or *unstructured*. For example, the tag vocabulary associated with a survey can be considered fixed and structured since the set of tags and the grouping of tags into coherent semantic categories (e.g., genres, instruments, emotions, usages) is predetermined by experts using domain knowledge [41, 43]. By contrast, social tagging communities produce a vocabulary that is extensible since any user can suggest any free-text token to describe music. This vocabulary is also unstructured since tags are not organized in any way. In general, we prefer an extensible vocabulary because a fixed vocabulary limits text-based retrieval to a small set of predetermined tags. In addition, a structured vocabulary is advantageous since the ontological relationships (e.g., genre hierarchies, families of instruments) between tags encode valuable semantic information that is useful for retrieval.

Finally, the accuracy with which tags are applied to songs is perhaps the most important point of comparison. Since there is no ideal ground truth and listeners do not always agree whether (or to what degree) a tag should be applied to a song (i.e., "the subjectivity problem" [31]), evaluating accuracy can be tricky. Intuitively, it is preferable to have trained musicologists, rather than untrained nonexperts, annotate a music corpus. It is also advantageous to have multiple individuals, rather than a single person, annotate each song. Last, individuals who are given incentives to provide good annotations (e.g., a high score in a game) may provide better annotations than unmotivated individuals.

8.3.1   Conducting a Survey

Perhaps the most well-known example of the music annotation survey is Pandora's[2] "Music Genome Project" [11, 46]. Pandora uses a team of approximately 50 expert music reviewers (each with a degree in music and 200 hours of training) to annotate songs using structured vocabularies of between 150 to 500 "musically objective" tags depending on the genre of the music [17]. Tags, such as "Afro-Latin Roots," "Electric Piano Riffs," and "Political Lyrics," can be considered objective since, according to Pandora, there is a high level of inter-reviewer agreement when annotating the same song. Between 2000 and 2010, Pandora annotated about 750,000 songs[3] [46]. Currently, each song takes between 20 to 30 minutes to annotate and approximately 15,000 new songs are annotated each month. While this labor-intensive approach results in high-quality annotations, Pandora must be very selective of which songs

---

[2]http://www.pandora.com.

[3]Statistic retrieved from http://blog.pandora.com/faq/contents/29.html on July 15, 2010.

they choose to annotate given that other companies like Apple iTunes and Gracenote maintain growing databases with tens of millions of songs.[4]

Pandora, as well as companies like Moodlogic[5] and Allmusic[6], have devoted considerable amounts of money, time, and human resources to annotate their music databases with high-quality tags. As such, they are unlikely to share this data with the Music IR research community. To remedy this problem, we have collected the CAL500 data set of annotated music [41]. This data set contains one song from 500 unique artists, each of which have been manually annotated by a minimum of three nonexpert reviewers using a structured vocabulary of 174 tags. While this is a small data set, it is strongly labeled, relies on multiple reviewers per song, is publicly available, and as such, has become a standard data set for training and evaluating tag-based music retrieval systems.

8.3.2 Harvesting Social Tags

Last.fm[7] is a music discovery Web site that allows users to contribute *social* tags through a text box in their audio player interface. This *crowd sourcing* approach has the potential to generate an enormous amount of semantic music information and benefits by involving millions of individuals in the annotation process (i.e., "wisdom of the crowds"). By the beginning of 2007, Last.fm's large base of 40 million monthly users had built up an unstructured vocabulary of 960,000 free-text tags and used it to annotate millions of songs [32]. Unlike Pandora and AMG Allmusic, Last.fm makes much of this data available through their public API[8]. While this data is a useful resource for the Music-IR community, Lamere [21, 22] points out a number of problems with social tags. First, there is often a sparsity of tags for new and obscure artists (cold-start problem / popularity bias). Second, most tags are used to annotate artists rather than individual songs. This is problematic since we are interested in retrieving semantically relevant songs from eclectic artists. Third, individuals use ad-hoc techniques when annotating music. This is reflected by use of polysemous tags (e.g., "progressive"), tags that are misspelled or have multiple spellings (e.g., "hip hop," "hip-hop"), tags used for self-organization (e.g., "seen live"), and tags that are nonsensical. Finally, the public interface allows for malicious behavior. For example, individuals have been known to target artists (e.g., Paris Hilton) with misleading tags (e.g., "brutal death metal").

---

[4]Apple reports the iTunes store contains over 13 million songs for purchase and Gracenote claims they they have music fingerprints for 28 million unique songs. http://www.apple.com/itunes/features/#purchasingmusic, http://www.gracenote.com/business_solutions/music_id/(accessed: July 15, 2010).

[5]http://en.wikipedia.org/wiki/MoodLogic.

[6]http://www.allmusic.com.

[7]www.last.fm.

[8]http://www.last.fm/api.

8.3.3   Playing Annotation Games

Another crowd sourcing approach involves using games to collect music tags. At the 2007 ISMIR conference, music annotation games were presented for the first time: ListenGame [43], Tag-a-Tune [24], and MajorMiner [28]. ListenGame and its successor Herd It [4] are real-time games where a large group of users is presented with a song and a list of tags. The players choose the best and worst tags for describing the song. When a large group of players agree on a tag, the song has a strong (positive or negative) association with the tag. This game, like a music survey, has the benefit of using a structured vocabulary of tags. It can be considered a strong labeling approach since it also collects information that reflects negative semantic associations between tags and songs. Like the ESPGame for image tagging [45], Tag-a-Tune is a two-player game where the players listen to a song and are asked to enter "free text" tags until they both enter the same tag. MajorMiner is similar in nature, except the tags entered by the player are compared against the database of previously collected tags in an offline manner. Like social tagging, the tags collected using both games result in a unstructured but extensible vocabulary.

A major problem with this game-based approach is that players will inevitably attempt to *game* the system. For example, the player may only contribute generic tags (e.g., "rock," "guitar") even if less common tags provide a better semantic description (e.g., "grunge," "distorted electric guitar"). Also, despite the recent academic interest in music annotation games, no game has achieved large-scale success. This reflects the fact that it is difficult to design a viral game for this inherently laborious task.

8.3.4   Mining Web Documents

Artist biographies, album reviews, and song reviews are another rich source of semantic information about music. There are a number of research-based music IR systems that collect such documents from the Internet by querying search engines [19], monitoring MP3 blogs [9], or crawling a music site [47]. In all cases, Levy and Sandler point out that such Web mined corpora can be *noisy* since some of the retrieved Web pages will be irrelevant and, in addition, much of the text content on relevant Web pages will be useless [25].

Most of the proposed Web mining systems use a set of one or more documents associated with a song and convert them into a single document vector (see Section 8.2.1) [20, 48]. This *vector space* representation is then useful for a number of music IR tasks such as calculating music similarity [48] and indexing content for a text-based music retrieval system [20]. More recently, Knees et al. [19] have proposed a promising new Web mining technique called *relevance scoring* as an alternative to the vector space approaches. Both relevance scoring and vector space approaches are subject to popularity bias since short-head songs are generally represented by more documents than long-tail songs.

8.3.5 Autotagging Audio Content

All previously described approaches require that a song be annotated by humans, and as such, are subject to the cold-start problem. Content-based audio analysis is an alternative approach that avoids this problem. Early work on this topic focused (and continues to focus) on music classification by genre, emotion, and instrumentation [44, 26, 15]. These classification systems effectively "tag" music with class labels (e.g., "blues," "sad," "guitar"). More recently, *autotagging* systems have been developed to annotate music with a larger, more diverse vocabulary of (nonmutually exclusive) tags [41, 29, 14, 36]. In Turnbull et al. [41], we describe a generative approach that learns a Gaussian mixture model (GMM) distribution over an audio feature space for each tag in the vocabulary. Bertin-Mahieux et al. use a discriminative approach by learning a boosted decision stump classifier for each tag [14]. Similarly, Mandel et al. [29] follow a discriminative approach but use a Support Vector Machine (SVM) classifier to improve performance. Finally, Sordo et al. present a nonparametric approach that uses a content-based measure of music similarity to propagate tags from annotated songs to similar songs that have not been annotated [36]. For a more complete introduction to this emerging area of research, please refer to the recent book chapter by Bertin-Mahieux, Eck, and Mandel [6].

8.3.6 Additional Remarks

In Table 8.1, we list some of the relative strengths and weaknesses for each of the five approaches that we have covered in this section. In an ideal world, having multiple human experts annotate each song with a large, extensible vocabulary of tags would likely lead to the most accurate music index. However, this is an extremely expensive and unrealistic idea. To illustrate this point, consider that Pandora has been annotating music for over a decade and yet has only annotated only about 5% of the music that is available for sale on the Apple iTunes store. Social tagging offers an improvement in terms of scalability by crowd-sourcing the task of music annotation; however, it suffers from strong popularity bias and inconsistent annotation behavior. Music annotation games also rely on crowd-sourcing but have the potential to provide more consistent annotations for a controlled set of songs. However, to date, none of the games have been successful at attracting and sustaining a large group of users. This suggests that designing a compelling game for an inherently laborious task is a challenging undertaking.

Using text mining to extract semantic information from Web documents has potential since it does not involve direct human annotation. However, the text mining process is error prone and, in general, Web documents only exist

Approach	Strengths	Weaknesses
Survey	Custom-tailored vocabulary High-quality annotations Strong labeling Pick which songs to annotate	Small, predetermined vocabulary High cost for human labor Scalability issues Commercial data is not public
Social Tags	Collective wisdom of crowds Unlimited vocabulary Provides social context	Ad-hoc annotation behavior Strong popularity bias Weak labeling
Game	Collective wisdom of crowds Entertaining incentives produce high-quality annotations Pick which songs to annotate	Problems with "gaming" the system Difficult to create viral game Often based on short audio clips No large-scale success to date
Web Documents	Large, publicly available corpus of relevant documents No direct human involvement Provides social context	Noisy annotations from text-mining Sparse/missing in long-tail Weak labeling
Autotags	Not affected by cold-start problem No direct human involvement Strong labeling	Computationally intensive Limited by training data Based solely on audio content

Table 8.1
Strengths and Weaknesses of Music Information Data Sources

for more popular artists and songs. Content-based autotagging does not suffer from popularity bias but needs to be improved in terms of accuracy, especially for more nuanced and detail-oriented tags. The qualitative comparison of the five data sources is Shown in Table 8.2.

We should also mention that, in addition to the five data sources we have explored in this section, there are other sources of semantic music information. In Kim et al. [18], we show that we can use preference information (i.e., collaborative filtering) to first calculate artist similarity and then propagate tags from known artists to unknown artists. Despite the fact that preference information does explicitly represent semantic information, we showed that it was more useful for music indexing than tag propagation when calculating artist similarity based on audio content, Web documents, or social tags.

Images and (music) videos represent yet another source of music information. In early work [27], we showed that we can successfully index artists with music genre tags based on promotional photographs and album cover artwork. Other researchers have been exploring how to annotate consumer videos with tags based on the multimodal analysis of both the video and audio tracks [10].

	Survey (CAL500)	Social Tags (Last.fm)	Game (ListenGame)	Web Docs (WRS)	Autotags (SML)
			Scalability and Cost		
Financial	Expensive ~ \$3 per song	**Minimal** **public** **API**	Moderate design, deploy, promote game	**Minimal** **fully** **automated**	**Minimal** **fully** **automated**
Human Labor	Expensive 18 min per song	Moderate crowd-sourced	Moderate crowd-sourced	**Minimal** **none**	**Minimal** **training** **data**
Computation	**Minimal** **Database**	**Minimal** **Database**	Moderate Game Server, Database	Moderate Webcrawler, Text Mining	Expensive Computer Auditon
			Quality		
Popularity Bias	Decent can pick songs	Poor strong bias	Decent can pick songs	Poor strong bias	**Great** **only requires** **audio track**
Labeling Setup	**Strong**	Weak	**Strong**	Weak	**Strong**
Vocabulary	**Structured,** **Fixed**	Unstructured, **Extensible**	**Structured,** **Extensible**	N/A	N/A
Accuracy	**Great** **Focused** **Experts**	Good Ad-hoc Community	Good Competitive Community	Decent Noisy Text mining	Decent content-based limitations

Table 8.2
Qualitative Comparison of Data Sources (**Bold font** indicates a positive attribute of the data source.)

8.4   Comparing Sources of Music Information

In this section, we describe one music indexing system for each data source. Each is based on systems that have been recently developed within the music-IR research community [41, 19, 43]. Each produces a $|T| \times |S|$ tag-song music index matrix $\mathbf{X}$ where $|T|$ is the size of our tag vocabulary and $|S|$ is the number of songs in our corpus. Each element $x_{t,s} = [\mathbf{X}]_{t,s}$ in the matrix represents the strength of semantic association between tag t and song s.

We set $x_{t,s} = 0$ if the relationship between tag t and song s is missing (i.e., unknown). If the matrix $\mathbf{X}$ has many such values, then we refer to the matrix as *sparse*, otherwise we refer to it as *dense*. Missing data results from both weak labeling and the cold-start problem. Sparsity is reflected by the *tag density* of a matrix, which is defined as the percentage of nonzero values in the matrix.

Our goal is to find a tagging system that is able to accurately retrieve (i.e., rank-order) songs for a diverse vocabulary of tags (e.g., emotions, genres, instruments, usages). We quantitatively evaluate music retrieval performance of system a by comparing the matrix $\mathbf{X}^a$ against the CAL500 matrix $\mathbf{X}^{\text{CAL500}}$ (see Section 8.3.1). The $\mathbf{X}^{\text{CAL500}}$ matrix is a binary matrix where $x_{t,s} = 1$ if 80% or more of the individuals annotate song s with tag t, and 0 otherwise (see Section V.a of Turnbull et al. [41] for details). For the experiments reported in this section, we use a subset of 109 of the original 174 tags.[9] We assume that the subset of 87 songs from the Magnatunes [12] collection that are included in the CAL500 data set are representative of long-tail music. As such, we can use this subset to gauge how the various music indexing approaches are affected by popularity bias.

Each system is compared to the CAL500 data set using a number of standard information retrieval (IR) evaluation metrics [30]: area under the receiver operating characteristic curve (AUC), mean average precision (MAP), and Top-10 precision (10-Prec). A receiver operating characteristic curve (ROC) is a plot of the true positive rate as a function of the false positive rate as we move down this ranked list of songs. The area under the ROC curve (AUC) is found by integrating the ROC curve and is upper-bounded by 1.0. A random ranking of songs will produce an expected AUC score of 0.5. Average precision (AP) is found by moving down our ranked list of test songs and averaging the precisions at every point where we correctly identify a relevant song. Top-10 precision is the precision after we have retrieved the top 10 songs for a given tag. This metric is designed to reflect the 10 items that would be displayed on the first results page of a standard Internet search engine.

Each value reported in Table 8.3 is the mean of a metric after averaging over all 109 tags in our vocabulary. That is, for each tag, we rank-order our 500 song data set and calculate the value of the metric using CAL500 data as our ground truth. We then compute the average of the metric using the 109 values from the 109 rankings.

8.4.1   Social Tags: Last.fm

For each of our 500 songs, we attempt to collect two lists of social tags from the Last.fm Audioscrobbler Web site. One list is related specifically to the song and the other list is related to the artist. For the song list, each tag has a score ($x_{t,s}^{\text{Last.fm\_Song}}$) that ranges from 0 (low) to 100 (high) and is a secret function (i.e., trade secret of Last.fm) of both the number and diversity of users who have annotated song s with tag t. For the artist list, the tag score ($x_{t,s}^{\text{Last.fm\_Artist}}$) is again a secret function that ranges between 0 and 100, and reflects both tags that have been used to annotate the artist or songs by the artist. We found one or more tags for 393 and 472 of our songs and artists, respectively. This included at least one occurrence of 71 and 78 of the 109

---

[9]We have merged genre-best tags with genre tags, removed instrument-solo tags, removed some redundant emotion tags, and pruned other tags that are used to annotate less than 2% of the songs. For a complete list of tags, see: http://cosmal.ucsd.edu/cal.

Approach	Songs	Density	AUC	AP	10-Prec
Survey (CAL500)	All Songs	1.00	1.00	1.00	0.97
Ground Truth	Long Tail	1.00	1.00	1.00	0.57
Baseline	All Songs	1.00	0.50	0.15	0.13
Random	Long Tail	1.00	0.50	0.18	0.12
Social Tags	All Songs	0.23	0.62	0.28	0.37
Last.fm	Long Tail	0.03	0.54	0.24	0.19
Game ListenGame[†]	All Songs	0.37	0.65	0.28	0.32
Web Documents	All Songs	0.67	0.66	0.29	0.37
SS-WRS	Long Tail	0.25	0.56	0.25	0.18
Autotags	All Songs	1.00	0.69	0.29	0.33
SML	Long Tail	1.00	0.70	0.34	0.27
Rank-based	All Songs	1.00	0.74	0.32	0.38
Interleaving (RBI)	Long Tail	1.00	0.71	0.33	0.28

Table 8.3

Quantitative Comparison of Data Sources (Each approach is compared using all *CAL500* songs and a subset of 87 more obscure *long-tail* songs from the Magnatune data set. *Tag Density* represents the proportion of song-tag pairs that have a nonempty value. The three evaluation metrics [*AUC, AP, 10 Precision*] are found by averaging over 109 tag queries. [†]Note that ListenGame is evaluated using half of the CAL500 songs and that the results do not reflect the realistic effect of the popularity bias [see Section 8.4.2].)

tags in our vocabulary, respectively. While this suggests decent coverage, tag densities of 4.6% and 11.8%, respectively, indicate that the tag-song matrices, $\mathbf{X}^{\text{Last.fm\_Song}}$ and $\mathbf{X}^{\text{Last.fm\_Artist}}$, are sparse even when we consider mostly short-head songs. When evaluated for music retrieval, these sparse tag-song matrices produce AUC of 0.57 and 0.58, respectively.

To remedy this problem, we create a single Last.fm tag-song matrix by leveraging the Last.fm data in three ways. First, we match tags to their synonyms.[10] For example, a song is considered to be annotated with "down tempo" if it has instead been annotated with "slow beat." Second, we allow wildcard matches for each tag. That is, if a tag appears as a substring in another tag, we consider it to be a wildcard match. For example, "blues" matches with "delta electric blues," "blues blues blues," "rhythm & blues." Although synonyms and wildcard matches add noise, they increase the respective densities to 8.6% and 18.9% and AUC performance to 0.59 and 0.59. Third, we combine the song and artist tag-song matrices in one tag-song matrix:

$$\mathbf{X}^{\text{Last.fm}} = \mathbf{X}^{\text{Last.fm\_Song}} + \mathbf{X}^{\text{Last.fm\_Artist}}$$

This results in a single tag-song matrix that has a density of 23% and AUC of

---

[10]Synonyms are determined by the author using a thesaurus and by manually exploring Last.fm's vocabulary of tag.

0.62. Of the 109 tags 95 are represented at least once in this matrix. However, the density for the Magnatune (e.g., long-tail) songs is only 3% and so the Magnatune matrix produces retrieval results that are not much better than random.

8.4.2   Games: ListenGame

In Turnbull et al. [43], we describe a music annotation game called ListenGame in which a community of players listens to a song and is presented with a set of tags. Each player is asked to vote for the single *best* tag and single *worst* tag to describe the music. From the game, we obtain the tag-song matrix $\mathbf{X}^{\mathrm{Game}}$ by:

$$[\mathbf{X}^{\mathrm{Game}}]_{t,s} = \#(\text{best votes}) - \#(\text{worst votes})$$

when song s and tag t are presented to the players.

During a two-week pilot study, 16,500 annotations (best and worst votes) were collected for a random subset of 250 CAL500 songs. Each of the 27,250 song-tag pairs were presented to users an average of 1.8 times. Although this represents a very small sample size, the mean AUC for the subset of 250 songs averaged over the 109-tag vocabulary is 0.65. Long-tail and short-head results do not accurately reflect the real-world effect of popularity bias since all songs were selected for annotation with equal probability. As such, the "long tail" results have been omitted from Table 8.3.

8.4.3   Web Documents: Weight-Based Relevance Scoring

In order to extract tags from a corpus of Web documents, we adapt the relevance scoring (RS) algorithm that has recently been proposed by Knees et al. [19]. They have shown this method to be superior to algorithms based on vector space representations. To generate tags for a set of songs, the RS works as follows:

1. **Collect Document Corpus:** For each song, repeatedly query a search engine with each song title, artist name, or album title. Collect web documents in search results. Retain the (many-to-many) mapping between songs and documents.

2. **Tag Songs:** For each tag:

    (a) Use the tag as a query string to find the relevant documents, each with an associated *relevance weight* (defined below) from the corpus.

    (b) For each song, sum the relevance scores for all the documents that are related to the song.

We modify this algorithm in two ways. First, the relevance score in Knees et al. [19] is inversely proportional to the rank of the relevant document. We use a weight-based approach to relevance scoring (WRS). The relevance weight of a document given a tag can be a function of the number of times the tag appears in the document (tag-frequency), the number of documents with the tag (document frequency), the number of total words in the document, the number of words or documents in the corpus, etc. For our system, the relevance weights are determined by the MySQL match function.[11]

We calculate an entry of the tag-song matrix $\mathbf{X}^{\text{WRS}}$ as,

$$\mathbf{X}_{t,s}^{\text{WRS}} = \sum_{d \in D_t} w_{d,t} I_{d,s}$$

where D_t is the set of relevant documents for tag t, $w_{d,t}$ is the relevance weight for document d and tag t, and $I_{d,s}$ is an indicator variable that is 1 if document d was found when querying the search engine with song s (in Step 1) and 0 otherwise. We find that weight-based RS (WRS) produces a small increase in performance over rank-based RS (RRS) (AUC of 0.66 versus 0.65). In addition, we believe that WRS will scale better since the relevance weights are independent of the number of documents in our corpus.

The second modification is that we use *site-specific* queries when creating our corpus of Web documents (Step 1). That is, Knees et al. [19] collect the top 100 documents returned by Google when given queries of the form:

- "<artist name>" music
- "<artist name>" "<album name>" music review
- "<artist name> " "<song name>" music review

for each song in the data set. Based on an informal study of the top 100 Web pages returned by nonsite-specific queries, we find that many pages contain information that is only slightly relevant (e.g., music commerce site, ticket resellers, noisy discussion boards, generic biographical information). By searching music-specific sites, we are more likely to find detailed music reviews and in-depth artist biographies. In addition, the Web pages at sites like Pandora and Allmusic specifically contain useful tags in addition to natural language content [39, 34].

We use site-specific queries by appending the substring 'site:<music site url>' to the three query templates, where <music site url> is the url for a music Web site that is known to have high quality information about songs, albums or artists. These sites include allmusic.com, amazon.com, bbc.co.uk, billboard.com, epinions.com, musicomh.com, pandora.com, pitchforkmedia.com, rollingstone.com, wikipedia.org. For these 10 music sites and one nonsite-specific query, we collect and store the top 10 pages returned by the Google search engine. This results in a maximum of 33 queries and a maximum of 330 pages per song. On average, we are only able to collect 150 Web pages per song since some of the long-tail songs are not well represented by these music sites.

Our *site-specific weight-based relevance scoring* (SS-WRS) approach produces a relatively dense tag-song matrix (46%) compared with the approach

---

[11] http://dev.mysql.com/doc/refman/5.0/en/fulltext-natural-language.html.

involving Last.fm tags. However, like the Last.fm approach, the density of the tag-song matrix is greatly reduced (25%) when we consider only long-tail songs.

8.4.4 Autotagging: Supervised Multiclass Labeling

In Turnbull et al. [41], we use a supervised multiclass labeling (SML) model to automatically annotate songs with a diverse set of tags based on audio content analysis. The SML model is parameterized by one Gaussian mixture model (GMM) distribution over an audio feature space for each tag in the vocabulary. The parameters for the set of GMMs are trained using annotated training data. Given a novel audio track, audio features are extracted and their likelihood is evaluated using each of the GMMs. The result is a vector of probabilities that, when normalized, can be interpreted as the parameters of a multinomial distribution over the tag vocabulary. This *semantic multinomial* distribution represents a compact and interpretable index for a song where the large parameter values correspond to the most likely tags.

Using 10-fold cross validation, we can estimate a semantic multinomial for each of the CAL500 songs. By stacking the 50 test set multinomials from each of the 10 folds, we can construct a strongly labeled tag-song matrix $\mathbf{X}^{\text{SML}}$ that is based purely on the audio content. As such, this tag-song matrix is dense and not affected by the cold-start problem.

8.4.5 Summary

Comparing systems using a two-tailed, paired t-test ($N = 109$, $\alpha = 0.05$) on the AUC metric, we find that all pairs of the four systems are significantly different, with the exception of game and Web documents.[12] If we compare the systems using the other two metrics (average precision and top 10 precision), we no longer find statistically significant differences. It is interesting that social tags and Web documents (0.37) have slightly better top 10 precision than autotags (0.33). This reflects the fact that for some of the more common individual tags, we find that social tags and Web documents have exceptional precision at low recall levels. For both Web documents and social tags, we find significant improvement in retrieval performance of short-head songs over long-tail songs. However, as expected, there is no difference for autotags. This confirms the intuition that systems based on Web documents and social tags are influenced by popularity bias, whereas content-based autotagging systems are not.

---

[12]Note that when we compare each system with the Game system, we compare both systems using the reduced set of 250 songs.

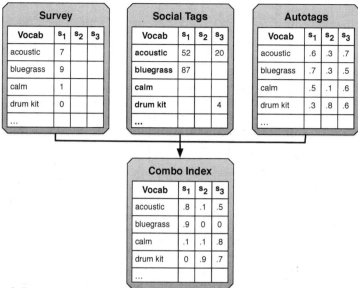

Figure 8.3
Creating a combined music index from multiple data sources.

8.5   Combining Sources of Music Information

Given the various strengths and weaknesses of each data source, it seems intuitive that we can benefit if we combine information from multiple data sources. To motivate this section, we will begin with the illustrative example shown in Figure 8.3.

Let's suppose that we are considering three data sources (Survey, Social Tags, Autotags), three songs (s_1, s_2, s_3), and four tags ("acoustic," "bluegrass," "calm," and "drum kit.") The first song is a favorite song by a well-known bluegrass band, the second song is a hard rock song by the local house band in your neighborhood pub, and the third is the first single from an up-and-coming folk-rock singer/songwriter. If we look at the tag-song matrix of the survey, we see that the song from the established bluegrass band has been annotated, but the other two songs have not yet been annotated and it is unlikely that the local band's song will ever be annotated.

It is a similar story for social tags information in that we can observe aspects of popularity bias and the cold-start problem. In addition, certain tags like "calm" and "drum kit" are rarely provided by the users in a social tagging context despite the fact that these tags might be useful for music retrieval. Autotags, on the other hand, provide us with a fully dense tag-song matrix. However, while auto tagging systems have been shown to produce decent annotations of music, their performance is not as accurate (or decisive) as human annotators especially for specific tags ("horns," "voice," "trumpet" [29]) that are poorly modeled using current state-of-the-art techniques.

In addition to data sparsity and accuracy problems, we also must consider how to *rescale* the values from each data source. For example, survey data often takes the form of rating data on an N-point scale (e.g., 10-point Likert scale), social tag data may be a raw count of the number of users who have applied a tag to a song, and an autotagger often outputs a probabilistic estimate for the predicted strength of semantic association. Two common rescaling techniques are *0-1 normalization* and *z-score standardization*. To 0-1 normalize the tag-song matrix X for a data source, we will find the minimum x_t^{min} and maximum x_t^{min} separately for each tag t. We then replace the value for $x_{t,s}$ with $x_{t,s}'$:

$$x_{t,s}' = \frac{x_{t,s} - x_t^{min}}{x_t^{max} - x_t^{min}} \tag{8.2}$$

so that each new value lies in the range $[0, 1]$. Z-score normalization is calculated by first finding the mean $\bar{x}_t$ and standard deviation s_t for each tag and then calculating the z-score:

$$x_{t,s}' = \frac{x_{t,s} - \bar{x}_t}{s_t}. \tag{8.3}$$

By definition, a z-score is the number of standard deviations a value is away from the mean. Technically, they range between $[-\infty, \infty]$ but in practice are almost always between $[-3, 3]$. While we have described rescaling approaches that require separate calculations for each tag, it may be reasonable to rescale the entire tag-song matrix without individually considering each tag.

Even after rescaling, combining scores from multiple data sources can be particularly challenging since our confidence in each score often depends not only on the data source, but also on the specific tag and song for each data source. For example, our autotagging algorithm may be particularly bad at predicting accurate scores for the tag "drum kit," or the social tags for the third song in Figure 8.3 may be inaccurate because only a small number of users have contributed tags to this song. This will require us to focus on the problems caused by sparse or inaccurate information.

In this section, we describe some of the algorithms that have be can used to combine multiple data sources. We roughly categorize these algorithms into ad-hoc approaches that are generally simple to implement, quick to compute, and tend to perform surprisingly well. The second kind of algorithm automatically *learns* how to combine data sources based on training data. These algorithms generally involve more computation but produce better performance results.

8.5.1   Ad-Hoc Combination Approaches

In this section, we describe two types of algorithms, Fixed Combination Rules and Rank-based Interleaving, that can be used to quickly combine multiple tag-song matrices. However, both require that we make various heuristic decisions that can have a significant impact on retrieval performance [38].

Fixed Combination Rules: A fixed combination rule produces an output score $x'_{t,s}$ as a simple function of the input scores $x^{DS}_{s,t}$'s where DS represents one of our data sources [38]:

$$x^{FCR}_{s,t} = f(x^{\text{SocialTag}}_{t,s}, x^{\text{Game}}_{t,s}, x^{\text{WebDocs}}_{t,s}, x^{\text{Autotag}}_{t,s}). \tag{8.4}$$

The most common functions are *max*, *min*, *median*, *sum*, and *product*. As discussed above, it is important to rescale raw tag-song matrices before applying a fixed combination function.

Rank-Based Interleaving (RBI): RBI is an algorithmic approach where, for each tag, we interleave the top-ranked songs according to each of the data sources [40]. The values in the combined tag-song matrix are the ranks of this combined rank ordering.

More formally, for a given tag, each data source produces a row in the tag-song matrix that defines a rank-ordering over our set of songs.[13] The RBI algorithm re-ranks songs by their best rank under any of the tag-song matrices. That is, for a single tag t and each data source DS, consider rank orderings, each denoted by $\mathbf{r}^{DS}_t$ and ordered by the values in row t of tag-song matrix $\mathbf{X}^{DS}$, such that

$$r^{DS}_{t,s} = \begin{cases} |S| - 1 & \text{if } s \text{ is the top ranked song} \\ |S| - 2 & \text{if } s \text{ is the 2nd ranked song} \\ \dots & \\ 0 & \text{if } s \text{ is the lowest ranked song} \\ & \text{or } x^{DS}_{t,s} = 0 \end{cases}$$

where $x^{DS}_{t,s} = 0$ implies that there is either no semantic association between tag t and song s or that the data is missing due to weak labeling. We then represent each song by its best rank according to each of the data sources:

$$x^{\text{RBI}}_{t,s} = \max(r^{\text{SocialTag}}_{t,s}, r^{\text{Game}}_{t,s}, r^{\text{WebDocs}}_{t,s}, r^{\text{Autotag}}_{t,s}) \tag{8.5}$$

for all songs s and all tags t. We note that, since RBI only uses the rank-orderings induced by song-tag matrices and not the actual values of these matrices, we need not rescale the song-tag matrices when using RBI. We also note that, as with fixed combination rules, we can imagine using other functions (e.g., average, median) when combining the ranks for our data sources.

The results of RBI on the experiment performed in Section 8.4.5 are reported in the last row of Table 8.3. We observe a significant increase in performance (AUC 0.74) over the best single approach (autotags with AUC = 0.69). This suggests that even a simple approach like RBI can be used to improve tag-based music retrieval.

---

[13] We will assume that we will randomly break ties if two song have the same score for a tag.

8.5.2   Learned Combination Approaches

A disadvantage of the ad-hoc approaches is that each data source is effectively given equal weight. That is, these approaches do not take advantage of the fact that one data source may be more or less reliable than the others. If we have annotated training data, we can *learn* which data sources to put more weight on and which data sources to put less weight on.

Regression: One approach is to formulate the data source combination problem as a *regression* problem. Like a fixed combination rule in Equation 8.4, we will combine the data sources using function of the input scores, but the parameters of the function will be determined using training data. We will briefly describe some common regression models but refer the reader to a basic textbook on statistics or machine learning for details on parameter estimation [7].

Linear regression involves learning scalar weights ($w's$) for a linear combination of the data source scores:

$$x_{t,s}^{LinR} = w_t^0 + \sum_{DS \in \{Sources\}} w_t^{DS} x_{t,s}^{DS} \qquad (8.6)$$

where w_t^0 is an (optional) offset value and DS represents one of our data sources (e.g., social tags, Web documents, autotags). This model is useful when we are trying to predict real-valued relevance scores.

If we would like to predict values in the range $[0, 1]$, a logistic regression model may be more appropriate. This is the case if, for example, $x_{t,s}$ is meant to represent the probability that tag t applies to song s. The functional form for logistic regression is:

$$x_{t,s}^{LogR} = \frac{1}{1 + \exp(-(w_t^0 + \sum_{DS \in \{Sources\}} w_t^{DS} x_{t,s}^{DS}))} \qquad (8.7)$$

where *exp* is the exponential function.

One potential drawback of the linear and logistic regression models described above is that we independently learn one function for each tag. One could imagine that there is useful information that can be shared *across* tags. For example, if Social Tags are good at predicting a large number of genre tags, we would expect $w_{t,s}^{SocialTag}$ to be large for many of these tags. In [38], we explore more complex *Bayesian hierarchical* regression models that simultaneously learn all regression models for all tags. However, our results show that the simple independent linear model performs as well as these more complicated models for the task of tag-based music indexing.

Calibrated Score Averaging (CSA): Using training data, we can learn a piecewise-constant function $g(\cdot)$ that *calibrates* scores such that $g(x_{t,s}) \approx P(t|x_{t,s})$. This allows us to compare data sources in terms of calibrated posterior probabilities rather than incomparable scores.

We use isotonic regression [35] to estimate a function g for each data source. More specifically, we use the pair-adjacent violators (PAV) algorithm [2, 13] to learn the isotonic (i.e., nondecreasing) function that produces the best fit in terms of minimum mean-squared error. To learn this function g for tag t, we start with a (low to high score) rank-ordered training set $s^{(1)}, s^{(2)}, ..., s^{(i)}, ..., s^{(|S|)}$ of $|S|$ songs where $x_{t,(i-1)} < x_{t,(i)}$. We initialize g to be equal to the sequence of binary training labels; i.e., $g(x_{t,s}) = 1$ if training song s is positively associated with tag t, and 0 otherwise. If the training data is perfectly ordered, then g is isotonic and we are done. Otherwise, there exists an i where there is a *pair-adjacent violation* such that $g(x_{t,(i-1)}) > g(x_{t,(i)})$. To remedy this violation, we update $g(x_{t,(i-1)})$ and $g(x_{t,(i)})$ so that they both become $[g(x_{t,(i-1)}) + g(x_{t,(i)})]/2$. We repeat this process until we have eliminated every pair-adjacent violation. At this point, g is isotonic and we combine it with the corresponding scores $x_{t,s}$ to produce a stepwise function that maps scores to approximate probabilities.

For example, if we have seven training songs with relevance scores equal to $(1, 2, 4, 5, 6, 7, 9)$ and ground truth labels equal to $(0, 1, 0, 1, 1, 0, 1)$, then $g(x) = 0$ for $x < 2$, $g(x) = 1/2$ for $2 \leq r < 6$, $g(x) = 2/3$ for $6 \leq r < 9$, and $g(x) = 1$ for $9 \leq r$. We can use Dümbgen's [13] linear-time $O(|S|)$ implementation of the PAV algorithm to compute g. For missing data from a weakly labeled data source, we use the training data to estimate a calibrated score when $x_{t,s} = 0$:

$$P(t|x_{t,s} = 0) = \frac{\#(\text{relevant songs with } x_{t,s} = 0)}{\#(\text{songs with } x_{t,s} = 0)}. \tag{8.8}$$

Once we have learned a calibration function for each data source, we convert raw scores for a novel song to approximate posterior probabilities. We can combine these posterior probabilities by using a fixed combination rule described in section 8.5.1. Of all these combination rules, we find that the arithmetic average produces the best empirical tag-based retrieval results.

RankBoost: RankBoost is a boosting algorithm that is designed to combine multiple rank-orderings of data [16]. The algorithm produces a *strong* ranking function H that is a weighted combination of *weak* ranking functions h_t. Each weak ranking function is defined by a data source, a threshold, and a default value for missing data. For a given song, the weak ranking function is an indicator function that outputs 1 if the score for the associated data source is greater than the threshold or if the score is missing and the default value is set to 1. Otherwise, it outputs 0. During training, RankBoost iteratively builds an ensemble of weak ranking functions and associated weights. At each iteration, the algorithm selects the weak learner (and associated weight) that maximally reduces the *rank loss* of a training data set given the current ensemble. For implementation details, please refer to Freund et al. [16].

Kernel Combination SVM: A support vector machine (SVM) is a popular binary classifier that learns the parameters of a separating hyperplane (i.e., a decision boundary) between, in our case, a set for relevant training songs and set of irrelevant training songs for a given tag. That is, a SVM learns a linear decision function that defines the distance of a new song, s, from the hyperplane boundary between the relevant and irrelevant training example songs:

$$x_{t,s}^{SVM} = \sum_{i=1}^{|S|} \alpha_i K(s, i) + b, \qquad (8.9)$$

where b is the offset of the decision boundary and a_i is a learned weight for each training example. In practice, a_i is nonzero for only a small number of training songs (i.e., the support vectors). The function $K(\cdot, \cdot)$ is a *kernel* function that measures the similarity between pairs of songs.

For our training data, we can compute an $|S| \times |S|$-dimensional kernel matrix $\mathbf{K}^{DS}$ matrix for each data source. For example, for social-tag data source, we might compute a radial basis function (RBF) kernel with entries:

$$[\mathbf{K}^{\text{SocialTag}}]_{i,j} = K^{\text{SocialTag}}(i, j) = \exp(-\frac{\|\mathbf{x}_i - \mathbf{x}_j\|^2}{2\sigma^2}), \qquad (8.10)$$

where $K(i, j)$ represents the similarity between $\mathbf{x}_i$ and $\mathbf{x}_j$, the column vectors of tag-song matrix $\mathbf{X}^{\text{SocialTag}}$ corresponding to songs i and j. The hyperparameter σ is estimated using cross validation.

To combine data sources, we can compute a single kernel matrix $\mathbf{K}$ that is a linear combination of the kernel matrices for each of the individual data sources:

$$\mathbf{K} = \sum_{DS \in \{sources\}} \mu^{DS} \mathbf{K}^{DS}, \quad \text{where } \mu^{DS} > 0 \qquad (8.11)$$

where the μ^{DS}'s are learned weights. Both the α_i's in Equation 8.9 and the μ^{DS}'s in Equation 8.11 can be efficiently learned together using convex optimization (see Barrington et al. [5] and Lanckriet et al. [23] for details).

Approach	AUC	MAP
Social Tags	0.623	0.431
Web Documents	0.625	0.413
Autotags	0.731	0.473
Kernel Combo (KC)	0.756	0.529
RankBoost (RB)	0.760	0.531
Calib. Score Avg. (CSA)	**0.763**	**0.538**

Table 8.4
Evaluation of Combination Approaches (The performance differences between single source and multiple source algorithms are significant [one-tailed, paired t-test over the vocabulary with $\alpha = 0.05$].)

8.5.3 Comparison

In this section, we compare three individual data sources with three learned combination approaches for the task of tag-based music retrieval. As in Section 8.4, we experiment on the CAL500 data set but use a smaller subset of 72 tags by requiring that each tag be associated with at least 20 songs and removing some tags that we deemed to be redundant or overly subjective. We evaluate the rankings using two metrics, mean AUC and mean average precision (AP), which are computed using 10-fold cross validation.

As shown in Table 8.4, we observe that all three of our combination approaches produce significantly improved tag-based music retrieval results. We also produce qualitative search results in Table 8.5 to provide context for our

Synthesized Song Texture	Acoustic Song Texture
0.80 / 0.71	0.73 / 0.76
Tricky—*Christiansands* (m)	Robert Johnson—*Sweet Home Chicago*
Propellerheads—*Take California*	Neil Young—*Western Hero*
Aphex Twin—*Come to Daddy*	Cat Power—*He War* (m)
New Order—*Blue Monday*	John Lennon—*Imagine*
Massive Attack—*Risingson*	Ani DiFranco—*Crime for Crime*
Female Vocals	**Male Vocals**
0.95 / 0.90	0.71 / 0.82
Billie Holiday—*God Bless the Child*	The Who—*Bargain*
Andrews Sisters—*Boogie Woogie Bugle Boy*	Bush—*Comedown*
Alanis Morissette—*Thank U*	AC/DC—*Dirty Deeds Done Dirt Cheap*
Shakira—*The One*	Bobby Brown—*My Prerogative*
Alicia Keys—*Fallin'*	Nine Inch Nails—*Head Like a Hole*
Jazz	**Blues**
0.96 / 0.82	0.84 / 0.45
Billie Holiday—*God Bless the Child*	B.B. King—*Sweet Little Angel*
Thelonious Monk—*Epistrophy*	Canned Heat—*On the Road Again* (m)
Lambert, Hendricks & Ross—*Gimme That Wine*	Cream—*Tales of Brave Ulysses* (m)
Stan Getz—*Corcovado*	Muddy Waters—*Mannish Boy*
Norah Jones—*Don't Know Why*	Chuck Berry—*Roll Over Beethoven* (m)
Calming	**Aggressive**
0.81 / 0.66	0.84 / 0.51
Crosby, Stills & Nash—*Guinnevere*	Pizzle—*What's Wrong with My Foot?*
Carpenters—*Rainy Days and Mondays*	Rage Against the Machine—*Maggie's Farm*
Cowboy Junkies—*Postcard Blues*	Aphex Twin—*Come to Daddy*
Tim Hardin—*Don't Make Promises*	Black Flag—*Six Pack*
Norah Jones—*Don't Know Why*	Nine Inch Nails—*Head Like a Hole*

Table 8.5

Tag-Based Music Retrieval Examples for Calibrated Score Averaging (CSA) (The top ranked songs for each of the first five folds [during 10-fold cross validation] for eight representative tags. In each box, the tag is listed first [in bold]. The second row is the area under the ROC curve (AUC) and the average precision [AP] for the tag [averaged over 10-fold cross validation]. Each artist-song pair is the top ranked song for the tag and is followed by "[m]" if it is considered misclassified, according to the ground truth. Note that some of the misclassified songs may actually be representative of the tag [e.g., see "Blues"].)

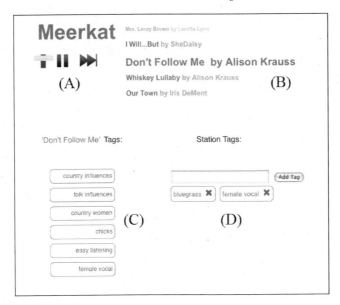

Figure 8.4
Screenshot of Meerkat: (A) The volume control, play/pause toggle, and fast-forward button allow the user to control the music playback. (B) Past songs (in a gray font) are displayed above the current song (in a red font) while future songs (also in a gray font) are below. (C) A list of tags associated with the current song provides context for the music. A user can click on a tag if he or she wishes to add it to the radio station. (D) The user selects tags to control the radio station. A user can also add tags using the text box.

evaluation metrics. The best performance is achieved using CSA, though the performance is not significantly better than RankBoost or KC-SVM.

8.6   Meerkat: A Semantic Music Discovery Engine

We conclude this chapter with the description of a research-based semantic music discovery engine called *Meerkat* [33]. Meerkat is a personalized Internet radio player that functions much like Pandora[14] and Slacker.[15] However, Meerkat has been explicitly designed to use semantic tags, rather than artist or song similarity, as the primary mechanism for generating personalized playlists. That is, a user might start a playlist by asking for "bluegrass"

---

[14]http://www.pandora.com.
[15]http://www.slacker.com.

music. Bluegrass music starts playing and the user has the opportunity to control the music by adding additional tags like "female vocals." (See Figure 8.4 for a screenshot of the user interface.)

Meerkat's backend uses four sources of music information (Web documents, social tags, autotags, and preference information) to index music with a large set of semantic tags [38]. These data sources are quickly combined using linear regression (see Equation 8.6.) The current music corpus consists of 10,870 songs that are representative of 18 genres (e.g., "rock," "electronic," "classical") and 180 subgenres (e.g., "grunge," "trance," "romantic period opera") of music [37]. Our vocabulary of tags consists of hundreds of genres and subgenres, dozens of emotions ("happy"), dozens of instruments ("distorted electric guitar"), hundreds of acoustic characteristics ("dominant bass riff"), and thousands of free-text social tags (from Last.fm). Based on the user-specified list of tags, the backend returns a ranked list (i.e., a playlist) of semantically relevant songs which have been ordered according to cosine similarity (see Equation 8.1).

Last.fm's MultiTag Radio[16] and the AMG Tapestry Demo[17] are two examples of commercial systems that enable tag-based playlist generation. However, both systems only allow users to seed a station with tags whereas Meerkat allows for the dynamic modification of a station as a user adds or removes tags over time. Also, each commercial system relies only on one data source to power their music discovery engines.[18]

Glossary

AP: Average precision. A useful IR evaluation metric that is found by moving down our ranked list of songs and averaging the precisions at every point where we correctly identify a relevant song.

AUC: Area under the Receiver Operating Character curve. A useful IR evaluation metric that we can use to evaluate the quality of rank-ordered list of relevant and irrelevant *documents*. AUC has an expected value of 0.5 when we randomly rank documents and 1.0 when we have a perfect ranking of documents (i.e., all relevant documents ranked before all irrelevant documents).

cold-start problem: A problem when a new (or less popular) *document* (e.g., a song) cannot be retrieved because it has not been indexed. This problem is often related to the *popularity bias*.

---

[16]http://www.last.fm/listen/.

[17]http://www.amgtapestry.com/radio/.

[18]Last.fm uses social tags and AMG relies on expert human surveys when calculating tag-based music relevance.

corpus: A set of documents (e.g., songs).

crowd-sourcing: Engaging a large group of individuals to solve a big problem by breaking it down into smaller, more manageable tasks.

document: A generic data item like a Web page, song, or artist.

information retrieval: A field of study that focuses on techniques for efficiently accessing data.

long tail: The long-tail/short-head metaphor refers to a plot of the rank of a document based on its popularity versus the popularity of the document. This curve tends to be shaped like a power law probability distribution (i.e., $y = 1/x$) where most of the mass is attributed to a small number of popular documents (the short-head) and the remaining mass is distributed over a large number of unpopular songs (the long-tail) [1, 22].

popularity bias: The idea that more popular documents (e.g., "Hey Jude" by The Beatles) receive more attention and thus will likely be more thoroughly annotated in a crowd sourcing environment.

precision: Percentage of retrieved documents that are relevant.

recall: Percentage of relevant documents that are retrieved.

semantic music discovery engine: A computer-based tool that helps people find music through the use of *tags* as well as other forms of music information.

short head: See definition for *long tail*.

social tag: A *tag* that is applied to a *document* in social networking framework (e.g., Last.fm, Digg, Flicker).

strongly labeled data: Data where every document has a known positive or negative association with each tag.

subjectivity problem: When two or more annotators disagree whether (or to what degree) a term applies to a document.

tag: A short text-based token like "bluegrass" or "distorted electric guitar."

term: A label that can be used to describe (i.e., index) a document. For music indexing, a term is often referred to as a *tag*.

weakly labeled data: Data where the absence of an association between a document and a tag does not necessarily mean that they are not related to one another.

vocabulary: A set of tags or terms.

Acknowledgments

Parts of this book chapter also appear as parts of various conference papers. The material found in Sections 8.3 and 8.4 was originally developed in *Five Approaches to Collecting Tags for Music* [40] with my coauthors Luke Barrington and Gert Lanckriet. They, as well as Merhdad Yazdani, significantly contributed to *Combining Audio Content and Social Context for Semantic Music Discovery* [42], much of which is described in Section 8.4. The material on regression in Section 8.4 is summarized from my work with Brian Tomasik in *Using Regression to Combine Data Sources for Semantic Music Discovery*. Finally, the Meerkat Music Discovery Engine that is described in Section 8.6 was designed and developed in large part by Ashley Oudenne, Damien O'Malley, Youngmoo Kim, Brian Tomasik, Douglas Woos, Derek Tingle, Joon Hee Kim, Malcolm Augat, Margaret Ladlow, and Richard Wicentowski. A user study that evaluates Meerkat is provided in *Meerkat: Exploring Semantic Music Discovery Using Personalized Radio* [33].

Bibliography

[1] C. Anderson. *The Long Tail: Why the Future of Business Is Selling Less of More.* Hyperion, New York, 2006.

[2] M. Ayer, H.D. Brunk, G.M. Ewing, W.T. Reid, and E. Silverman. An empirical distribution function for sampling with incomplete information. *Annals of Mathematical Statistics*, 1955.

[3] R. Baeza-Yates and B. Ribeiro-Neto. *Modern Information Retrieval.* Addison-Wesley, Upper Saddle River, New Jersey, 2008.

[4] L. Barrington, D. Turnbull, D. O'Malley, and G. Lanckriet. User-centered design of a social game to tag music. *ACM KDD Workshop on Human Computation*, 2009.

[5] L. Barrington, M. Yazdani, D. Turnbull, and G. Lanckriet. Combining feature kernels for semantic music retrieval. *ISMIR*, 2008.

[6] T. Bertin-Mahieux, D. Eck, and M. Mandel. Automatic tagging of audio: The state-of-the-art. *Machine Audition: Principles, Algorithms and Systems*, 2010.

[7] C. Bishop. *Pattern Recognition and Machine Learning.* Springer, New York, 2007.

[8] G. Carneiro, A. B. Chan, P.J. Moreno, and N. Vasconcelos. Supervised learning of semantic classes for image annotation and retrieval. *IEEE PAMI*, 29(3):394–410, 2007.

[9] O. Celma, P. Cano, and P. Herrera. Search sounds: An audio crawler focused on Weblogs. *ISMIR*, 2006.

[10] S.F. Chang, D. Ellis, W. Jiang, K. Lee, A. Yanagawa, A. Loui, and J. Luo. Large-scale multimodal semantic concept detection for consumer video. *ACM Multimedia Workshop on Multimedia Information Retrieval*, 2007.

[11] S. Clifford. Pandora's long strange trip. *Inc.com*, 2007.

[12] J.S. Downie. Music information retrieval evaluation exchange (MIREX), 2005.

[13] L. Dümbgen. Isotonic regression software (MATLAB), http://staff.unibe.ch/duembgen/software/#Isotone.

[14] D. Eck, P. Lamere, T. Bertin-Mahieux, and S. Green. Automatic generation of social tags for music recommendation. *Neural Information Processing Systems Conference (NIPS)*, 2007.

[15] S. Essid, G. Richard, and B. David. Inferring efficient hierarchical taxonomies for music information retrieval tasks: Application to musical instruments. *ISMIR*, 2005.

[16] Y. Freund, R. Iyer, R. Schapire, and Y. Singer. An efficient boosting algorithm for combining preferences. *JMLR*, 4:933–969, 2003.

[17] W. Glaser, T. Westergren, J. Stearns, and J. Kraft. Consumer item matching method and system. U.S. Patent Number 7003515, 2006.

[18] J.H. Kim, B. Tomasik, and D. Turnbull. Using artist similarity to propagate semantic information. *ISMIR*, 2009.

[19] P. Knees, T. Pohle, M. Schedl, D. Schnitzer, and K. Seyerlehner. A document-centered approach to a natural language music search engine. *ECIR*, 2008.

[20] P. Knees, T. Pohle, M. Schedl, and G. Widmer. A music search engine built upon audio-based and Web-based similarity measures. *ACM SIGIR*, 2007.

[21] P. Lamere. Social tagging and music information retrieval. *JNMR*, 2008.

[22] P. Lamere and O. Celma. Music recommendation tutorial notes. *ISMIR Tutorial*, September 2007.

[23] G. Lanckriet, N. Cristianini, P. Bartlett, L. El Ghaoui, and M. Jordan. Learning the kernel matrix with semidefinite programming. *JMLR*, 5:27–72, 2004.

[24] E. Law, L. von Ahn, and R. Dannenberg. Tagatune: A game for music and sound annotation. *ISMIR*, 2007.

[25] M. Levy and M. Sandler. A semantic space for music derived from social tags. *ISMIR*, 2007.

[26] T. Li and G. Tzanetakis. Factors in automatic musical genre classification of audio signals. *IEEE WASPAA*, 2003.

[27] J. Lībeks and D. Turnbull. Exploring "artist image" using content-based analysis of promotional photos. *International Computer Music Conference*, 2010.

[28] M. Mandel and D. Ellis. A Web-based game for collecting music metadata. *ISMIR*, 2007.

[29] M. Mandel and D. Ellis. Multiple-instance learning for music information retrieval. *ISMIR*, 2008.

[30] C.D. Manning, P. Raghavan, and H. Schtze. *Introduction to Information Retrieval*. Cambridge University Press, Cambridge, Massachusetts, 2008.

[31] C. McKay and I. Fujinaga. Musical genre classification: Is it worth pursuing and how can it be improved? *ISMIR*, 2006.

[32] F. Miller, M. Stiksel, and R. Jones. Last.fm in numbers. *Last.fm press material*, February 2008.

[33] A. Oudenne, Y. Kim, and D. Turnbull. Meerkat: Exploring semantic music discovery using personalized radio. *ACM International Conference on Multimedia Information Retrieval*, 2010.

[34] J. Reed and C. H. Lee. A study on attribute-based taxonomy for music information retrieval. *ISMIR*, 2007.

[35] T. Robertson, F. Wright, and R. Dykstra. *Order Restricted Statistical Inference*. Wiley and Sons, Hoboken, New Jersey, 1988.

[36] M. Sordo, C. Lauier, and O. Celma. Annotating music collections: How content-based similarity helps to propagate labels. *ISMIR*, 2007.

[37] D. Tingle, Y. Kim, and D.Turnbull. Exploring automatic music annotation with "acoustically objective" tags. *ACM International Conference on Multimedia Information Retrieval*, 2010.

[38] B. Tomasik, J. Kim, M. Ladlow, M. Augat, D. Tingle, R. Wicentowski, and D. Turnbull. Using regression to combine data sources for semantic music discovery. *ISMIR*, 2009.

[39] D. Turnbull, L. Barrington, and G. Lanckriet. Modelling music and words using a multi-class naïve bayes approach. *ISMIR*, pages 254–259, 2006.

[40] D. Turnbull, L. Barrington, and G. Lanckriet. Five approaches to collecting tags for music. *ISMIR*, 2008.

[41] D. Turnbull, L. Barrington, D. Torres, and G. Lanckriet. Semantic annotation and retrieval of music and sound effects. *IEEE TASLP*, 16(2), 2008.

[42] D. Turnbull, L. Barrington, M. Yazdani, and G. Lanckriet. Combining audio content and social context for semantic music discovery. In *ACM SIGIR*, 2009.

[43] D. Turnbull, R. Liu, L. Barrington, and G. Lanckriet. Using games to collect semantic information about music. *ISMIR '07*, 2007.

[44] G. Tzanetakis and P.R. Cook. Musical genre classification of audio signals. *IEEE Transaction on Speech and Audio Processing*, 10(5):293–302, 2002.

[45] L. von Ahn and L. Dabbish. Labeling images with a computer game. *ACM CHI*, 2004.

[46] T. Westergren. Personal notes from Pandora get-together in San Diego, California, March 2007.

[47] B. Whitman and D. Ellis. Automatic record reviews. *ISMIR*, pages 470–477, 2004.

[48] B. Whitman and S. Lawrence. Inferring descriptions and similarity for music from community metadata. *ICMC*, 2002.

9

Human Computation for Music Classification

Edith Law

Carnegie Mellon University

CONTENTS

9.1 Introduction

One of the key challenges in music information retrieval is the need to quickly and accurately index the ever growing collection of music on the Web. There has been an influx of recent research on machine learning methods for automatically classifying music by semantic tags, including Support Vector Machines [12, 13], Gaussian Mixture Models [18, 19], Boosting [2], Logistic Regression [1], and other probabilistic models [6]. The majority of these methods are supervised learning methods, requiring a large amount of labeled music as training data, which has traditionally been difficult and costly to obtain. Today, the shortage of labeled music data is no longer a problem. There is now a proliferation of online music Web sites that millions of users visit daily, generating an unprecedented amount of useful information about each piece of music. For example, Last.fm, a collaborative tagging Web site, collect on the order of 2 million tags per month [8]. Without prompting, human users are performing meaningful computation each day, mapping music to tags.

Collaborative tagging Web sites operate on the premise of an open call—anyone can freely label any pieces of music. In contrast, there are computational systems that exert finer control over how computation (in this case, the mapping from music to tags) is carried out, in order to optimize the efficiency and accuracy of the computation. These so-called *human computation systems* take many forms. For example, Amazon Mechanical Turk is a human computation market where workers perform explicit tasks (e.g., annotating music) in return for small monetary payments. In this chapter, we focus on the use of human computation games, or *games with a purpose* (GWAP), to elicit participation from humans to perform computation.

Human computation games are multiplayer online games with an underlying mechanism (a set of rules) that incentivize players to volunteer data in a truthful manner. These games typically enjoy high throughput—by designing a game that is entertaining, players are willing to spend a huge amount of time playing the game, contributing massive amounts of data as a by-product. Conceptualized in 2003 [21], human computation games are designed to tackle difficult artificial intelligence (AI) problems that humans find easy but computers still cannot solve. The very first human computation game—the ESP Game [21]—was designed to tackle the image labeling problem. In this game, two players are asked to provide tags for the same image. If any of their tags match, players are rewarded and that tag becomes a label for that image. The ESP Game operates under the *output-agreement* mechanism in which players are rewarded for agreeing with each other. Since players do not have any knowledge of their partner, the best strategy is to think of tags that are likely to be entered by *any* person, essentially motivating players to enter tags that *do* describe the image. The ESP Game is hugely successful: millions of image tags have been collected, and ultimately used to power image search on the Web [4]. Since its deployment, many human computation games have been built to tackle other AI problems, including image ranking, music classification, semantic role analysis, translation, Web search, knowledge extraction, and so forth.

Despite being powerful applications capable of engaging millions of workers to perform useful computation, the effectiveness (i.e., in terms of efficiency and accuracy) of a human computation system depends critically on how well it meets two objectives—the objectives of the human worker (i.e., to earn money, to be entertained, and so on) and the computation objectives. In the case of music tagging, one of the computation objectives is to collect labeled data that is useful towards training supervised music tagging algorithms. Specifically, the ideal data set should have a set of *distinct* labels that are each associated with *roughly equal* and a *large number* of examples. Game players, on the other hand, are mainly interested in playing a *fun* game that is easy to understand, and neither too trivial nor too difficult. To be successful, a human computation game designed to collect tags for music must strike a balance between these two objectives—if players are unmotivated or have difficulty playing the game, then computation would be inefficient, requiring a long time to complete. On the other hand, designing the game around the needs of the players can compromise data quality—the resulting data set can be extremely noisy or unbalanced, making it necessary to invent new learning and evaluation methods before one can make full use of the collected data.

TagATune is a prime example of a human computation game with such design tradeoffs. In this chapter, we will describe the rationale behind the design of TagATune, the positive and negative consequences of this design, and techniques for mitigating each of the challenges.

9.2 TagATune: A Music Tagging Game

Our first attempt at building a music tagging game was a straightforward adaptation of the ESP Game for music. In the prototype game [11], two players are given the same music clip and asked to enter tags to describe the music. Similar to the ESP Game, players are rewarded if their tags match. From the pilot study, it was quickly realized that players had difficulty matching on a tag—36% of the time players opted to pass after failing a few unsuccessful attempts. To mitigate this problem, an additional design choice in the prototype game was to specify the *kinds* of tags (e.g., genre, mood, speed) we want the players to enter during a particular round, in the hopes of improving the chances of agreement. This design did not fare well—even under constraints, there is enough variation in the way music is described that matching remains difficult. For example, there are many ways to express the same mood of a piece of music (e.g., soothing, calm, calming, serene, meditative, peaceful, quiet, soft, relaxing, tranquil). For audio clips, this problem is even more severe, as players try to express what they hear in *phrases*, e.g., "cars on a street" versus "traffic." In their attempt to match, players chose tags that are common and general (e.g., "music"). This problem of tags being *uninformative* is characteristic of games that use the output agreement mechanism [16], and has inspired research on new game mechanisms [5] that can generate more specific tags.

9.2.1 Input-Agreement Mechanism

There were several music-related human computation games built using the output-agreement mechanism, including MajorMiner [14], the Listen Game [20], Herd It, and MoodSwings [7]. While distinct in the details of their designs, all of these games use agreement as an indicator of output accuracy. Among these games, the ones for music tagging address the problem of matching by having players match against a database of previously collected tags [14], or

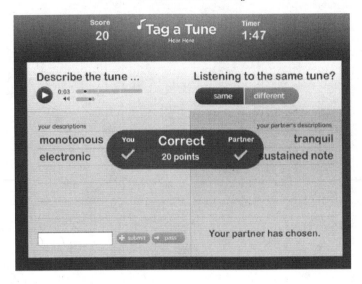

Figure 9.1
A TagATune screenshot.

by having players select from a small set of preselected tags [20]. There are disadvantages to such design approaches: having people play by themselves eliminates the social aspects of online game, and limiting players to predefined set of tags may make the game significantly less fun and useful.

The matching problem implies that a game mechanism for tagging music must make use of other means of verifying the correctness (or trustworthiness) of players' output. In the final design of TagATune [10], we took a completely different design approach—instead of requiring two players to match, the game gives each of the players a music clip and asked them to exchange tags with each other, then decide whether the music clips are the same or different. Figure 9.1 shows the interface of TagATune.

TagATune leverages the fact that players must be truthful to each other in order to successfully guess whether the music clips are the same or different. The game belongs to a class of mechanisms called *function computation* mechanism, in which players are motivated to exchange their private information (e.g., tags) with each other in order to jointly compute a function (e.g., whether the two pieces of music are the same or different). The TagATune mechanism is a specific instance of the *function computation* mechanism called *input-agreement* [10], where the function to compute is "1 if the inputs (i.e., music clips) are the same, 0 otherwise." The input-agreement mechanism can be generalized to handle other types of input data (e.g., images, video, or text) that has high description entropy.

The open communication protocol introduced by the input-agreement mechanism has two natural consequences. By allowing players to freely communicate using an *open vocabulary* that poses no restriction on what tags can be entered, the resulting data set is noisy—there is a huge number of imbalanced classes, each corresponding to a tag that can be misspelled, redundant (i.e., synonymous), or irrelevant to content. On the other hand, because players are allowed to see each other's outputs, TagATune can also serve as a platform for collecting human evaluations of machine-generated music tags, simply by having music tagging algorithms (also referred to as *music taggers*) pose as human players. In the following two sections, we will explore these two aspects of TagATune.

9.2.2 Fun Game, Noisy Data

As reported by Law and von Ahnl [10], over a seven-month period, there were a total of 49,088 games played by more than 14,224 unique players. The number of games each person played ranged from 1 to 6,286, or equivalently, between 3 minutes and 420 *hours* of game play. There were 512,770 tags collected, of which 108,558 were verified by at least two players, and 70,908 were unique. The average number of tags generated by the game per minute is four. The final design of TagATune was a huge improvement over the prototype version—the game allows players to openly communicate with each other, without requiring them to match. As evidence of lowered player frustration, players only passed on 0.50% of the total number of rounds, as opposed to the 36% pass rate in the output-agreement based prototype game.

In a Mechanical Turk evaluation of the tags collected by TagATune for 20 randomly selected songs, it was found that on average five to six out of seven tags are deemed to be accurate descriptions of the music clip [10]. The small error (the one to two out of seven tags that do not describe the music) can be attributed to the existence of easily filterable junk words that we decided to present to the judges in the experiment (such as "same," "diff," and so on). As a human computation game, TagATune enjoyed great success in collecting accurate music tags quickly. The collected tags were made into one of the largest tagged music data sets (called *magnatagatune*) made publicly available to the research community.

Table 9.1 shows examples of different types of tags collected using the TagATune game. The *Common Tags*, sampled from the 50 most frequently used tags, typically describe the genre, instrumentation, speed, and mood of the music. Among the most frequent tags are *Negation Tags*, which are tags that describe what is *not* in the music, for example, "not classical" or "no piano." Negative tags are natural consequences of the open communication protocol, and are unique to the input-agreement mechanism. For example, when a player enters "singing," the partner might respond with "no singing" in order to assert his beliefs that the music clips are different.

Found among the least frequently used tags are *compound tags, transcription tags, misspelled tags,* and *communication tags*. Compound tags are descriptive and accurate tags that consist of two or more words, for example,

Common Tags	Negation Tags
rock	no piano
guitar	no guitar
singing	no lyrics
soft	not classical
light violin	not English
Compound Tags	**Transcription Tags**
epileptic seizure music	fill me up ...
big guitar riff intro	rain on my parade
burly man song	fa la la la
quiet string instrument	the highest of sunny days
ballet extravaganza	you can run, you can fall
Misspelled Tags	**Communication Tags**
dreums	nice to meet you
sofr	you're alright at this :)
wman vocal	I love this song
violene	gonna say diff
orcherstras	more info please

Table 9.1
Characteristics of the Tags Collected from the TagATune Game

"helicopter sound," "Halloween," "cookie monster vocals," "wedding reception music." Transcription tags are phrases that contain lyrics from the song. Both types of tags are useful for describing and retrieving music, but difficult to process into well-defined labels for training music tagging algorithms. Misspelled tags and communication tags are found among the least frequent tags, but sometimes also among the most frequent. For example, "same," "diff," "yes," and "no" are often used by players to communicate their final decision of whether the music clips are the same or not.

9.2.3   A Platform for Collecting Human Evaluation

While open communication leads to more noisy data, it also has a positive effect—allowing players to observe each other's tags enables TagATune to function as a new platform for evaluating music tagging algorithms. The key idea is to have human players play the game against a music tagger, instead of another human player. The extent to which players can make correct guesses using the tags generated by the music tagger becomes an implicit measure of the algorithm's performance.

This new method of soliciting human evaluation addresses some of the problems associated with current methods of evaluation. The conventional way to determine whether an algorithm is producing accurate music tags is to compute the level of agreement between the output generated by the algorithm and the ground truth set. Agreement-based metrics, for example, accuracy, precision, F-measure and ROC curve, have been long-time workhorses of evaluation, accelerating the development of new algorithms by providing an automated way to gauge performance. The most serious drawback to using agreement-based metrics is that ground-truth sets are never fully comprehensive. First, there are exponentially many sets of suitable tags for a piece of music—creating all possible sets of tags and then choosing the best set of tags as the ground truth is difficult, if not impossible. Second, tags that are appropriate for a given piece of music can simply be missing in the ground truth set because they are less salient, or worded differently (e.g., baroque versus 17th century classical). Furthermore, because an exhaustive set of negative tags is impossible to specify, when a tag is missing, it is impossible to know whether it is in fact inappropriate for a particular piece of music.

Agreement-based metrics also impose restrictions on the type of algorithms that can be evaluated. To be evaluated, tags generated by the algorithms must belong to the ground truth set. This means that audio tagging algorithms that are not trained on the ground truth set, for example, those that use text corpora or knowledge bases to generate tags, cannot be evaluated using agreement-based metrics. Finally, to be useful, tags generated by audio tagging algorithms must, from the perspective of the end user, accurately describe the music. However, because we do not yet fully understand the cognitive processes underlying the representation and categorization of music, it is often difficult to know what makes a tag "accurate" and what kinds of inaccuracies are tolerable. For example, it may be less disconcerting for users to receive a folk song when a country song is sought, than to receive a sad, mellow song when a happy, up-beat song is sought. Ideally, an evaluation metric should measure the quality of the algorithm by implicitly or explicitly capturing the users' differential tolerance of incorrect tags generated by the algorithms.

9.2.3.1 The TagATune Metric

The TagATune metric for measuring the performance of tagging algorithms is simple. Suppose a set of algorithms $\mathcal{A} = \{a_i, \ldots, a_{|\mathcal{A}|}\}$ and a test set $\mathcal{S} = \{s_j, \ldots, s_{|\mathcal{S}|}\}$ of music clips. During each round of the game, a particular algorithm i is given a clip j from the test set and asked to generate a set of tags for that clip. To be a valid evaluation, we only use rounds where the clips given to the human player and the algorithm bot are the same. This is because if the clips are different, an algorithm can output the wrong tags for a clip and actually *help* the players guess correctly that the clips are different.

A human player's guess is denoted as $G = \{0, 1\}$ and the ground truth is denoted as $GT = \{0, 1\}$, where 0 means that the clips are the same and 1 means that the clips are different. The performance P of an algorithm i on clip j under TagATune metric is as follows:

$$P_{i,j} = \frac{1}{N} \sum_{n}^{N} \delta(G_{n,j} = GT_j) \tag{9.1}$$

where N represents the number of players who were presented with the tags generated by algorithm i on clip j, and $\delta(G_{n,j} = GT_j)$ is a Kronecker delta function which returns 1 if, for clip j, the guess from player n and the ground truth are the same, 0 otherwise. The overall score for an algorithm is averaged over the test set S:

$$P_i = \frac{1}{S} \sum_{j}^{S} P_{i,j} \tag{9.2}$$

9.2.3.2 MIREX Special TagATune Evaluation

To test out the feasibility of TagATune as an evaluation platform, we organized a "Special TagATune Evaluation" benchmarking competition as an off-season Music Information Retrieval Evaluation Exchange (MIREX) task. Participating algorithms were asked to provide two different types of outputs:

1. A binary decision on whether each tag is relevant to each clip.

2. A real-valued estimate of the "affinity" of the clip for each tag. A larger affinity score means that a tag is more likely to be applicable to the clip.

There were five submissions to the competition, which we will refer to as Mandel, Manzagol, Marsyas, Zhi, and LabX[1] from this point on. A sixth algorithm we are using for comparison is called AggregateBot, which serves tags from a vocabulary pool of 146 tags collected by TagATune since deployment, 91 of which overlap with the 160 tags used for training the algorithms. The AggregateBot essentially mimics the aggregate behavior of human players from previous games. The inclusion of AggregateBot demonstrates the utility of TagATune in evaluating algorithms that have different tag vocabulary.

We trained the participating algorithms on a subset of the TagATune data set. The training and test sets comprise 16,289 and 100 music clips respectively. The test set was limited to 100 clips for both the human evaluation using TagATune and evaluation using the conventional agreement-based metrics, in order to facilitate direct comparisons of their results. Each clip is 29 seconds long, and the set of clips are associated with 6,622 tracks, 517 albums, and 270 artists. The data set is split such that the clips in the training and test sets do not belong to the same artists. Genres include

---

[1] The LabX submission was identified as having a bug which negatively impacted its performance, hence, the name of the participating laboratory has been obfuscated. Since LabX essentially behaves like an algorithm that randomly assigns tags, its performance establishes a lower bound for the TagATune metric.

Algorithm	TagATune
AggregateBot	**93.00%**
Mandel	**70.10%**
Marsyas	68.60%
Manzagol	67.50%
Zhi	60.90%
LabX	26.80%

Table 9.2
Evaluation Statistics under the TagATune versus Agreement-Based Metrics

Classical, New Age, Electronica, Rock, Pop, World, Jazz, Blues, Metal, Punk, and so forth. The tags used in the experiments are each associated with more than 50 clips, where each clip is associated only with tags that have been verified by more than two players independently.

Algorithm Ranking

While there are many benchmarking competitions for algorithms, little is said about the level of performance that is acceptable for real-world applications. In our case, the performance of the AggregateBot serves as an upper ceiling for performance—if the algorithms can achieve the same level of performance as the AggregateBot under the TagATune metric, they are essentially achieving *human level performance* at playing this game. In our experiment, it was shown (Table 9.2) that human players can correctly guess that the music are the same 93% of the times when paired against the AggregateBot, while only approximately 70% of the times when paired against an algorithm. In other words, human performance (i.e., AggregateBot) is significantly better than the performance of all the music tagging algorithms used in the competition.

Game Statistics

In a TagATune round, the game selects a clip from the test set and serves the tags generated by a particular algorithm for that clip. For each of the 100 clips in the test set and for each algorithm, 10 unique players were elicited (unknowingly) by the game to provide evaluation judgments. This totals to 5,000 judgments, collected over a one-month period, involving approximately 2,272 games and 657 unique players.

How many tags were actually evaluated by players?

One complication with using TagATune for evaluation is that players are motivated to end the round as soon as they believe that they have enough information to guess correctly. Not surprisingly, it was found that players reviewed only a small portion (i.e., two to five tags) of the tags before guessing. In essence, the TagATune metric is similar to the precision@N metric, which only consider the accuracy of the top N tags.

Is the TagATune metric correlated with precision?

Results show that there are generally more true positive tags (that are reviewed by players before guessing) in rounds where players successfully guessed the answer than in rounds where they failed. Additionally, players reviewed, on average, fewer number of tags in the failed rounds, suggesting that guesses are made more hastily when the music taggers produce the wrong tags.

Can we detect which tag(s) cause players to make incorrect guesses?

Our hypothesis is that in a failed round, the last tag reviewed before guessing is usually an incorrect tag that causes players to make a mistake. To test this hypothesis, we evaluate the correctness of the last tag a player reviewed before making an incorrect guess.

Table 9.3 shows the percentage of times that the last tag is actually wrong in failed rounds, which is above 75% for all algorithms. In contrast, the probability of the last tag being wrong is much lower in successful rounds, showing that using game statistics alone, one can detect problematic tags that cause most players to make the wrong guess in the game. This trend does not hold for LabX, possibly because players were left guessing randomly due to the lack of information (since this algorithm generated only one tag per clip).

9.2.3.3 Strength and Weaknesses

One drawback of using TagATune for evaluation is that players do not expect their human partners to make mistakes in tagging, that is, they have very little tolerance for a partner that appears "stupid" or wrong. Unfortunately, the music tagging algorithms that are posing as human players *do* make mistakes—for example, they often generate tags that are contradictory

System	Failed Round	Success Round
Mandel	86.15%	49.00%
Marsyas	80.49%	45.00%
Manzagol	76.92%	33.33%
Zhi	84.38%	70.10%
LabX	100.0%	95.77%

Table 9.3
Percentage of the Time That the Last Tag Displayed before Guessing Is Wrong in a Failed Round versus Success Round

Algorithm	Generated	Contradictory or Redundant
Mandel	36.47	16.23
Marsyas	9.03	3.47
Manzagol	2.82	0.55
Zhi	14.0	5.04
LabX	1.0	0.00

Table 9.4
Average Number of Tags Generated by Algorithms and Contradictory/Redundant Ones among the Generated Tags

(e.g., *slow* followed by *fast*, or *guitar* followed by *no guitar*) or redundant (e.g., *string, violins, violin*). Table 9.4 provides a summary of the number of tags generated (on average) by each algorithm for the clips in the test set, and how many of those are contradictory or redundant. To ensure that the game remains fun despite the mistakes of the music taggers, we adopted two preventive measures. First, we filtered out tags (with lower affinity scores) that are contradictory or redundant. Second, we limited the use of music taggers to only a very small portion of the game.

Our work has shown that TagATune is a feasible and cost-effective platform for collecting a large number of evaluations from human users. We were able to get more than 100 hours worth of evaluation work done by 657 unique human judges, without incurring any cost. However, as an evaluation platform, the TagATune mechanism is not efficient—for example, not every tag is evaluated, and not every round of the game is used for evaluation. The design of new game mechanisms that are more suitable for evaluation is one of our future research directions.

9.3 Learning to Tag Using TagATune Data

The open vocabulary tags collected by TagATune are extremely noisy, requiring substantial postprocessing before they can be used to train music tagging algorithms. For example, after filtering out unreliable tags (ones that were entered by only a single player) and rare tags (those that are associated with very few music clips), the magnatagatune data set consists of only a very small subset (i.e., 188 tags) of the 10,000+ tags that were collected by TagATune. In contrast, most data sets used for training music classification algorithms [2, 18, 12, 13, 19, 1, 6] typically consist of labels that are considered to be devoid of errors and belonging to a small, fixed vocabulary. In this section, we will describe a particular learning method (based on Topic Models) that can be trained on an open vocabulary data set [9].

9.3.1   A Brief Introduction to Topic Models

A topic model [3, 17] is a hierarchical probabilistic model that describes the process of how the constituents of an entity, such as the words of a document or the tags of a music clip, are generated from a set of hidden topics. In our case, a topic is a distribution over music tags, and each music clip is associated with a set of topics with different probabilities.

Consider the example (shown in Figure 9.2(a)) of a topic model learned over the music tags collected by TagATune. Figure 9.2(b) and Figure 9.2(c)

1	indian drums sitar eastern drum jazz tribal oriental
2	slow quiet classical soft solo classic low calm
3	ambient slow synth new_age electronic weird quiet soft
4	classical violin strings harpsichord cello classic violins orchestra
5	opera female woman vocal choir singing female_vocal choral
6	flute classical flutes oboe classic slow organ clarinet
7	guitar strings slow classical harp country solo classical_guitar
8	electronic beat fast drums synth dance beats electro
9	rock guitar loud metal drums hard_rock fast heavy
10	male man male_vocal pop vocal male_voice vocals singing

(a) Topic Model

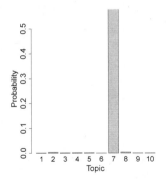

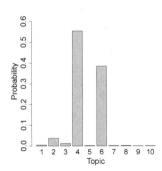

(b) sitar, twang, country, steel_guitar, bluegrass, yeah, fast, guitar, western, strum, twangy, yep, blues, strings

(c) horn, classical, violins, flute, getting_louder, woodwinds, flute, chamber, orchestra, wind_instrument, violin, quiet, odd, strings

Figure 9.2

An example of a topic model learned over music tags, and the representation of two music clips by topic distribution.

show the topic distributions for two very distinct music clips and the ground-truth tags associated with them (in the caption), with one music clip associated with only topic 7 (the "classical/country/guitar" topic), and the other music clip associated with a mix of topic 2 (the "soft/calm" topic), topic 4 (the "violin/cello" topic), and topic 6 (the "flute/oboe" topic).

A widely used method in topic modeling is a *latent variable model* called the latent dirichlet allocation (LDA) [3]. In our adaptation of LDA, the music tags collected by TagATune are the observed variables, and the latent variables to be inferred are (i) the topic distribution for each music clip (such as the ones shown in Figure 9.2(b) and 9.2(c)), (ii) the probability of tags in each topic, and (iii) the topics responsible for generating the observed tags. LDA is a *generative* model; it describes the process of how players of TagATune, using a topic structure that they have in mind, could have generated the tags that we observed in the game—given a music clip, the player first selects a topic according to the topic distribution for that clip, then generates a tag according to the tag distribution of the chosen topic. Our goal, then, is to discover these latent topic structures and use them to automatically tag new music.

9.3.2 Leveraging Topic Models for Music Tagging

The main idea behind the algorithm introduced by Law et al. [9] is to organize noisy tags into well-behaved labels using topic modeling, and learn to predict tags accurately using a mixture of topic labels. This technique is scalable (i.e., makes full use of an arbitrarily large set of noisy labels) and efficient (i.e., the training time remains reasonably short as the tag vocabulary grows).

Our music tagging algorithm (referred to as *Topic Method*) consists of two phases. During the training phase, we use the music tags collected by TagATune to induce a topic model, using which we can infer the topic distribution of each music clip. Using these topic distributions as labels, we then train a classifier g to map audio features to topic distributions. During the inference phase, we use the learned classifier to predict the topic distribution of an previously unseen music clip. Based on this predicted topic distribution, each tag can be given a relevance score for a given music clip c, by multiplying the probability of that tag in each topic and the probability of that topic in c, summing over all topics.

There are two desiderata in the design of a music tagging algorithm that make use of an open vocabulary data set—the algorithm should be (a) capable of generating tags that are reasonably accurate and useful toward indexing music for retrieval, and (b) scalable. We compared the performance of our algorithm to binary classification (i.e., training a classifier for each tag), which we referred to as the *Tag Method*, as well as to a random baseline. Our experiments address the feasibility, annotation and retrieval performance and efficiency of our proposed algorithm, as well as highlight some challenges regarding the evaluation of music taggers against an open vocabulary ground-truth set.

9.3.2.1   Experimental Results

Feasibility

The first question is whether the topic models derived from the music tags collected from TagATune are semantically meaningful. Table 9.5 shows examples of learned topic models with 10, 20, and 30 topics. In general, the topics reflect meaningful groupings of tags, such as synonyms (e.g., "soft/quiet," or "talking/speaking"), misspellings (e.g., "harpsichord/harpsicord" or "cello/chello"), or associations (e.g., "jungle/bongos/fast/percussion"). As the number of topics increases, we also observe the emergence of new topics as well as refined topics (e.g., topics 6 and 24 in the 30-topic model all describe male vocals, but for different genres of music).

The second feasibility question is how well topic distributions can be predicted from audio features. Table 9.6 summarizes the performance results, in terms of accuracy, average rank of the most relevant topic, and KL divergence between the ground truth and predicted topic distributions. Although the performance degrades as the number of topics increases, all models significantly outperform the random baseline. Even with 50 topics, the average rank of the most relevant topic is still around 3, suggesting that the classifier is quite capable of predicting the most relevant topic. This is crucial, as the most appropriate tags for a music clip are likely to be found in the most relevant topics.

Annotation and Retrieval Performance

The second set of questions concerns the quality of the generated tags, both in terms of their accuracy as well as their utility in music retrieval. We evaluated the precision, recall and F-1 measure for each tag, averaged over all the tags that are generated by our algorithm. Results (in Table 9.7) show that the Topic Method significantly outperforms the Tag Method under these metrics.

However, as it has been pointed out by previous work [6, 19], these metrics do not take into account the fact that the algorithms can achieve high scores just by omitting tags that are less common and more difficult to learn. In fact, when we analyze the same set of results under the omission-penalizing metrics [9], the tag method is performing better (Table 9.8). This is because the Tag Method omits much fewer tags than the Topic Method.

10 Topics	
1	indian drums sitar eastern drum jazz tribal oriental middle_eastern beat
2	slow quiet classical soft solo classic low calm silence strings
3	ambient slow synth new_age electronic weird quiet soft dark spacey
4	classical violin strings harpsichord cello classic violins orchestra fast slow
5	opera female woman vocal choir singing female_vocal choral vocals female_voice
6	flute classical flutes oboe classic slow organ clarinet pipe wind
7	guitar strings slow classical harp country solo classical_guitar acoustic banjo
8	electronic beat fast drums synth dance beats electro pop modern
9	rock guitar loud metal drums hard_rock fast heavy electric_guitar male
10	male man male_vocal pop vocal male_voice vocals singing male_vocals guitar

20 Topics	
1	male man male_vocal vocal male_voice pop vocals singing male_vocals voice
2	ambient slow dark weird drone water synth quiet low birds
3	ambient synth slow new_age electronic soft weird instrumental spacey organ
4	guitar slow strings classical harp solo classical_guitar acoustic soft spanish
5	drums drum beat beats tribal percussion indian fast jungle bongos
6	choir choral opera chant chorus vocal vocals singing voices chanting
7	flute classical flutes oboe slow classic clarinet pipe wind woodwind
8	violin classical strings cello violins classic orchestra slow string solo
9	electronic beat synth fast dance drums beats electro trance electric
10	guitar country blues irish folk banjo fiddle celtic harmonica clapping
11	jazz jazzy drums sax bass funky guitar funk trumpet beat
12	opera female woman classical vocal singing female_opera female_voice voice
13	indian sitar eastern oriental strings middle_eastern foreign arabic india guitar
14	rock guitar loud metal hard_rock drums heavy fast electric_guitar heavy_metal
15	slow soft quiet sad calm solo classical mellow low bass
16	classical solo harp classic slow fast soft quiet strings light
17	female woman vocal female_vocal singing female_voice vocals female_vocals pop voice
18	harpsichord classical harpsicord baroque strings classic organ harp medieval harps
19	fast loud upbeat quick fast_beat very_fast fast_paced dance happy fast_tempo
20	quiet slow soft classical silence silent low very_quiet strings ambient

30 Topics	
1	ambient slow synth new_age electronic soft bells spacey instrumental quiet
2	cello violin classical strings solo slow classic string violins chello
3	electronic synth beat electro weird ambient electric modern drums new_age
4	rock guitar loud metal hard_rock drums heavy fast electric_guitar heavy_metal
5	ambient slow dark bass low drone deep synth weird quiet
6	rap hip_hop male man male_vocal vocals beat voice vocal male_vocals
7	indian sitar eastern oriental middle_eastern strings arabic india guitar foreign
8	classical oboe orchestra flute classic strings clarinet horns horn violin
9	quiet slow soft classical silence silent low very_quiet calm ambient
10	talking weird voice electronic loud voices beat speaking strange male_voice
11	female woman female_vocal vocal female_voice pop singing female_vocals vocals voice
12	classical violin strings violins classic orchestra slow string cello baroque
13	classical solo fast classic slow soft quiet light clasical instrumental
14	violin irish fiddle celtic folk strings clapping medieval country violins
15	vocal vocals singing foreign female voices women choir woman voice
16	harp strings guitar dulcimer classical slow string sitar plucking oriental
17	organ classical solo slow classic keyboard accordian new_age soft modern
18	opera female woman classical vocal singing female_opera female_voice operatic
19	guitar country blues banjo folk harmonica bluegrass twangy acoustic fast
20	flute classical flutes pipe slow wind woodwind classic soft wind_instrument
21	drums drum beat beats tribal percussion indian fast jungle bongos
22	ambient water weird slow birds wind quiet new_age dark drone
23	harpsichord classical harpsicord baroque strings classic harp medieval harps guitar
24	male man male_vocal vocal male_voice pop singing vocals male_vocals rock
25	choir choral opera chant chorus vocal male chanting vocals singing
26	fast loud quick upbeat very_fast fast_paced fast_beat happy fast_tempo fast_guitar
27	jazz jazzy drums sax funky guitar funk bass pop reggae
28	beat fast electronic dance drums beats synth electro trance upbeat
29	slow soft quiet sad solo calm mellow classical very_slow low
30	guitar classical slow strings solo classical_guitar acoustic soft harp spanish

Table 9.5

Topic Model with 10, 20, and 30 Topics

No. of Topics	Method	Accuracy	KL Divergence	Average Rank
10	LDA (Distribution)	0.6972 (0.0054)	0.9982 (0.0056)	0.6676 (0.0161)
	Random	0.1002 (0.0335)	2.5906 (0.0084)	4.4287 (0.5155)
20	LDA	0.6370 (0.0060)	1.3233 (0.0142)	1.1375 (0.0328)
	Random	0.0335 (0.0094)	2.9056 (0.0081)	10.2867 (0.4676)
30	LDA	0.5684 (0.0040)	1.6010 (0.01461)	1.7240 (0.04437)
	Random	0.03408 (0.00898)	2.9964 (0.0076)	15.0050 (0.7400)
40	LDA	0.5174 (0.0047)	1.8499 (0.0077)	2.3378 (0.04211)
	Random	0.02925 (0.0082)	3.0817 (0.0055)	19.5751 (1.3955)
50	LDA	0.4943 (0.0029)	2.008 (0.0062)	3.0993 (0.0522)
	Random	0.0223 (0.0084)	3.1258 (0.0074)	25.0087 (0.9001)

Table 9.6
Results Showing How Well Topic Distribution or the Best Topic Can Be Predicted from Audio Features (The metrics used include accuracy and average rank of the top topic, and KL divergence between the *ground truth* and predicted topic distributions.)

Model	Precision	Recall	F-1
10 Topics	0.3159 (0.0117)	**0.4360** (0.0243)	**0.3359** (0.0159)
20 Topics	0.3390 (0.0049)	0.3517 (0.0087)	0.2970 (0.0087)
30 Topics	0.3617 (0.0113)	0.3268 (0.0053)	0.2885 (0.0036)
40 Topics	0.3720 (0.0078)	0.3171 (0.0065)	0.2848 (0.0064)
50 Topics	**0.3846** (0.0113)	0.3138 (0.0077)	0.2834 (0.0073)
Tag Method	0.3176 (0.0037)	0.2141 (0.0038)	0.2131 (0.0034)
Random	0.0132 (0.0004)	0.01174 (0.0003)	0.0115 (0.0003)

Table 9.7
Annotation Performance

Specifically, as shown in Figure 9.3, the Tag Method gains in outputting rarer tags (such as "meditation," "didgeridoo," etc.) that the Topic Method omits. For more common tags (such as "opera," "drums," "oboe," etc.), we found that the Topic Method and Tag Method are very similar in their performance, with the 50-topic model slightly outperforming the 10-topic model.

The quality of tags can also be measured by how well they facilitate music retrieval. Given a search query, each music clip can be rank ordered by the KL divergence between the boolean vector representing the query and the probability distribution over the tags generated by our algorithm for that clip. Retrieval performance is measured using the mean average precision (MAP) [15] metric, which computes precision (the number of retrieved music clips whose ground-truth tags include the search query) while placing more weight on the higher-ranked clips. Table 9.9 shows that the retrieval performance of the Topic Method (with 50 topics) is indistinguishable from the Tag method, and both methods significantly outperform the random baseline.

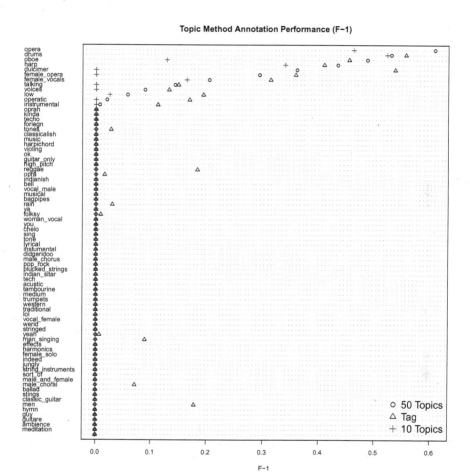

Figure 9.3
Detailed performance of the algorithms under the F-1 measure.

Model	Precision	Recall	F-1	No. Tags Outputted
10 Topics	0.0321 (0.0004)	0.0361 (0.0008)	0.0278 (0.0004)	70.8 (2.3)
Upper Bound	0.1068 (0.0024)	0.0829 (0.0027)	0.0829 (0.0027)	854
20 Topics	0.0465 (0.0004)	0.0431 (0.0007)	0.0364 (0.0008)	104.6 (1.4)
Upper Bound	0.1425 (0.0013)	0.1225 (0.0016)	0.1225 (0.0016)	854
30 Topics	0.0558 (0.0022)	0.0462 (0.0005)	0.0407 (0.0007)	120.6 (2.0)
Upper Bound	0.1598 (0.0021)	0.1412 (0.0022)	0.1412 (0.0022)	854
40 Topics	0.0597 (0.0015)	0.0470 (0.0005)	0.0422 (0.0006)	126.6 (1.9)
Upper Bound	0.1664 (0.0020)	0.1482 (0.0022)	0.1482 (0.0022)	854
50 Topics	0.0618 (0.0021)	0.0467 (0.0002)	0.0422 (0.0003)	127.2 (2.8)
Upper Bound	0.1670 (0.0031)	0.1489 (0.0033)	0.1489 (0.0033)	854
Tag Method	**0.0990** (0.0015)	**0.0592** (0.0006)	**0.0589** (0.0009)	**236.2** (3.5)
Upper Bound	0.2878 (0.0040)	0.2766 (0.0041)	0.27658 (0.0041)	854
Random	0.0131 (0.0004)	0.0115 (0.0003)	0.0119 (0.0003)	853.4 (0.49)

Table 9.8
Annotation Performance under the Omission-Penalizing Metrics (Note that these metrics are upper bounded by a quantity that depends on the number of tags outputted by the algorithm.)

Model	Average Mean Precision
10 Topics	0.2572 (0.0054)
20 Topics	0.2964 (0.0120)
30 Topics	0.3042 (0.0182)
40 Topics	0.3140 (0.0119)
50 Topics	**0.3236** (0.0111)
Tag Method	0.3117 (0.0114)
Random	0.1363 (0.0061)

Table 9.9
Retrieval Performance, in Terms of Average Mean Precision

Efficiency

Scalability is one of the main motivations behind the use of topic models for music tagging—it is much faster to train a classifier to predict 50 topic classes than 834 tag classes. Table 9.10 shows an estimate of the training time of the different models.

Model	Training Time (in minutes)
10 Topics	18.25
20 Topics	19.00
30 Topics	45.25
40 Topics	42.50
50 Topics	48.00
Tag Method	845.5

Table 9.10
Annotation Performance, in Terms of Training Time

While the training time does increase with the number of topics, training time does reach a plateau. More importantly, the Topic Method is approximately 94% times faster to train than the Tag Method, confirming our intuition that Topic Method will be significantly more scalable as the size of the tag vocabulary grows.

Human Evaluation

The performance metrics we used to evaluate tags assume the existence of a perfect, comprehensive ground-truth set. This is far from true, especially for data sets that contain open vocabulary labels. There are often missing tags (e.g., the music tagger can generate tags that a music clip has never been associated with) or vocabulary mismatch (e.g., the music tagger predicts "serene," but the music clip has only been tagged "calm"). In fact, in several experiments [9] where we asked Mechanical Turk workers to evaluate the annotation and retrieval performance of the algorithms, it was found that comparisons against an open vocabulary ground-truth set grossly underestimate performance. Out of the tags that turkers consider relevant, on average, approximately 50% of them are missing from the ground-truth sets [9].

Another interesting finding is that in the evaluation of tags, the sum of parts does not equal whole. For example, when the Topic Method and Tag Method are evaluated by having workers judge the relevance of each tag, the performance of the two algorithms are quite similar. However, when asked "which list of tags do you prefer the most," the Tag Method was strongly favored (winning 6.20 out of 10 votes, as opposed to the 3.34 votes for the Topic Method and 0.46 votes for the Random Method). This shows that human evaluations can be invaluable if we want to measure the performance of music taggers the same way humans perceive them. As we described in the previous section, TagATune is a viable solution for collecting human evaluations of music tags in an economical and timely fashion.

9.4 Conclusion

In this chapter, we described the design of a human computation game called TagATune, which is shown to be capable of eliciting the help of the crowd to annotate music. While the game is made more fun by allowing players to openly communicate with each other, the data collected via TagATune is also more noisy, necessitating the development of new learning algorithms and evaluation methods. The key take-away message is simple—without motivated human volunteers, there will be no human computation systems. Therefore, a human computation system must be designed to solve the computational problem at hand *without* failing to meet the users' needs.

Acknowledgments

This chapter is based on three articles—"Input Agreement: A New Mechanism for Data Collection Using Human Computation Games" [10] coauthored with Luis von Ahn; "Learning to Tag from Open Vocabulary Labels" [9] coauthored with Burr Settles and Tom Mitchell; and "Evaluation of Algorithms Using Games: The Case of Music Tagging" coauthored with Kris West, Michael Mandel, Mert Bay, and Stephen Downie. All authors have made substantial contributions to the results obtained in this chapter.

Bibliography

[1] J. Bergstra, A. Lacoste, and D. Eck. Predicting genre labels for artists using FreeDB. *ISMIR*, pages 85–88, 2006.

[2] T. Bertin-Mahieux, D. Eck, F. Maillet, and P. Lamere. Autotagger: A model for predicting social tags from acoustic features on large music databases. *TASLP*, 37(2):115–135, 2008.

[3] D. Blei, A. Ng, and M. Jordan. Latent dirichlet allocation. *Journal of Machine Learning Research*, 3:993–1022, 2003.

[4] The Google Image Labeler, http://images.google.com/imagelabeler/.

[5] C.J. Ho, T.H. Chang, J.C. Lee, J. Hsu, and K.T. Chen. Kisskissban: A competitive human computation game for image annotation. In *Proceedings of the KDD Human Computation Workshop (HCOMP)*, pages 11–14, 2009.

[6] M. Hoffman, D. Blei, and P. Cook. Easy as CBA: A simple probabilistic model for tagging music. *ISMIR*, pages 369–374, 2009.

[7] Y.E. Kim, E. Schmidt, and L. Emelle. Moodswings: A collaborative game for music mood label collection. *ISMIR*, pages 231–236, 2008.

[8] P. Lamere. Social tagging and music information retrieval. *Journal of New Music Research*, 37(2):101–114, 2008.

[9] E. Law, B. Settles, and T. Mitchell. Learning to tag from open vocabulary labels. In *ECML*, pages 211–226. Springer, New York, 2010.

[10] E. Law and L. von Ahn. Input agreement: A new mechanism for collecting data using human computation games. *CHI*, pages 1197–1206, 2009.

[11] E. Law, L. von Ahn, R. Dannenberg, and M. Crawford. Tagatune: A game for music and sound annotation. *ISMIR*, 2007.

[12] T. Li, M. Ogihara, and Q. Li. A comparative study on content-based music genre classification. *SIGIR*, pages 282–289, 2003.

[13] M. Mandel and D. Ellis. Song-level features and support vector machines for music classification. *ISMIR*, 2005.

[14] M. Mandel and D. Ellis. A Web-based game for collecting music metadata. *Journal of New Music Research*, 37(2):151–165, 2009.

[15] Performance metrics for information retrieval, http://en.wikipedia.org/wiki/Information_retrieval.

[16] S. Robertson, M. Vojnovic, and I. Weber. Rethinking the ESP game. In *Proceedings of the 27th International Conference on Human Factors in Computing Systems (CHI)*, pages 3937–3942, 2009.

[17] M. Steyvers and T. Griffiths. Probabilistic topic models. In T. Landauer, D.S. McNamara, S. Dennis, and W. Kintsch, editors, *Handbook of Latent Semantic Analysis*. Erlbaum, Hillsdale, New Jersey, 2007.

[18] D. Turnbull, L. Barrington, D. Torres, and G. Lanckriet. Toward musical query-by-semantic description using the CAL500 data set. *SIGIR*, pages 439–446, 2007.

[19] D. Turnbull, L. Barrington, D. Torres, and G. Lanckriet. Semantic annotation and retrieval of music and sound effects. *TASLP*, 16(2):467–476, February 2008.

[20] D. Turnbull, R. Liu, L. Barrington, and G. Lanckriet. A game-based approach for collecting semantic annotations of music. *ISMIR*, pages 535–538, 2007.

[21] L. von Ahn and L. Dabbish. Labeling images with a computer game. *CHI*, pages 319–326, 2004.

Part IV

Advanced Topics

10

Hit Song Science

François Pachet

Sony CSL

CONTENTS

Hit Song Science is an emerging field of investigation that aims at predicting the success of songs before they are released on the market. This chapter defines the context and goals of *Hit Song Science* (HSS) from the viewpoint of music information retrieval. In the first part, we stress the complexity of the mechanisms underlying individual and social music preference from an experimental psychology viewpoint. In the second part, we describe current attempts at modeling and predicting music hits in a feature oriented view of popularity and, finally, draw conclusions on the current status of this emerging but fascinating field of research.

10.1   An Inextricable Maze?

Can someone predict whether your recently produced song will become a hit? Any pop song composer would probably laugh at this question and respond: How could someone predict the success of what took so much craft, pain, and immeasurable creativity to produce? I myself do not even have a clue!

This question raises a recurring fantasy in our culture: wouldn't it be thrilling to understand the "laws of attraction" that explain how this sort of preference system of music in human beings works, to the point of being able to predict the success of a song or any other cultural artifact before it is even released? This fantasy is elaborated in detail in Malcom Gladwell's story "The Formula" [14]. In this fiction, a—needless to say, fake—system is able to predict the success of movies by analyzing their script automatically. The system is even smart enough to propose modifications of the script to increase the success of the movie, with a quantified estimation of the impact in revenues. In the introduction, Gladwell begins by describing the reasons why we like a movie or not as resulting from a combination of small details. He writes:

> Each one of those ... narrative details has complicated emotional associations, and it is the subtle combination of all these associations that makes us laugh or choke up when we remember a certain movie... Of course, the optimal combination of all those elements is a mystery. [14]

This process is also true for music: what makes us like a song or not probably has to do with a complex combination of micro-emotions, themselves related to our personal history, to the specifics of the song and to many other elusive elements that escape our direct understanding. In spite of the many claims that writing hit songs is just a matter of technique (see, for example, Blume [5]), it is likely that, as the highly successful Hollywood screenwriter William Goldman said: "Nobody knows anything" [15].

Or is this the case? However daring, Hit Song Science attempts to challenge this assumption by precisely undertaking the task of making these kinds of predictions. Several companies now claim to be able to automatically analyze songs in order to predict their success (HSS, PlatiniumBlue) and to sell their results to record labels. Unfortunately, the exact mechanisms behind these predictions are not disclosed, and no reproducible data is provided to check the accuracy of these predictions. At the same time, the very existence of these services shows that hit prediction is taken seriously by the music industry.

Considering this hit song prediction fantasy from a scientific viewpoint raises issues in several disciplines, interrelated in complex ways that involve the following issues: (1) the psychology of music listening and the effects of repeated exposure, (2) the paradoxical nature of the Western media broadcasting system, radios in particular, and (3) the social influence human beings exert and receive from each other. Before describing the specific Music Information Retrieval (MIR) approach to Hit Song Science, each of these issues is first addressed.

10.1.1 Music Psychology and the Exposure Effect

Surprisingly, the question of "Why we like or not a particular song?" has received little attention from music psychology. Although music preference is recognized as a central aspect of modern identities, the field is "still in its infancy" [30]. The issue of *liking* per se is indeed difficult to study directly, and music psychologists have traditionally focused on less elusive, more directly measurable phenomena such as memorization, recognition or learning.

In our context, a central issue in trying to explain music hits is *exposure*, that is, the simple fact of listening to a musical piece. What is the effect of exposure on preference or liking? Studies on exposure show that there is indeed an impact of repeated exposure on liking, but also that this impact is far from simple. Parameters such as the context, type of music or listening conditions (focused or incidental), seem to influence the nature of this impact, and many contradictory results have been published.

The popular idea that repeated exposure tends to increase liking was put forward early [21] and was confirmed experimentally in a wide variety of contexts and musical genres [27]. The so-called *mere exposure* effect, akin to the *familiarity principle*, or *perceptual fluency*, is considered by many psychologists to be a robust principle, pervading many facets of music listening.

However, as noted by Schellenberg [29], this increase in liking may be restricted to musically impoverished or highly controlled stimuli. Indeed, other studies have shown a more subtle effect of repeated exposure. The study by Siu-Lan et al. [31] showed different effects of exposure on intact and patchwork compositions. An inverted U-curve phenomena was observed in particular by Szpunar et al. [33] and Schellenberg [30], itself explained in large part by the "two factor model" of Berlyne [3]. In this model, two forces compete to build up liking: (1) the *arousal* potential of the stimulus (the music), which decreases with repeated listening, thereby increasing liking (with the habituation to this arousal potential), and (2) *familiarity*, which tends to create boredom. These two forces combined produce typical inverted U-shapes that have been observed in many studies of preference. This model is itself related to the famous "Wundt curve" [36]. The Wundt curve describes the typical experience of arousal as being optimal when achieving a compromise between repetition/boredom and surprise (Figure 10.1). Interestingly, reaching such

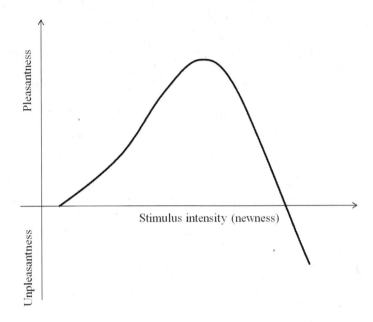

Figure 10.1
The Wundt curve describes the optimal "hedonic value" as the combination
of two conflicting forces.

an optimal compromise in practice is at the root of the psychology of flow
developed by Csíkszentmihályi [8].

Yet, other studies [35] show in contrast a *polarization* effect, whereby repeated exposure does not influence initial likings but makes them stronger,
both positively or negatively. Finally, Loui et al. [19] studied exposure effects
by considering exotic musical temperaments, to study the relation between
learning and preference. They showed that passive exposure to melodies built
in an entirely new musical system led to learning and generalization, as well
as increased preference for repeated melodies. This work emphasizes the importance of learning in music preference.

These psychological experiments show that a relation between exposure
and liking exists, but that this relation is complex and still not well understood, in particular for rich, emotionally meaningful pieces. It is therefore
impossible to simply consider, from a psychological point of view, that repeated exposure necessarily increases liking: it all depends on a variety of
factors.

10.1.2 The Broadcaster/Listener Entanglement

If the relation between exposition and preference is unclear in socially neutral contexts, the issue becomes even more confusing when considering the tangled interplay between the preference engine of individuals and the editorial choices of broadcasters, radio programmers, in particular.

Indeed, exposure is largely dependent upon the editorial strategies of programmers in the field of radio (in a broad definition of the term). Again, it is often said that it suffices to play a tune often enough to make it a hit, and that therefore hits are basically built by music marketing. However, the influence of radios on musical taste is paradoxical. On one hand, mass media (radio, television, etc.) want to broadcast songs that most people will like, in the hope of increasing their audience. Yet, what these media broadcasts actually influence, in turn, is the taste of audiences by means of repeated, forced exposition.

One process by which radios, for instance, maximize their audiences is so-called *radio testing*, which is performed regularly by various companies (e.g., musicresearch.com). Radio testing consists in playing songs to a selected panel that is representative of the radio audience, and then asking the listeners to rank songs. Eventually, only songs having received top ranks are kept and programmed. This radio testing phenomenon is more and more prevalent in Western society [17], but, strangely, has received so far little attention from researchers. Montgomery and Moe [22] exhibit a dynamic relationship between radio airplay and album sales, with vector autoregressive models, in an attempt to better inform marketing managers. Montgomery and Moe also stress the unpredictable evolution of this relationship as audiences may progressively evaluate these airplays critically by considering them as forms of advertisements.

This situation creates a paradox, also stressed by Montgomery and Moe [22]: "Not only is a radio station able to influence the public, but the public can also affect what is aired on the radio. Increased album sales may lead radio stations to play an album more." In turn, these songs are repeatedly exposed to a larger population with effects that are not completely clear, as seen above. As a result, if it is clear that radios do have an impact on musical taste, it is, again, difficult to assess exactly which one.

10.1.3 Social Influence

This is not the whole story. The situation is further complicated by the social influence that we all exert on one other. Knowing that a song is a hit, or at least preferred by others in our community, influences our liking. This phenomenon has been studied by Salganik et al. [28] in a remarkable experiment, which consisted in studying preferences in two groups of people: in the first group (*independent*), users had to rank individually, unknown songs. In the second group (*social influence*), users had the same task, with additional information about what the other users of the group ranked. This information regarding the preferences of others had itself two strength levels.

The comparison between these two groups showed two interesting facts: (1) In the independent group, the distribution of preference was not uniform, showing that there are indeed songs that are statistically preferred to others, independently of social influence. This preference can only come from the songs themselves and can be considered as an indicator of their intrinsic quality. (2) The strength of the "social signal" increases the unpredictability of hits, that is, the more information about what others like, the less replicable are the experiments in terms of which songs become hits. This unpredictability, well-studied in network analysis, is referred to as the *cumulative advantage* effect. In the social influence group, hits are much more popular than in the independent group, but they are also different for each experiment, with the same initial conditions. One interesting argument put forward in this study is that the determination of which songs will become hits, in the social influence condition, eventually depend on "early arriving individuals" [34], in other words on initial conditions, which are themselves essentially random.

Under all of these conditions (unknown effects of repeated exposure, complex interplay between broadcasters and listeners, and the effects of social influence), is it reasonable even to attempt to program computers to predict hits in the first place?

10.1.4   Modeling the Life Span of Hits

Recognizing the importance of social pressure and the rich-get-richer effect, some works have attempted to predict hits using only social information, regardless of the intrinsic characteristics of songs.

For instance, Chon et al. [7] attempt to predict the popularity and life span of a jazz album given its entry position in the charts. This work used only charts information from Billboard, an American magazine maintaining music charts on a weekly basis. Analysis of the distribution of hits over time showed that the life span of a song tended to increase with its starting position in the charts. This result was interpreted as an encouragement to record labels to market albums before the sales, since the higher the starting position is, the longer it will stay in the charts. However, such a technique does not seem to be sufficient to yield more accurate predictions.

In the same vein, Bischoff et al. [4] attempted to identify critical early-stage effects of cumulative advantage. More precisely, this work posits that the success of a hit depends only on two factors: (1) its initial observed popularity after one week, as well as (2) contextual information such as the album, the general popularity of the artist, and the popularity of other tracks in the album. Similarly, this approach does not use any information concerning the actual content of the songs. Initial popularity and contextual information are converted into a 18 feature vector, and standard machine-learning techniques are then used to train and test a predictor (as described in detail in the next section). Like in the previous work, ground-truth data is taken from Billboard. This experiment was conducted on a database of 210,350 tracks, performed by 37,585 unique artists. The results yield an improvement of 28% in AUC (area under ROC) compared to the work of Dhanaraj and Logan [9] described below.

These works show that there are patterns in the way social pressure generates hits. However, requiring initial popularity data, and being independent of both the characteristics of the songs and the listeners, they don't tell us much about why we like or not a given song. The following approaches take the opposite stance, trying explicitly to identify the features of songs that make them popular, regardless of social pressure effects.

10.2 In Search of the Features of Popularity

Several MIR researchers have recently attempted to consider hit prediction from a candid viewpoint. Like mathematicians trying to predict forecast or evolutions of financial markets, Hit Song Science has emerged as a field of predictive studies. Starting from the observation of the nonuniform distribution of popularity [12], the goal is to understand better the relation between intrinsic characteristics of songs (ignored in the preceding approaches) and their popularity, regardless of the complex and poorly understood mechanisms of human appreciation and social pressure at work.

In this context, popularity is considered as a *feature* of a song, and the problem, then, is to map this feature to other features that can be measured objectively. In other words, MIR sees Hit Song Science as yet another "feature problem," like genre or instrument classification.

It is important to stress the hypothesis at stake in this view, in light of the three difficulties described in the previous section. The attempt to directly model popularity with objective features ignores the difficulties that experimental psychology encounters in explaining the exposure effect. However, the yet unclear nature of human music habituation mechanisms does not imply that a predictor cannot be built. Of course, even if successful, such a predictor would probably not say much about the mysteries of the exposure effect.

The radio entanglement problem is related to the social influence issue: the final distributions of hits in a human community depend on random initial conditions which are not under control, from the choice of the members in the panel to the preferences of the "early arriving individuals." This intrinsic unpredictability in the hit distribution seems at first glance to threaten the whole Hit Song Science enterprise. An answer to this criticism consists in considering Hit Song Science as an idealistic attempt to determine the "objective causes" of individual music preference, independently of the effects of social influence.

Even if it is not realistic for individuals to listen and rate songs independently of each other, such an attempt is an important and sound approach for two reasons. First, if the works of Salganik et al. [28] aim at stressing the importance of social influence, they also show, in passing, that individual preferences do exist and are consistent and grounded. The difference between the nonuniform distribution of the independent group and a random one can precisely be seen as what *remains of individual preference*, once social influence is ignored. Because these remains of individuality are not random, it is worth trying to model them. Second, the search for the causes of our aesthetic experience, even partial ones, is a legitimate goal of cognitive science and should also be a goal of modern musicology. The remainder of this chapter focuses on this MIR view of hits, and more precisely on the following problem: under the absence of social pressure, which features of songs are able to explain their popularity?

10.2.1 Features: The Case of Birds

Before reviewing works that specifically address music features, we review here a fascinating and successful case of feature-based hit song prediction in a less complex area than human music: *bird songs*. Researchers in animal behavior have long been interested in the phenomenon of bird song production and its role in the mating process. In several bird species, male birds produce songs primarily to attract females. The issue of what makes a bird song more attractive than others has received particular attention in the recent years. This question echoes the Hit Song Science question (What are the features of popularity?), but in a simpler context, where social pressure is considered to be less significant.

Various results have indeed shown that specific features of songs can account for their popularity. For instance, great reed warbler females (*Acrocephalus arundinaceus*) were shown to prefer long over short songs in the wild [2].

More interestingly, the study by Draganoiu et al. [10] focused on the case of the domesticated canary (*Serinus canaria*). In this species, male bird songs have a specific phrase structure. Two features of these phrases were shown to significantly increase liking: (1) frequency bandwidth and (2) trill rate. However, it was also shown that these two features are somehow contradictory: a trade-off is observed in real phrases, due to the specific motor constraints of the bird vocal track.

The breakthrough experiment by Draganoiu et al. [10] consisted in synthesizing artificial phrases optimizing these two features in an unrealistic way, that is "beyond the limits of vocal production." The exposition of these artificial phrases to bird females showed unequivocally that females preferred these artificial phrases to the natural ones (see Figure 10.2). An interesting interpretation for this preference is that the production of "difficult" phrases

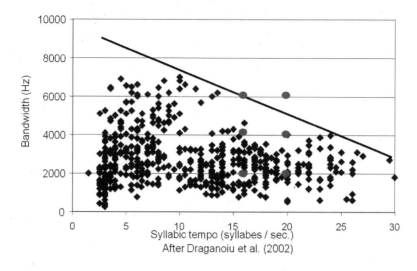

Figure 10.2
The distribution of canary phrases, in a bandwidth/tempo space, representing the natural trade-off between bandwidth and syllabic tempo. Circles represent the phrases used for the experiment. The artificial top right phrases optimizing the two features in unrealistic ways were the most successful [10].

maximizing both bandwidth and syllable rate may be a reliable indicator of male physical or behavioral qualities.

This evolutionary argument emphasizes the role of virtuosity in music appreciation. In popular music, virtuosity is explicitly present in specific genres (e.g., *shredding* in hard-rock, or melodic-harmonic virtuosity in bebop). However, it is probably a marginal ingredient of most popular styles (pop, rock), although virtuosity is still a largely understudied phenomenon. To which extent can these works be transposed to popular music?

10.2.2 The Ground-Truth Issue

In the MIR view of Hit Song Science, the nonuniform distribution of preferences is taken as *ground-truth* data. The problem is then to find a set of song features that can be mapped to song popularity. Once the mapping is discovered, the prediction process from a given, arbitrary new item (a song or a movie scenario) can be automated.

Considering the preceding arguments, a suitable database to conduct the experiment should ideally contain preference data which results from non-"socially contaminated" rankings. Such rankings can be obtained as in the experiment by Salganik et al. [28]. However, this process works only for a set of carefully chosen, unknown songs by unknown artists. In practice, there is no database containing "normal songs" associated with such pure preference data. The experiments given in the following sections are based on databases of known music, with socially determined preference data, as described below. The impact of this approximation is not clear, and this approach *by default* could clearly be improved in future works.

10.2.3   Audio and Lyrics Features: The Initial Claim

The first attempt to model directly popularity in a feature-oriented view of music preference is probably the study by Dhanaraj and Logan [9]. This study consisted in applying the traditional machine-learning scheme, ubiquitous in MIR research, to the prediction of popularity. The features consisted both in traditional audio features Mel-frequency cepstral coefficients (MFCCs) extracted and aggregated in a traditional manner, as well as features extracted from the lyrics. The lyrics features were obtained by extracting an eight-dimensional vector representing the closeness of the lyrics to a set of eight semantic *clusters*, analyzed in a preliminary stage using a nonsupervised learning scheme.

The experiment was performed on a 1,700 song database, using Support Vector Machines (SVMs), and a boosting technique [13]. The conclusion of this study is that the resulting classifiers using audio or lyric information do perform better than random in a significant way, although the combination of audio and lyric features do not improve the accuracy of the prediction. However, a subsequent study described below showed contradictory results.

10.3   A Large-Scale Study

The studies by Pachet and Roy [23, 25] describe a larger-scale and more complete experiment designed initially to assess to which extent high-level music descriptors could be inferred automatically using audio features. A part of this study was devoted to the specific issue of popularity, seen as a particular high-level descriptor among many others. This experiment used a 32,000 song database of popular music titles, associated to fine-grained human metadata, in the spirit of the Pandora effort (http://www.pandora.com) as well as popularity data, obtained from published charts data like in the preceding approaches. To ensure that the experiment was not biased, three sets of different features were used: a generic acoustic set *à la* MPEG-7, a specific acoustic set using proprietary algorithms, and a set of high-level metadata produced by humans. These feature sets are described in the next sections.

10.3.1 Generic Audio Features

The first feature set was related to the so-called bag-of-frame (BOF) approach. The BOF approach owes its success to its simplicity and generality, as it can be, and has been, used for virtually all possible global descriptor problems. The BOF approach consists in modeling the audio signal as the statistical distribution of audio features computed on individual, short segments. Technically, the signal is segmented into successive, possibly overlapping frames, from which a feature vector is computed. The features are then aggregated together using various statistical methods, varying from computing the means/variance of the features across all frames to more complex modeling such as Gaussian Mixture Models (GMM). In a supervised classification context, these aggregated features are used to train a classifier. The BOF approach can be parameterized in many ways: frame length and overlap, choice of features and feature vector dimension, choice of statistical reduction methods (statistical moments or Gaussian Mixture Models), and choice of the classifier (decision trees, Support Vector Machines, GMM classifiers, etc.). Many articles in the Music Information Retrieval (MIR) literature report experiments with variations on BOF parameters on several audio classification problems [1, 11, 20, 26]. Although perfect results are rarely reported, these works demonstrate that the BOF approach is relevant for modeling a wide range of global music descriptors.

The generic feature set considered here consisted of 49 audio features taken from the MPEG-7 audio standard [18]. This set includes spectral characteristics (Spectral Centroid, Kurtosis and Skewness, High-Frequency Centroids, Mel-frequency cepstrum coefficients), temporal (Zero-Crossing Rate, Inter-Quartile Range), and harmonic (chroma). These features were intentionally chosen for their generality, that is they did not contain specific musical information nor used musically *ad hoc* algorithms. Various experiments (reported by Pachet and Roy [25]) were performed to yield the optimal BOF parameters for this feature set: localization and duration of the signal, statistical aggregation operators used to reduce dimensionality, frame size and overlap. The best trade-off between accuracy and computation time was achieved with the following parameters: 2,048 sample frames (at 44,100 Hz) with a 50% overlap computed on a two-minute signal extracted from the middle part of the title. The aggregated features were the two first statistical moments of this distribution (mean and variance) yielding eventually a feature vector of dimension 98 (49 means + 49 variances).

10.3.2 Specific Audio Features

The specific approach consisted in training the same (SVM) classifier with a set of "black-box" acoustic features developed especially for popular music analysis tasks by Sony Corporation [32]. These proprietary features have been used in commercial applications such as hard disk based Hi-Fi systems. Altogether, the specific feature set also yielded a feature vector of dimension

98, to guarantee a fair comparison with the generic feature set. As opposed to the generic set, the specific set did not use the BOF approach: each feature was computed on the whole signal, possibly integrating specific musical information. For instance, one feature described the proportion of perfect cadences (i.e., resolutions in the main tonality) in the whole title. Another one represented the proportion of percussive sounds to harmonic sounds.

10.3.3 Human Features

The last feature set considered was a set of human-generated features. We used the 632 Boolean labels provided by a manually annotated database (see the following section) to train the classifiers. This was not directly comparable to the 98 audio features as these labels were Boolean (and not floating point values). However, these features were good candidates for carrying high-level and precise musical information that are typically not well learned from features extracted from the acoustic signal.

10.3.4 The HiFind Database

10.3.4.1 A Controlled Categorization Process

Several databases of annotated music have been proposed in the MIR community, such as the RWC database [16], the various databases created for the MIREX tests [6]. However, none of them had the scale and number of labels needed to conduct this experiment. For this study the authors used a music and metadata database provided by the defunct HiFind Company. This database was a part of an effort to create and maintain a large repository of fine-grained musical metadata to be used in various music distribution systems, such as playlist generation, recommendation, or advanced music browsing. The HiFind labels were binary (0/1 valued) for each song. They were grouped in 16 categories, representing a specific dimension of music: Style, Genre, Musical setup, Main instruments, Variant, Dynamics, Tempo, Era/Epoch, Metric, Country, Situation, Mood, Character, Language, Rhythm, and Popularity. Labels described a large range of musical information: objective information such as the "presence of acoustic guitar," or the "tempo range" of the song, as well as more subjective characteristics such as "style," "character" or "mood" of the song. The Popularity category contained three (Boolean) labels: *low, medium,* and *high,* representing the popularity of the title, as observed from hit charts and records of music history. These three labels were mutually exclusive.

The HiFind categorization process was highly controlled. Each title was listened to entirely by one categorizer. Labels to be set to true were selected using an *ad hoc* categorization software. Label categories were considered in some specific order. Within a category, some rules could apply that prevented specific combinations of labels to be selected. The time taken, for a trained categorizer, to categorize a single title was about six minutes. Categorized titles were then considered by a categorization supervisor, who checked consistency and coherence to ensure that the description ontologies were well understood and utilized consistently across the categorization team. Although errors and inconsistencies could be made during this process, the process nevertheless guaranteed a relative good "quality" and consistency of the metadata, as opposed for instance to collaborative tagging approaches with no supervision. As a consequence, the metadata produced was very precise (up to 948 labels per title), a precision difficult to achieve with collaborative tagging approaches.

The total number of titles considered in this study was 32,978, and the number of labels 632. Acoustic signals were given in the form of a wma file at 128 kbps. This database was used both for training and testing classifiers, as described in Section 10.3.5.3.

10.3.4.2   Assessing Classifiers

To avoid the problems inherent to the sole use of precision or recall, a traditional approach is to use F-measure to assess the performance of classifiers. For a given label, the recall R is the proportion of positive examples (i.e., the titles that are true for this label) that were correctly predicted. The precision P is the proportion of the predicted positive examples that were correct. When the proportion of positive examples is high compared to that of negative examples, the precision will usually be artificially very high and the recall very low, regardless of the actual quality of the classifier. The F-measure addresses this issue and is defined as:

$$F = 2 \times \frac{R \times P}{R + P}$$

However, in this specific case, the authors had to cope with a particularly unbalanced two class (True and False) database. Therefore, the mean value of the F-measure for each class (True and False) could be artificially good. To avoid this bias, the performances of classifiers were assessed with the more demanding *min F-measure*, defined as the minimum value of the F-measure for the positive and negative cases. A min-F-measure near 1 for a given label really means that the two classes (True and False) are well predicted.

10.3.5   Experiment

10.3.5.1   Design

The HiFind database was split in two "balanced" parts, Train and Test, so that Train contained approximately the same proportion of examples and counterexamples for each label as Test. This state was obtained by performing repeated random splits until a balanced partition was observed.

Three classifiers were then trained, one for each feature set (generic, specific, and human). These classifiers all used an SVM algorithm with a radial-basis function (RBF) kernel. Each classifier, for a given label, was trained on a maximally "balanced" subset of Train, that is, the largest subset of Train with the same number of "True" and "False" titles for this label (popularity: Low, Medium, and High). In practice, the size of these individual train databases varied from 20 to 16,320. This train database size somehow represented the "grounding" of the corresponding label. The classifiers were then tested on the whole Test base. Note that the Test base was usually not balanced with regards to a particular label, which justified the use of the min-F-measure to assess the performance of each classifier.

10.3.5.2 Random Oracles

To assess the performance of classifiers, these were compared to that of random oracles defined as follows: given a label with p positive examples (and therefore $N - p$ negative ones, with N the size of the test set), this oracle returns true with a probability $\frac{p}{N}$. By definition, the min-F-measure of a random oracle only depends on the proportion of positive and negative examples in the test database.

For instance, for a label with balanced positive and negative examples, the random oracle defined as above has a min-F-measure of 50%. A label with 200 positive examples (and therefore around 16,000 negative examples) leads to a random oracle with a min-F-measure of 2.3%. So the performance of the random oracle was a good indicator of the size of the train set and could therefore be used for comparing classifiers as described below.

10.3.5.3 Evaluation of Acoustic Classifiers

The comparison of the performance of acoustic classifiers with random oracles showed that the classifiers did indeed learn something about many of the HiFind labels. More than 450, out of 632 labels, were better learned with the acoustic classifiers than with random oracle. Table 10.1 indicates, for each feature set, the distribution of the relative performance of acoustic classifiers with regards to random oracles.

Table 10.1 also shows that around 130 to 150 labels lead to low-performance classifiers, that is, acoustic classifiers that did not perform significantly better than a random oracle (the last row Table 10.1); approximately half of the labels led to classifiers that improve over the performance of a random classifier by less than 10; the rest (top rows) clearly outperformed a random oracle, that is, they were well modeled by acoustic classifiers.

It is interesting to see that the performance of these acoustic classifiers varied from 0% for both feature sets to 74% for the generic features and 76% for the specific ones. The statistical distribution of the performance was close to a power law distribution, as illustrated by the log-log graph of Figure 10.3.

These acoustic classifiers learned aspects of human musical categorization with a varying degree of success. The problem, as outlined below, is that popularity stands at the bottom line of this scale.

Improvement	Specific	Generic
50	8	0
40	12	15
30	43	20
20	111	79
10	330	360
0	128	158

Table 10.1

Number of Labels for Which an Acoustic Classifier Improves over a Random Classifier by a Certain Amount (Column "Improvement" reads as follows: there are 111 labels for which a specific acoustic classifier outperforms a random classifier by +20 [in min-F-measure].)

Not surprisingly, it could be observed that specific features performed always better than the generic ones (see Figure 10.4). Since the classifiers were both based on the same SVM/kernel, the difference in performance could only come from the actual features considered.

Last, the relationship between the performance and the size of the training set was studied. The trend lines in Figure 10.5 show that the performance of acoustic classifiers increase with the training data set size, regardless of the feature set. This was consistent with the acknowledged fact that machine-learning algorithms require large numbers of training samples, especially for high-dimensional feature sets.

These experiments showed that acoustic classifiers definitely learned musical information, with varying degrees of performance. It also showed that the subjective nature of the label did not seem to influence their capacity to be learned by audio features. For instance, the label "Mood nostalgic" was learned with 48% (specific features), and 43% (generic features), to be compared to the 6% of the random oracle. Similarly, label "Situation evening mood" was learned with 62% and 56% respectively, against 36% for random. Since *a priori* high-level features of songs could be learned with some success, why not popularity?

10.3.5.4 Inference from Human Data

This double feature experiment was complemented by another experiment with classifier trained using all the HiFind labels but the Popularity ones. Some pairs of HiFind labels were perfectly well correlated so this scheme worked obviously perfectly for those, but this result was not necessarily meaningful in general (e.g., to infer the country from the language). The same Train / Test procedure described above applied with the 629 nonpopularity labels as input yielded the following result (min-F-measure): 41% (Popularity-Low), 37% (Popularity-Medium), and 3% (Popularity-High).

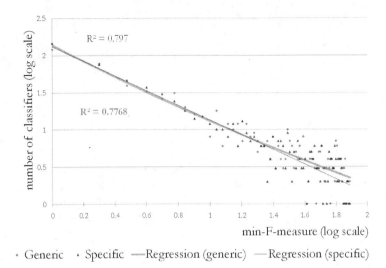

Figure 10.3
Log-log graph of the distribution of the performance of acoustic classifiers for both feature sets.

10.3.6 Summary

The results concerning the Popularity labels are summarized in Table 10.2. These results show clearly that the Popularity category was not well-modeled by acoustic classifiers: its mean performance was ranked fourth out of 16 categories considered, but with the second lowest maximum value among categories.

Although these results appear to be not so bad at least for the "Low" label, the comparison with the corresponding random classifiers shows that popularity is in fact not learned. Incidentally, the performance was not improved with the *correction scheme*, a method that exploits inter-relations between labels to correct the results [25]. Interestingly, human features (all HiFind labels) did not show either any significant improvement over random classifiers.

A last experiment was conducted with *a priori* irrelevant information: the letters of the song title, that is, a feature vector of size 26, containing the number of occurrences of each letter in the song title. The performances of the corresponding classifiers were respectively 32%, 28%, and 3%. (For the low-, medium-, and high-popularity labels, see Table 10.2.) This shows that even dumb classifiers can slightly improve the performance of random classifiers (by 5% in this case for the medium- and low-popularity labels). Obviously, this information does not teach us anything about the nature of hits and can be considered as some sort of noise.

These results suggest that there are no significant statistical patterns

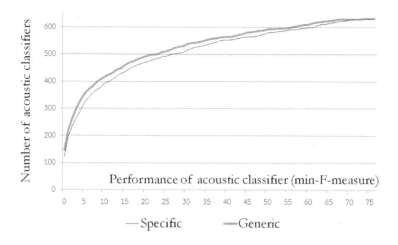

Figure 10.4
Cumulated distribution of the performance of acoustic classifiers for the generic and specific feature sets.

concerning popularity using any of the considered features sets (audio or humans). This large-scale evaluation, using the best machine-learning techniques available to date, contradicts the initial claims of Hit Song Science, that is that the popularity of a music title could be learned effectively from well-identified features of music titles. A possible explanation is that these early claims were likely based on spurious data or on biased experiments. This experiment was all the more convincing that other subjective labels could be learned reasonably well using the features sets described here (e.g., the "mood nostalgic" label).

The question remains: Do these experiments definitely dismiss the Hit Song Science project?

10.4   Discussion

The experiments described above show that current feature-oriented approaches to hit song prediction are essentially not working. This negative result does not mean, however, that popularity could not be learned from the analysis of a music signal or from other features. It rather suggests that the features used commonly for music analysis are not informative enough to grasp anything related to subjective aesthetic judgments.

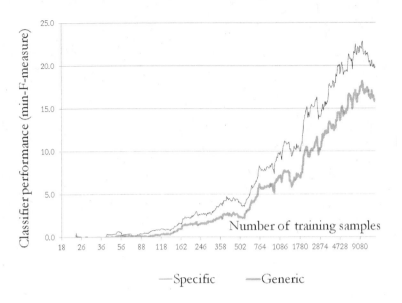

Figure 10.5
The relative performance of the 632 acoustic classifiers (i.e., the difference between the min-F-measures of the classifier and of the corresponding random oracle) for specific and generic features, as a function of the training database size. The performance of the acoustic classifiers increases with the size of the training database.

A natural way forward is to consider other feature sets. A promising approach is the use of *feature generation* techniques, which have been shown to outperform manually designed features for various audio classification tasks [24]. However, more work remains to be done to understand the features of subjectivity for even simpler musical objects such as sounds or monophonic melodies. Concerning the problem of social pressure, an interesting approach is to use music produced with exotic musical temperaments, an approach described by Loui et al. [19] to study the effects of exposure on musical learning and preference. This approach cannot be used on existing music, but has the great advantage of avoiding the biases of social pressure.

These negative results cast serious doubts on the predictive power of commercial Hit Song Science systems. Therefore, notwithstanding the limitations of current feature-based approaches, the arguments of social pressure effects are crippling: Hit Song Science cannot be considered, in its current state, as a reliable approach to the prediction of hits, because of the chaotic way individual preferences are mingled and propagated.

In spite of these negative results, we think that the main scientific interest of Hit Song Science, from a MIR viewpoint, lies precisely in the feature

Popularity	Generic Features	Specific Features	Corrected Specific	Human Features	Dummy Features	Random Oracle
Low	36	35	31	41	32	27
Medium	36	34	38	37	28	22
High	4	3	3	3	3	3

Table 10.2
The Performance (Min-F-Measures) of the Various Classifiers for the Three Popularity Labels (No significant improvement on the random oracle is observed.)

questions: Are there features of popularity, for an individual or for a community, and, if yes, what are they? From this perspective, Hit Song Science is a fascinating enterprise for understanding more what we like, and hence, what we are. The studies presented here have only started to scratch the surface of these questions: Hit Song Science is not yet a science but a wide open field.

Bibliography

[1] J.-J. Aucouturier and F. Pachet. Improving timbre similarity: How high is the sky? *Journal of Negative Results in Speech and Audio Sciences*, 1(1), 2004.

[2] S. Bensch and D. Hasselquist. Evidence for active female choice in a polygynous warbler. *Animal Behavior*, 44:301–311, 1991.

[3] D.E. Berlyne. Novelty, complexity and hedonic value. *Perception and Psychophysics*, 8(5A):279–286, 1970.

[4] K. Bischoff, C. Firan, M. Georgescu, W. Nejdl, and R. Paiu. Social knowledge-driven music hit prediction. In R. Huang, Q. Yang, J. Pei, J. Gama, X. Meng, and X. Li, editors, *Advanced Data Mining and Applications*, volume 5678 of *Lecture Notes in Computer Science*, pages 43–54. Springer, Berlin/Heidelberg, 2009.

[5] J. Blume. *6 Steps to Songwriting Success: The Comprehensive Guide to Writing and Marketing Hit Songs*. Billboard Books, New York, 2004.

[6] P. Cano, E. Gómez, F. Gouyon, P. Herrera, M. Koppenberger, B. Ong, X. Serra, S. Streich, and N. Wack. ISMIR 2004 audio description contest. Technical Report MTG-TR-2006-02, University Pompeu Fabra, Italy, 2006.

[7] S.H. Chon, M. Slaney, and J. Berger. Predicting success from music sales data: A statistical and adaptive approach. In *AMCMM '06: Proceedings of the 1st ACM Workshop on Audio and Music Computing for Multimedia*, pages 83–88, ACM Press, New York, 2006.

[8] M. Csíkszentmihályi. *Beyond Boredom and Anxiety.* Jossey-Bass, San Francisco, 1975.

[9] R. Dhanaraj and B. Logan. Automatic prediction of hit songs. In *Proceedings of the International Conference on Music Information Retrieval (ISMIR)*, pages 488–491, 2005.

[10] T.I. Draganoiu, L. Nagle, and M. Kreutzer. Directional female preference for an exaggerated male trait in canary (*Serinus canaria*) song. *Proceedings of the Royal Society of London B*, 269:2525–2531, 2002.

[11] S. Essid, G. Richard, and B. David. Instrument recognition in polyphonic music based on automatic taxonomies. *IEEE Transactions on Audio, Speech, and Language Processing*, 14(1):68–80, 2006.

[12] R.H. Frank and P.J. Cook. *The Winner-Take-All Society.* Free Press, New York, 1995.

[13] Y. Freund and R. E. Schapire. A decision-theoretic generalization of online learning and an application to boosting. *Journal of Computer and System Sciences*, 55(1):119–139, 1997.

[14] M. Gladwell. The formula. *The New Yorker*, October 16, 2006.

[15] W. Goldman. *Adventures in the Screen Trade.* Warner Books, New York, 1983.

[16] M. Goto, H. Hashiguchi, T. Nishimura, and R. Oka. RWC Music Database: Popular, Classical, and Jazz music databases. In *Proceedings of the 3rd International Conference on Music Information Retrieval*, pages 287–288, 2002.

[17] M. Kelner. Heard the same song three times today? Blame the craze for testing tunes. *The Guardian*, May 19, 2008.

[18] H.G. Kim, N. Moreau, and T. Sikora. *MPEG-7 Audio and Beyond: Audio Content Indexing and Retrieval.* Wiley & Sons, New York, 2005.

[19] P. Loui, D.L. Wessel, and C.L. Hudson Kam. Humans rapidly learn grammatical structure in a new musical scale. *Music Perception*, 25(5):377–388, June 2010.

[20] D. Lu, L. Liu, and H. Zhang. Automatic mood detection and tracking of music audio signals. *IEEE Transactions on Audio, Speech, and Language Processing*, 14(1):5–18, January 2006.

[21] M. Meyer. Experimental studies in the psychology of music. *American Journal of Psychology*, 14:456–478, 1903.

[22] A.L. Montgomery and W.W. Moe. Should record companies pay for radio airplay? Investigating the relationship between album sales and radio airplay. Technical Report, Marketing Dept., The Wharton School, University of Pennsylvania, June 2000.

[23] F. Pachet and P. Roy. Hit song science is not yet a science. In J. P. Bello, E. Chew, and D. Turnbull, editors, *Proceedings of the 9th International Conference on Music Information Retrieval (ISMIR)*, pages 355–360, Philadelphia, 2008.

[24] F. Pachet and P. Roy. Analytical features: A knowledge-based approach to audio feature generation. *EURASIP Journal on Audio, Speech, and Music Processing*, 1:1–23, February 2009.

[25] F. Pachet and P. Roy. Improving multilabel analysis of music titles: A large-scale validation of the correction approach. *IEEE Transactions on Audio, Speech, and Language Processing*, 17(2):335–343, 2009.

[26] E. Pampalk, A. Flexer, and G. Widmer. Improvements of audio-based music similarity and genre classification. In T. Crawford and M. Sandler, editors, *In Proceedings of the International Conference on Music Information Retrieval (ISMIR)*, pages 628–633, London, 2005.

[27] I. Peretz, D. Gaudreau, and Bonnel A.-M. Exposure effects on music preference and recognition. *Memory and Cognition*, 26(5):884–902, 1998.

[28] M.J. Salganik, P.S. Dodds, and D.J. Watts. Experimental study of inequality and unpredictability in an artificial cultural market. *Science*, 311(5762):854–856, 2006.

[29] E.G. Schellenberg. The role of exposure in emotional responses to music. *Behavioral and Brain Sciences*, 31:594–595, 2008.

[30] E.G. Schellenberg, I. Peretz, and S. Vieillard. Liking for happy- and sad-sounding music: Effects of exposure. *Cognition and Emotion*, 22(2):218–237, 2008.

[31] T. Siu-Lan, M.P. Spackman, and C.L. Peaslee. The effects of repeated exposure on liking and judgments of musical unity of intact and patchwork compositions. *Music Perception*, 23(5):407–421, 2006.

[32] Sony. 12-tone analysis technology, http://www.sony.net/SonyInfo/technology/technology/theme/12tonealalysis_01.html, 2010.

[33] K.K. Szpunar, E.G. Schellenberg, and P. Pliner. Liking and memory for musical stimuli as a function of exposure. *Journal of Experimental Psychology: Learning, Memory, and Cognition*, 30(2):370–381, 2004.

[34] D.J. Watts. Is Justin Timberlake a product of cumulative advantage? *New York Times*, April 15, 2007.

[35] C.V.O. Witvliet and S.R. Vrana. Play it again Sam: Repeated exposure to emotionally evocative music polarizes liking and smiling responses, and influences other affective reports, facial EMG, and heart rate. *Cognition and Emotion*, 21(1):3–25, 2007.

[36] W. Wundt. *Outlines of Psychology*. Englemann, Leipzig, 1897.

11

Symbolic Data Mining in Musicology

Ian Knopke

British Broadcasting Corporation, Future Media and Technology

Frauke Jürgensen

University of Aberdeen

CONTENTS

11.1 Introduction

Symbolic data mining in musicology is concerned with extracting musical information from symbolic representations of music: that is, music that is represented by sequences of symbols over time. Representing music symbolically is certainly not a new practice; the majority of Western music exists primarily in a symbolic notation system evolved over more than a thousand years, and there are many examples of symbolic musical notation in other cultures. However, in the majority of cases the symbols used in this kind of study come from common practice musical notation, in the Western European tradition, within the last half millennium. Usually, these are the familiar notes and rests, but sometimes other representations such as chord symbols, neumes, tablature, or various kinds of performance markings are used.

There are significant differences between working with symbolic data and audio recordings. One view is that a score is an "ideal" version of music that exists prior to a performance, and recordings are simply instantiations. This is a common viewpoint but is rather limited and does not usually deal well with situations such as transcriptions made from recordings. From a data mining viewpoint, another perhaps more useful way to view symbolic music is as a kind of filtered, normalized recording where many details such as timbre, pitch variation, or performance aspects that usually complicate audio analysis are simply not present. Something as simple as finding the notes in a melody can be a difficult task to accomplish with precision from a recording, and working with polyphony is extremely difficult in most cases. With symbolic music data these are usually the starting point for analysis. In effect, the symbolic music analysis is often close to what an audio analysis would hope to achieve if the present state of our existing transcription techniques were more flexible, and these considerations can greatly influence the kinds of analysis that can be undertaken. Many types of musical analysis regarding traditional musicological concepts such as melody, harmonic structure, or aspects of form are easier to investigate in the symbolic domain, and usually more successfully.

Data mining is used for many different types of knowledge discovery, including predictive and exploratory tasks. Data mining in musicology is mostly concerned with trying to answer specific musicological questions that may be difficult to answer using other means. Here we attempt to summarize lessons we have learned from our own experiences in working in this area [8, 9, 3, 12]. After a discussion of the role of the computer in this work in Section 11.2 and some methodological aspects of this kind of data mining (Section 11.3), in Section 11.4 we present a case study involving the determination of double leading tones and tuning in a large corpus of 15th-century music.

11.2   The Role of the Computer

One might wonder if these kinds of investigations should not be left to existing methodologies such as musicology or music theory. In fact, why involve a computer at all? The most obvious reason of course is that computers save time. Traditionally, musical analysis is a task accomplished primarily through manual examination and analysis. A music analyst or theoretician often begins an analysis by making marks directly on a music score. While many types of tasks are possible, the majority involve, for example, tasks such as frequency counts or the comparison of recurring features. These results are then tabulated and presented to the reader in either a summarized form, or as a set of examples, and this usually forms the basis for further interpretation. The initial stages of this process can be extremely time consuming; the computer can be an excellent tool for overcoming these limitations.

Another consideration is that traditional musical analysis tends to concentrate on what we might term the "extraordinary" attributes of a musical work; that is, the features of a piece of music or collection that are most interesting to a particular analyst. While it is common practice to overlook the statistical flaw in this methodology, this is also partially a byproduct of the amount of work involved. Usually, this results in only a few pieces receiving concentrated analysis, often a single composition, or alternately a limited analysis of a larger group of works. With a large collection of pieces the analyst may only be able to accommodate surface details of each piece, or concentrate on a single feature, such as the opening themes of a group of sonatas.

This leads to another consequence. The kinds of computer methods being explored in this field make it possible to analyze large groups of pieces in detail. This is especially important as this is an area that has rarely been dealt with in music theory, again a consequence of the productivity of single researchers. A study on the scale that is possible with computers would previously have taken many years to complete, and possibly a researcher's entire career! Current trends in research funding do not tend to favor the sort of long-term study necessary for this kind of work. It is now possible to analyze enormous collections in minutes (assuming a suitable methodology) and it is not difficult to imagine comparisons across complete collections of multiple composers, styles, or time periods, within the boundaries of current research resources.

One additional advantage of symbolic data mining, and probably the least obvious, is that it brings to the work a level of consistency that can be difficult for a single researcher to achieve. A primary difference between computers and people is that people learn and develop while undertaking a task, and computers do not. A human proceeding with an analysis will begin to adapt to it, and analysis projects spanning weeks or months tend to accentuate this process of adaptation. It is not uncommon for a musicologist to expand his opinions, methods, and even goals during the course of an analysis. In contrast, the computer does not do this, and treats each piece in a collection in exactly the same manner. In a very particular sense this makes the computer more objective than a human. However, this should not be confused with true objectivity; computational methods of analysis still contain the analytical biases of the the analyst, but at least these biases are applied equally across the entire data set.

11.3 Symbolic Data Mining Methodology

Data mining is typically described as having several stages, including preprocessing, data mining, and the validation and extraction of knowledge from the

results. This section discusses some of the issues that are specific to this type
of work in the data mining context.

11.3.1   Defining the Problem

One of the most important steps in the entire data mining process is deciding on what the goals are of the analysis and what it is likely to achieve.
Is it to find a way to predict melodies for hit songs, for instance, to learn
something about cadence usage patterns, or to discover recurring elements
of a composer's style across his entire corpus? Giving some consideration to
what is to be investigated is extremely important, as it can save considerable
effort and avoid difficulties that may occur later in the process. This is especially important if a preexisting data set is not available. Experiments that
require new data to be entered require some careful thought as to the choice
of encodings and formats. Also some musicological questions are simply too
difficult or ephemeral to answer in a data mining context (although a great
many can) and some forethought at this stage is recommended, and often not
given enough attention.

11.3.2   Encoding and Normalization

Symbolic data for musical data mining is available in many different formats.
Probably the most common is the Musical Instrument Digital Interface (MIDI)
format. MIDI is primarily designed to store keyboard performances and is acceptable for pitch and duration queries, but tends to represent notated scores
rather poorly. For instance, measures are not explicitly encoded, there is no
differentiation between flat and sharp accidentals, and rests must be inferred
from the absence of notes. However, this may be sufficient for some research
questions.

There are also a variety of more specific music formats, including Humdrum [7], MuseData [17], MusicXML [4], and GUIDO [5]. In our work we tend
to prefer using the Humdrum kern notation because it is easy to use, edit,
and parse, and has some visual affinity as a data format with score notation
(albeit vertically instead of horizontally)[10]. A recent addition is the Music
Encoding Initiative (MEI) format that is still under development and looks
extremely promising [16].

If an existing collection is not available, often the only choice is to create
the data from scratch. Encoding data directly from musical scores by hand can
be an extremely labor-intensive process, and can take considerable amounts
of time for proofreading. One method for creating collections more easily is to
use an optical music recognition (OMR) system. OMR systems can be faster
for adding data but are prone to error [2, 11]. They may not capture some of
the details of a score, although these systems keep improving. Symbolic music
information can also be entered by a skilled performer using a MIDI keyboard
or some other entry device, although this usually entails using the deficient

MIDI file format. In any case, all three methods usually require considerable editing, proofreading, and verification of the data against the original scores.

Some consideration must also be given to how this data is to be normalized. Keys, note names, accidentals, rhythms, and chord symbols can be encoded in different ways, and the choice of representation can be influenced by the problem under investigation. For instance, when comparing melodies, it is often beneficial to transpose them to a common key or mode instead of encoding them as found in the original scores.

The primary purpose of data mining is to discover knowledge from a data set by reducing it to something more manageable and exposing hidden information, and the cases we present here do not deviate from this idea. Most of our work has focused on answering specific musicological questions. Generally we are more concerned with discovering answers in preexisting groups of pieces, and less concerned with predicting specific outcomes. While there are many different types of classifiers used in data mining, the tools we have found most useful for summarization are histograms, scatter plots, decision trees, and basic clustering procedures. Also, the results of the extraction process will usually require additional filtering of some kind.

11.3.3 Musicological Interpretation

Music is a complex subject, and to discover worthwhile knowledge in this context requires a final interpretive stage. In our opinion, this is best done in conjunction with an informed researcher experienced in the music and literature of the domain under investigation, so that false trails, assumptions, and conclusions may be avoided. Additionally, this interpretive stage often suggests new directions and refinements of the entire process.

11.4 Case Study: The Buxheim Organ Book

To demonstrate the possibilities inherent in symbolic music data mining we discuss applications to two related problems in the area of 15th-century performance practice. A brief outline of each problem is followed by an analysis of the data-mining process, results, and musicological interpretation.

Both questions concern the same collection of music. The core of this collection consists of the Buxheim Organ Book, the largest extant manuscript of 15th-century keyboard music, and notated in Old German organ tablature. Many of the pieces are based on models that exist elsewhere in the conventional mensural notation of the period. The collection also contains the concordances of these pieces found in mensural manuscripts. Altogether, the Buxheim Organ Book consists of about 250 pieces, about 60 of which have

Buxheim files	268
Buxheim notes	155252
Model files	228
Model notes	76812
Earliest concordance	ca. 1400
Latest concordance	ca. 1520s
Known composers	25

Table 11.1
Buxheim Data Set

one or more concordances in nearly 100 manuscripts, for a total of about 500 files. Precise details of the data collection are given in Table 11.1.

11.4.1   Research Questions

Many principal research questions of interest to modern scholars of performance practice are concerned with the interpretation of pitch material, the precise details of which rely on performance conventions that have not been accurately preserved in written form. The case study discussed here addresses two important musicological questions.

Our first question concerns the "double leading tone" accidental. In the 15th century, performers relied on conventions for adding accidentals to notated music. One of the most common chromatic alterations was the raising of the leading tone at cadences. A simple definition of a cadence in this music is a progression between two voices of a sixth moving outwards to an octave. The repertoire under consideration is predominantly in three voices, and when two of these form a cadence, the third voice can often be found in parallel to the top voice, stepping from the fourth degree above the cadential note to the fifth degree (illustrated in Figure 11.1). A "double leading tone" is the raising of the fourth degree to create a type of leading tone to the fifth degree. This chromatic inflection is extremely rare in mensural notation. Our first task attempts to discover empirically the frequency of this inflection in the Buxheim Organ Book and the circumstances under which it is notated.

Our second question concerns possible keyboard tunings in the 15th century. While it is beyond the scope of this chapter to explore in detail all of the musicological aspects of this topic, as well as the controversies, the following summary is given as a background to understanding the problem.

The problem, simply stated, is as follows: purely tuned intervals do not fit neatly into an octave, to give purely tuned triads. For example, in the Pythagorean tunings that predominated until the late middle ages, the notes of a keyboard are generated by tuning a long chain of pure perfect fifths and transposing the pitches to end up within the same octave. Each successive fifth has a ratio of 3/2 against the previous frequency, so a chain beginning

basic cadence   parallel voice   double leading tone

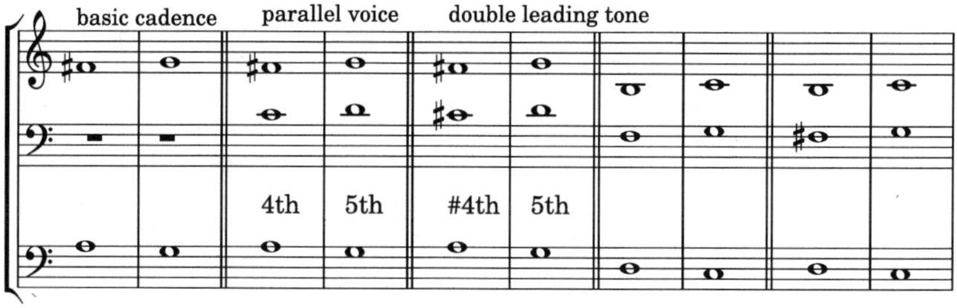

4th   5th   #4th   5th

Figure 11.1
Cadences to G and C, showing the double leading tone.

on "middle" C (261.63 Hz, according to the most common modern standard) would be followed by G (392.44 Hz), then D (588.66 Hz) and so on. At the end of the chain the B♯ (adjusted by octave to 260.74 Hz), which would fall on the same key as C, will be audibly different from the original C (261.63 Hz).

This small interval is known as the *syntonic comma*, and all keyboard temperaments are essentially clever attempts to conceal this inconsistency [15]. The specific way this is accomplished is known as a temperament, and depends greatly on the musical priorities of its inventor.[1]

The two main types of temperament systems in use during the mid-15th century are Pythagorean tunings and mean-tone tunings. In Pythagorean tunings all fifths except one are pure, and as a result most major thirds are far too large. Only very few of the triads will sound pure, depending on the placement of the bad, "wolf" fifth. In mean-tone tunings, some of the fifths are made smaller by fractions of the comma, in order to yield a greater number of good thirds. Although the most familiar mean-tone tuning is probably quarter-comma mean-tone, there are other variants.

The choice of tuning system was dependent on the musical priorities of the musician. While the mathematics of tuning had been well understood for many centuries, including the theoretical possibility of other tunings, in practice Pythagorean tuning dominated until the time period in question. In the case of earlier music the emphasis was on using perfect harmonic intervals, and the bad thirds of a Pythagorean temperament may not have been considered detrimental. However, as major thirds (and thus triads) become more musically significant as consonances, tuning systems were developed to make

---

[1]The difficulty of tunings extends beyond keyboard instruments, essentially concerning any instrument on which more than one pitch is fixed in place (including for example fretted stringed instruments of many types, as well as wind instruments with finger holes such as recorders). The reasons for investigating this problem extend beyond the practical (how can different types of instruments be used together?), into aesthetic and philosophical realms. For the purposes of this discussion, however, we are confining ourselves to the implications of a specific keyboard manuscript.

these available. One implication of this, of particular relevance here, is that music in which certain simultaneities are treated as consonances can be read to imply the use of a tuning system in which those simultaneities are especially good.

However, in order to evaluate the pitch content of a particular repertory, one needs to be fairly sure of the actual pitches that were meant to be performed. This is a particularly difficult problem in late medieval music, where unwritten performance conventions often led to alterations of notated pitches. There are two factors to consider here. First, tuning is more obviously a concern for instruments with fixed pitches such as keyboards, and these are the instruments where tablature notation was used. Second, the way in which pitches are notated in tablature suggests that scribes might have been more likely to notate the actual pitch intended. Certainly, there are far more explicitly notated accidentals in Buxheim than are found in mensural notation.

To what extent conclusions drawn from tablature, even from originally-vocal pieces that have been intabulated, can be retrofitted to vocal music, is controversial. Similarly, vocal performance, and indeed performance on all nonfixed-pitch instruments, allows for minute adjustments in tuning. This allows performers to produce perfectly good triads, even while they might be living inside an ostensibly Pythagorean conceptual framework. It is difficult to conceive of a productive way of examining the question, "will a composer use sonorities that are *theoretically* impure, when they can be *practically* pure?"[2] The evidence we have available to us, however, does invite us to examine keyboard performance practice more closely with regard to tunings.

Mark Lindley has been a principal scholar examining the music of this period for clues to tuning practices, and we have used his investigations as the inspiration for our Buxheim experiment. First, he gleaned descriptions of a particular variant of Pythagorean tuning, which he terms the F♯xB scheme, from predominantly Italian theoretical sources, including writers such as Ugolino, Prosdocimo, Zwolle, Gaffurio, and Aaron [13, pp. 4–5]. In this tuning, the white notes of the keyboard are tuned as a chain of fifths (F–B), and the black notes are generated by adding another chain of fifths to the bottom end of this chain (a chain of flats, from B♭ down to G♭). Since the Pythagorean diminished fourth (for example, D–G♭) is much closer to a pure major third than the Pythagorean third, major triads constructed of two white keys with a black key in the middle are virtually pure. For example, D–G♭–A, reinterpreted as D–F♯–A, is a very good triad, as are major triads on E and on A.

Lindley looked for evidence of this tuning in two of the principal tablature sources of the period, the Faenza Codex and the Buxheim Organ Book [13, pp. 38–43]. He regards the appearance of particularly prominent A- and D-major triads as indicative of this tuning practice, where examples of prominence might be protracted sonorities or thickened textures (such as temporary

---

[2] A good introduction to the debate can be found by Bent [1].

addition of an extra voice). In Buxheim, he identifies a group of pieces in the first eight (of nine) fascicles as likely candidates. He adds that "none of the arrangements bears Conrad Paumann's initials, none is derived from the transcriptions accompanying the 'Nürnberg 1452' version of his *Fundamentum organizandi* [in the Lochaimer Liederbuch]..., and none belongs to the most ornate category in the Buxheim repertory."[13, p. 42] Elsewhere he uses a Buxheim *Fundamentum* (keyboard exercise) to support his arguments that certain passages in Ramos's *Musica Practica* imply a mean-tone temperament, citing (along with Shohe Tanaka) Paumann's free use of diatonic triads [14, p. 50].

In effect, then, Lindley finds that two different temperament systems are implied by different groups of pieces in Buxheim: a group of early 15th-century chansons exhibiting F♯xB characteristics, and a group of Paumann-circle pieces suggesting mean-tone tuning.[3] This is not a problem, even if they were collected by one person to be played on one instrument (presumably with the mean-tone variant suggested by the later Paumann-circle pieces), since the F♯xB pieces would have been composed well before their incorporation into Buxheim, and would work just fine in a temperament designed to produce even more acceptable thirds than F♯xB.

The purpose of our experiment is both to test Lindley's theory empirically (as he does not provide any specific numbers), and to discover if any particular groupings of pieces or sonorities can be found that suggest changes in the tuning practices applicable to Buxheim. Other questions of interest, that could be addressed using data mining, include:

- Are Lindley's findings empirically borne out by our data?

- Are there clear groupings of pieces within Buxheim, that might suggest different tuning systems?

- If there are such groupings, do they bear any relation to factors such as possible date of composition or place of origin?

- Do originally mensural pieces that have been transcribed into Buxheim show compositional changes that might suggest a different tuning system?

11.4.2 Encoding and Normalization

To answer our two primary questions we mainly need to know about pitches and when in time these pitches occur, and require awareness of the separate voices. The principal features of pitch and rhythm are available from the late-medieval notation. In addition, we need to record information about the source manuscripts, (such as the date or place of copying, if known), and any unusual

---

[3] He does not give the exact pieces included in each group.

features that might make the notation more or less reliable (damage, obvious copying mistakes, and so on). This metadata is often useful in filtering the data collection. For example, results that consistently appear in conjunction with a "damaged page" flag might be discarded.

To represent pitches and rhythms, we used the `kern` representation associated with the Humdrum Toolkit [6]. Other features of the original, such as ligatures (symbols representing multiple notes in one), foliation, line-breaks, and areas of ambiguity, while not having immediate application, were seen as potentially useful. To accommodate this we constructed an additional Humdrum representation called `fol`, for "foliation."

Mensural notation, the top voice of tablature notation, and the lower voices of tablature notation all appear differently on a page. The top voice of the tablature, while on a staff, appears to have been notated in rhythmic values that are twice as fast as their mensural equivalents, and the lower tablature voices are written as letter symbols with separate rhythmic symbols above them (see Figure 11.2). In addition, mensural notation has signature accidentals, as well as "mensural signs" (time signatures). Neither of these exists in tablature notation. However, tablature notation has bar lines that are nonexistent in mensural notation. All of these types were transcribed into the same representation. A `kern` half note (minim) was used to represent a tablature minim and its mensural equivalent, a semibreve. `kern`'s bar lines were used to indicate divisions between mensural units in mensural notation, and the corresponding bar lines in tablature. Figure 11.2 shows how the beginning of the example tablature piece is represented in `kern`.

The encoding of the manuscripts was initially accomplished using direct manual entry from notation into `kern`, Later, to improve encoding speed, a MIDI keyboard was used to encode to MIDI format, with subsequent automated conversion into `kern` through a custom software solution. Initially, we investigated OMR, but were unable to find a system that could adequately handle manuscript music of this period. After data entry, each piece required several proofreading passes, as well as the addition of `fol` metadata. The collection took over a year of an expert's time to encode.

11.4.3   Extraction, Filtering, and Interpretation

11.4.3.1   Double Leading Tones

First, we had to devise an algorithm that would robustly identify all situations that could qualify as cadences. This was complicated by the frequent appearance of the "Landini ornament," which obscures the sixth-to-octave progression by turning the interval immediately preceding the octave into a fifth (see Figure 11.3). Then, situations with the parallel voice-leading described above had to be extracted, and the number of double-leading-tone inflections counted.

Once all double-leading tone cadences had been extracted, further filtering

```
!!!ONM: 124
!!!OTL: Fortune
!!!OTA: Gentil madonna
**kern   **kern   **kern   **fol
*tenor   *contra  *sup     *
2r       2r       8cc      66v1
.        .        8b       .
.        .        8a       .
.        .        8b       .
=1       =1       =1       .
1c       1g       2cc      .
.        .        2cc      .
4d       2G       4bM      .
4e       .        16cc     .
.        .        16b      .
.        .        16a      .
.        .        16b      .
=2       =2       =2       =
```

Figure 11.2
Piece No. 124 as it appears in Buxheim as tablature notation. (Bayerische
Staatsbibliothek München. Mus. Ms 3725, f. 66v.)

Figure 11.3
The "Landini" ornament.

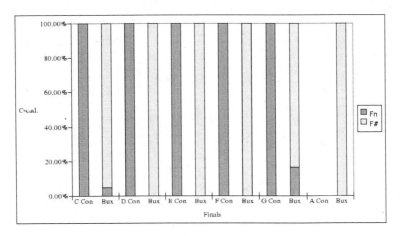

Figure 11.4
Double leading tones in cadences to C, grouped according to the final pitch
of the pieces.

was necessary. The pitch on which the cadence occurs is significant, because
certain accidentals are more common than others, and because this pitch could
have more or less significance, depending on its relationship to the "final" of
the piece (somewhat analogous to the tonic in later music). Figure 11.4 shows
a graph generated from the extraction results. The cadences shown are all
on the pitch C, making the double leading tone an F♯ (leading to the fifth
degree G). The X axis shows cadences in the mensural model concordances,
alternating with cadences in Buxheim. Each pair of columns represents pieces
based around the same pitch (C-pieces, D-pieces, etc.). This feature has been
separated out, to allow us to see if "final" in any way influences the tendency
for an indicated double leading tone.

The results shown in this graph are rather startling. It appears that in
the case of the mensural notation, the F♯ is *never* notated, whereas in the

tablature notation, it is notated over 90% of the time! The relationship of the cadential pitch to the final had no significant impact. The graph suggests that this particular chromatic alteration is mandatory, and supports the idea that tablature notation may be more reliable with regard to performed pitches than mensural notation.

Further investigation of double leading tone behavior at cadences to other pitches led to the observation that the intabulator appeared to be explicitly avoiding the pitch D♯, which would be the double leading tone in cadences to A. A possible explanation for this might be the use of a tuning system in which there are no good D♯s, those notes having been tuned give good E♭s instead. This piqued our interest, and suggested that further clues to tuning systems might indeed be found in Buxheim.

11.4.3.2 Keyboard Tuning

All of our tuning questions are addressed by the same core method: counting occurrences of various major triads in places where one might expect a good, pure sound to be favored. Three types of situations for prominent triads are: prolonged sonorities, thickenings of texture through the addition of extra voices, and prominent placement within the mensural unit (what would be called, in modern terms, a *downbeat*).

After the chords in these places had been counted, the results were filtered using metadata contained within the individual piece-files, to group pieces by characteristics such as concordant manuscripts, composers, or location within Buxheim.

For illustrative purposes, the following examples show only chords that occur at the beginning of mensural units. In our Buxheim collection these are trivial to locate, as the scribes used vertical lines in the manuscripts that are similar to modern bar lines. Rythmically prominent chords could easily be extracted by taking the chords following these lines, represented in kern using the bar-line symbol ("="). All triads that were not major triads were discarded, and these triads were then grouped roughly according to the tuning system they were thought to imply. Without going into too much detail, A minor, D major, and E major triads are in a "Pythagorean" group, whereas white-note major triads (C, F, G) and B♭ major are in a "Meantone" group.

Our first two graphs simply show the frequency of these triads in the Buxheim collection as compared with the concordant manuscripts. In Figure 11.5, downbeat triads in Buxheim pieces are shown as percentages of all downbeat triads in the Buxheim pieces. Only those pieces that have concordances in other manuscripts are shown. Our "Pythagorean" group of D major, A major, and E major is well represented, gathering about 15% of the occurrences in total, with D major and A major representing the largest groupings. By comparison, our "Meantone" group has nearly 45% of the total triads, with the E♭ major triads forming a small but significant group. G major and C major

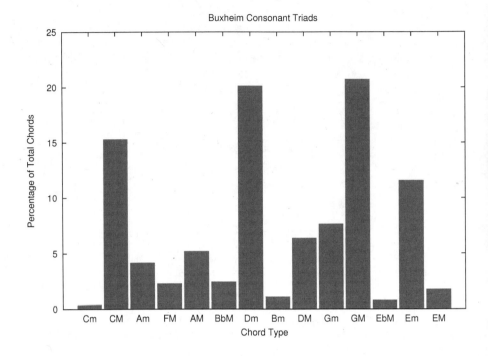

Figure 11.5
Buxheim consonant triads.

triads are the most prevalent, with B♭ major and F major triads represented
in distant third and fourth place.

Figure 11.6 shows a profile of downbeat triads in concordances. We are
including only mensural concordances here, to avoid contaminating our data
with accidentals from other keyboard manuscripts. Less than 2% of the triads
found here are in the "Pythagorean" group, and about 32% of the triads are
in the "Meantone" group. The order of triad frequency is the same as in the
intabulations, but the ratio has changed: G major now has the overwhelming
majority, with C major occurring much less prominently, and far closer to B♭
major and F major.

On the surface, it looks as though the Buxheim pieces support the presence
of a tuning that allows the "Pythagorean" triads, whereas the models would
seem to be sending us toward a tuning where these triads are not so good!
How can this be? We need to take into consideration the notational practices
of the mensural concordances: notated sharps (the F♯, C♯, and G♯ we are
looking for in our triads) are very rare in mensural manuscripts, and tend to
fall under the heading of conventional performer's alterations, as described
earlier for the double leading tones. We cannot therefore take the absence

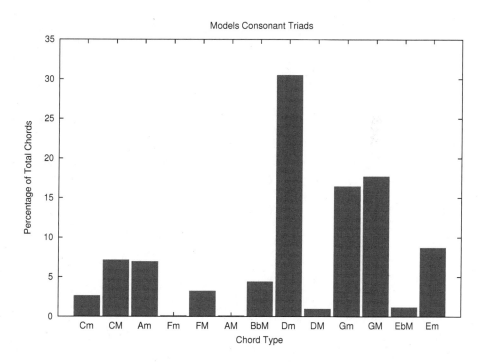

Figure 11.6
Model consonant triads.

of such sharps to indicate the absence of our "Pythagorean" triads from the mensural concordances *as they might have been performed*. The presence of these sharps in Buxheim, however, in fairly large numbers, *does* indicate that these triads must have sounded reasonably good on the instrument(s) available to the intabulator(s).

The differences among the "Meantone" triads invite some speculation: why does C major change so much in prevalence? Is it a particularly sweet triad on the Buxheim instrument(s), inviting the intabulators to use it more often?

Interpreting these graphs presents us with additional problems. We could use more context: we need to have some idea of how prevalent these sonorities are in the complete picture of downbeat sonorities. Perhaps it is not just the balance among triads that is different between Buxheim and the models, but also the extent to which complete triads, in general, are employed in exposed situations. The third voice (the contratenor) is often not as structurally significant in this repertoire as the other two (the superius and tenor), and so, it is the most subject to alterations. In Buxheim, many contratenors are not as faithfully intabulated as the other parts, and may indeed be missing. This could have several consequences on the chord profile: triads may be replaced by diads, or the contratenor may have been rewritten to accommodate more favored triads. A refinement to take this into consideration would be including diads in the total count against which chord percentages are calculated.

Figure 11.7 is similar to the previous two; however, it represents the triads found in Buxheim pieces in one concordant source only: the Lochaimer Liederbuch. It is unique, in that it is also in Old German organ tablature, and in that it seems to have a special link to Buxheim through repertoire connected with the noted blind organist Conrad Paumann, mentioned earlier. There is a far smaller variety of triads in use. The "Meantone" C major and G major are far less prevalent than in Buxheim, whereas the "Pythagorean" D major is more prevalent. This might suggest that the tuning most familiar in the performance context of this manuscript is more likely to have been Pythagorean. However, the sample is much smaller (only the few pieces that are concordant with Buxheim). Testing this further would require encoding a larger portion of the Lochaim manuscript for confirmation.

All these chord profiles cannot by themselves invite any conclusions, but they offer tantalizing thread-ends to pursue. This same methodology could also be used on later bodies of keyboard music for comparison, for some of which we have more external evidence about likely keyboard tunings. It also indicates the need for more traditional musicological investigation. These chord profiles must now be placed in the context of evidence gleaned not only from Lindley's theorists, but also archival records of instrument builders, for example.

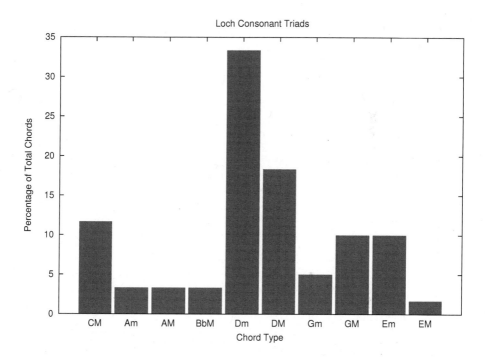

Figure 11.7
Lochaimer Liederbuch consonant triads.

11.5 Conclusion

The type of investigation we have described is not an end in itself; nor can it replace more traditional modes of enquiry. It can however provide a powerful tool that not only can quickly accomplish time-consuming traditional analysis tasks, but provide statistical pictures of a kind simply not feasible through analysis by hand. The use of data mining techniques shows great potential for musicological research.

Bibliography

[1] M. Bent. Diatonic *Ficta* revisited: Josquin's *Ave Maria* in context. *Music Theory Online*, 2(6), 1996.

[2] D. Byrd and M. Schindele. Prospects for improving optical music recognition with multiple recognizers. In *Proceedings of the International Conference on Music Information Retrieval*, pages 41–46, October 2006.

[3] B. Gingras and I. Knopke. Evaluation of voice-leading and harmonic rules of J.S. Bach's chorales. In *Proceedings of the Conference on Interdisciplinary Musicology*, pages 60–61, March 2005.

[4] M. Good. Representing music using XML. In *Proceedings of the International Symposium on Music Information Retrieval*, page n.p., 2000.

[5] H.H. Hoos, K. Renz, and M. Görg. GUIDO/MIR—An experimental musical information retrieval system based on GUIDO Music Notation. In *Proceedings of the International Symposium on Music Information Retrieval*, pages 41–50, 2001.

[6] D. Huron. *The Humdrum Toolkit: Reference Manual*. Center for Computer Assisted Research in the Humanities, Menlo Park, California, 1995.

[7] D. Huron. Music information processing using the Humdrum Toolkit: Concepts, examples, and lessons. *Computer Music Journal*, 26(1):15–30, 2002.

[8] F. Jürgensen and I. Knopke. A comparison of automated methods for the analysis of style in 15th-century song intabulations. In *Proceedings of the Conference on Interdisciplinary Musicology*, page n.p., 2004.

[9] F. Jürgensen and I. Knopke. Automated phrase parsing and structure analysis in 15th-century song intabulations. In *Proceedings of the Conference on Interdisciplinary Musicology*, pages 69–70, March 2005.

[10] I. Knopke. The perlhumdrum and perllilypond toolkits for symbolic music information retrieval. In *Proceedings of the Intenational Conference on Music Information Retrieval*, pages 147–152, 2008.

[11] I. Knopke and D. Byrd. Toward MUSICDIFF: A foundation for improved optical music recognition using multiple recognizers. In *Proceedings of the International Symposium on Music Information Retrieval*, pages 123–126, 2007.

[12] I. Knopke and F. Jürgensen. A system for identifying common melodic phrases in the masses of palestrina. *Journal of New Music Research*, 38(2):171–81, 2009.

[13] M. Lindley. Pythagorean intonation and the rise of the triad. *Royal Musical Association Research Chronicle*, 16:4–61, 1980.

[14] M. Lindley. An historical survey of meantone temperaments to 1620. *Early Keyboard Journal*, 8:1–29, 1990.

[15] G. Loy. *Musimathics: The Mathematical Foundations of Music. Vol. II. With a Foreword by John Chowning.* MIT Press, Cambridge, Massachusetts, 2007.

[16] P. Roland. Design patterns in XMLMusic representation. In *Proceedings of the International Symposium on Music Information Retrieval*, n.p., 2003.

[17] E. Selfridge-Field. The MuseData universe: A system of musical information. *Computing in Musicology*, 9:9–30, 1993–1994.

Index

0-1 normalization, 268
1/f proportions, 177
10-Prec, *see* Top-10 Precision

Aaron, P., 334
Accidental, 332
Acoustic features, 5
Acoustica.com, 85
Activity, 145
AdaBoost, 66
AdaBoost.RT, 158
ADRess, 122
Aesthetics, 175–177
Affect, 138
Affective Norms for English Words (ANEW), 140, 156
    Chinese, 151, 152
Affinity graph, 19
AGC, *see* Automatic gain control
AGMIS, *see* Automatically Generated Music Information System
Agostini, G., 113
Ahn, L. von, 300
AIS Algorithm, 11
Album cover, 221, 228, 260
All Music Guide, *see* Allmusic.com
Allinanchor, 224
Allmusic.com, 5, 27, 153, 220, 221, 231, 239, 257, 265
    Tapestry Demo, 275
Altavista, 235
Amazon.com, 235
AMG, *see* Allmusic.com
Anchor text, 224
ANEW, *see* Affective Norms for English Words
ANN, *see* Artificial Neural Network

API, 220, 233, 235
Apple, 257
A priori, 11
Area under the receiver operating characteristic curve, 262, 311
Aristo Music, 225
Aristotle, 176
Armonique, 196, 200
Armstrong, J.R., 168
Arnheim, R., 177
Arousal, 152, 175
Art of the Mix, 238, 239
Artificial Neural Network (ANN), 12, 14, 106, 111, 116, 119, 120, 174, 186, 191, 199
    feed-forward, 187
Artist classification, 24
Artist similarity, 234
Artist-to-genre classification, 236
ASF, *see* Audio spectrum flatness
Association mining, 11, 19
AUC, *see* Area under the receiver operating characteristic curve
Aucouturier, J.-J., 19, 173
Audio classification, 20
Audio compression, 53
Audio feature extraction, 43
Audio representation
    bio-inspired joint acoustic and modulation frequency, 66
    overcomplete, 53
    time-frequency, 44
Audio segmentation, 20
Audio spectrum flatness (ASF), 113
Audio-to-MIDI transcription, 197
Augat, M., 277
Autocorrelation, 63, 83